Geospatial Technology for Natural Resource Management

Scrivener Publishing
100 Cummings Center, Suite 541J
Beverly, MA 01915-6106

Publishers at Scrivener
Martin Scrivener (martin@scrivenerpublishing.com)
Phillip Carmical (pcarmical@scrivenerpublishing.com)

Geospatial Technology for Natural Resource Management

Edited by

Shruti Kanga

Gowhar Meraj

Suraj Kumar Singh

Majid Farooq

and

M. S. Nathawat

Scrivener Publishing

WILEY

This edition first published 2024 by John Wiley & Sons, Inc., 111 River Street, Hoboken, NJ 07030, USA and Scrivener Publishing LLC, 100 Cummings Center, Suite 541J, Beverly, MA 01915, USA

For more information about Scrivener publications please visit www.scrivenerpublishing.com.

Wiley Global Headquarters
111 River Street, Hoboken, NJ 07030, USA

For details of our global editorial offices, customer services, and more information about Wiley products visit us at www.wiley.com.

Library of Congress Cataloging-in-Publication Data

ISBN 978-1-394-16636-7

Front cover images supplied by Adobe Firefly
Cover design by Russell Richardson

Set in size of 11pt and Minion Pro by Manila Typesetting Company, Makati, Philippines

Printed in the USA

10 9 8 7 6 5 4 3 2 1

Contents

Preface

In today's rapidly evolving world, where discussions around the health of our planet have taken center stage, the significance of intertwining technological advancements with sustainability efforts cannot be understated. It is against this backdrop that *Geospatial Technology for Natural Resource Management* emerges, a collaborative effort by a distinguished panel of professionals who have dedicated significant parts of their careers to the fields of geography, technology, and environmental studies. This pivotal work unravels the intricate world of geo-informatics—a field that has been gaining momentum due to its potential to change the way we view and interact with our environment. The core tenet of the book is simple yet profound: for any sustainable development efforts to be truly effective, a comprehensive understanding of the Earth's diverse resources is indispensable. Beyond merely cataloging these resources, the book argues for a dynamic approach. This involves a cyclical evaluation, where resources are not only identified but are also assessed in terms of their current states, potential risks, and appropriate management interventions. The optimism around such a robust assessment process is largely fueled by technological innovations that have emerged over recent decades. Remote sensing (RS), for instance, is heralded in the book as a game changer. The sheer breadth and depth of data that RS can collect—spanning different geographical terrains and spectral ranges—are invaluable. It offers transformative insights, whether it is in predicting agricultural yields or devising strategies for disaster risk reduction. India, a country with an expansive and diverse geographical landscape, provides compelling case studies. The book highlights how the nation has leveraged RS, among other technologies, to gather and analyze data across multiple dimensions, becoming a benchmark in the global arena. But RS is not the only hero in this story. The global positioning system (GPS), often relegated to being a tool for navigation, emerges in a new light. The book enlightens readers on the broader capabilities of GPS, especially its role in continuously monitoring and documenting subtle shifts and patterns in the Earth's environment. When this vast pool of data is synthesized using geographic information systems (GIS)—a powerful tool that goes beyond presenting data to offering detailed spatial analyses—the possibilities are limitless. In drawing attention to these technological wonders, the editorial team does more than just present facts. They paint a vision of the future, a world where technology and environmental consciousness coalesce to address some of our most pressing challenges, from environmental degradation to unchecked urban expansion. At its core, *Geospatial Technology for Natural Resource Management* stands as a clarion call. It invites its readership, spanning academics, policymakers, practitioners, and even curious individuals, to immerse themselves in the vast potential of geo-informatics. It is not just an academic pursuit; it is a journey towards envisioning and working towards a more sustainable, harmonious world. The volume, in its essence, epitomizes the marriage of technology and environmental studies, making it an essential read for anyone invested in our planet's future.

Shruti Kanga
Gowhar Meraj
Suraj Kumar Singh
Majid Farooq
M. S. Nathawat
Editors

Acknowledgements

The journey of bringing *Geospatial Technology for Natural Resource Management* to life has been akin to crafting a detailed map, each contribution charting a course, and adding depth to the larger narrative. Most importantly, our heartfelt gratitude goes to the authors, who, from various regions of our planet, have brought together their unique experiences, insights, and expertise. Your willingness to share, discuss, and explore has transformed this volume from a dream into a tangible reality. The unsung heroes of any academic endeavor are the reviewers, and we wish to highlight their indispensable role. Their discerning eyes and commitment to excellence ensured that the quality of the study was not compromised. By meticulously navigating through each chapter, their feedback was instrumental in refining, redirecting, and reinforcing collective knowledge. Behind each of us stands a family that has championed our cause, offering unwavering support and countless sacrifices. Their belief in our vision, their patience through absences, and the celebration of our small victories have been the wind beneath our wings. To our mentors, colleagues, and the broader academic community, your engagement and encouragement have been invaluable. Your spirited discussions, shared resources, and enduring belief in the potential of this study have bolstered our determination and drive. We must not overlook the operational backbone: our administrative and logistical teams. Their diligence and dedication have been pivotal in ensuring that every "t" was crossed and every "i" dotted. To our readers, you are the reason this endeavor took shape. As you delve into these pages, our hope is that the passion and purpose with which this book was crafted resonates with you, offering both enlightenment and inspiration.

In essence, *Geospatial Technology for Natural Resource Management* is not just the product of its editors and contributors; it is a collective masterpiece shaped by every hand that touched it, every mind that engaged with it, and every heart that believed in it. For this symphony of collaboration, we remain profoundly grateful.

Shruti Kanga
Gowhar Meraj
Suraj Kumar Singh
Majid Farooq
M. S. Nathawat
Editors
Saturday, 28 October 2023

Introduction

In today's world, it is more important than ever to understand the relationship between the environment and technology. The opening chapters of *Geospatial Technology for Natural Resource Management* showcase the depth of this understanding. These chapters set the stage for the book's broader discussions about the marriage of technology and nature, the value of data in decision-making, and the urgent need for sustainable practices. Urban growth is not just about the rise of cities; it is about balancing growth with available resources. As cities expand, they face challenges such as sustaining the environment, using land wisely, finding energy sources, and ensuring water supply. This section of the book explores these challenges and emphasizes the importance of a balanced, sustainable approach to urban development. Water systems, from lakes and rivers to underground reserves, tell a story about our planet's health and changes. By examining these systems, we learn more about broader topics like climate change, changing patterns of snow, and the potential of groundwater. The chapters dedicated to this topic highlight the complexity of these water systems and emphasize their value and importance to the environment. As we continue in the 21st century, it becomes even more crucial to study our surroundings and understand how they are changing. Many of these changes come from human actions and they can influence both the environment and society. This collection of studies uses tools like geospatial techniques and surveys to better understand these changes. The aim is to offer insights and inspire everyone, from casual readers to policymakers, to work towards preserving and sustaining our world. To further guide the reader, individual prefaces will follow for each chapter. These provide a closer look into the specific topics and research methods of each contribution, offering a clear roadmap for the reader's journey through the book.

Chapter 1 sets the tone with a comprehensive study on groundwater in the Bhandara district of Central India, emphasizing its importance for agriculture and drinking. By correlating groundwater conditions with rainfall data spanning over four decades, the study paints a vivid picture of how the environment has shifted and what it means for the quality of groundwater. A highlight is the meticulous hydro-chemical characterization and the geospatial analysis techniques employed. In **Chapter 2**, the focus pivots to the dynamic realm of disaster response and recovery. The author dives into the world of technology, elucidating how innovations such as GIS, UAVs, and data analytics play crucial roles in enhancing disaster management capabilities. The chapter underscores the power of technology to not only respond to disasters but to predict and mitigate their impacts. **Chapter 3** ties in the global agenda of Sustainable Development Goals (SDGs) with the pressing concerns of natural and technological hazards. Khan's exploration into this synthesis is both timely and insightful. By identifying the intersections between specific SDGs and disaster risk reduction (DRR) strategies, the chapter offers a comprehensive roadmap for harmonizing global development objectives with hazard management. **Chapter 4** takes us on a journey to the Girna River Basin in Maharashtra, combining hydrological parameters with morphometric analysis. By employing remote sensing and GIS techniques, the study effectively characterizes the river basin's geomorphological features, shedding light on potential flood risks and the intricate dance between topography and water flow.

In **Chapter 5**, authors, present a vital study on the geospatial analysis of the effects of waste disposal on groundwater quality within a semi-rural community in Nigeria. As rapid urbanization continues to exert pressure on local ecosystems, understanding the relationship between waste disposal and water quality is crucial. Through laboratory investigations and surveys, the researchers provide quantitative data revealing a correlation between proximity to dumpsites and groundwater contamination. This chapter is a poignant reminder of the immediate threats posed to potable water sources by poor waste management practices and provides valuable insights for policy makers, environmentalists, and the general public. **Chapter 6** explores the impact of global construction material demand on natural resources, focusing on sustainable alternatives to sand in concrete. It evaluates the use of marble powder, glass fines, and fly ash as partial substitutes for natural sand. Results show that certain waste material combinations improve concrete properties and reduce environmental impact, highlighting the potential for sustainable construction practices. **Chapter 7** explores the dynamics of land use/land cover (LULC) in Maharashtra's Kolhapur District. The authors employ sophisticated remote sensing techniques and GIS methodologies to trace the watershed's changes over time. By offering a clear picture of the dominant land uses, such as agriculture and forest cover, the study provides critical data for local planners, environmentalists, and policymakers. The findings have vast implications for sustainable land management in the region.

Chapter 8 examines the use of village maps as tools for regional development planning. It focuses on generating geospatial data for Dungarpur Reelka, India, addressing deficiencies in education, health, and infrastructure. High-resolution images and spatial analysis were used to create thematic layers. The resulting maps could serve as models for other communities. **Chapter 9** pivots our attention to the geospatial analysis of land use and its dynamics. The authors present a comprehensive analysis of the land use changes in Varanasi, using sophisticated tools like GIS and landscape metrics. Their findings underscore the challenges faced by policymakers and the ripple effects of urban expansion on peri-urban areas. **Chapter 10** investigates the factors influencing online shopping behavior in North Bengal. It analyzes survey data from 700 respondents, identifying comfort, product variety, and cost savings as key drivers. The study highlights the importance of website quality, security, and customer support in enhancing consumer satisfaction and regional development. **Chapter 11** explores the relationship between protected areas (PAs) and tourism in India. It reviews the growth and management of PAs, emphasizing their role in biodiversity conservation and sustainable development. The study highlights the need for well-managed, connected, and financed PAs to address climate change and achieve conservation goals.

Chapter 12 immerses readers into the mesmerizing world of diatoms from the Renuka Lake, situated in the Indian Himalaya. By examining sediment cores from the lake, the authors intricately weave together the story of how these microscopic algae have responded to global warming and other climatic alterations over nearly two centuries. The chapter highlights the vulnerability of such freshwater systems to global phenomena, giving readers a nuanced understanding of regional responses to global challenges. Moving from the tranquil waters of a lake to the expansive territories of the *Beas Basin* in **Chapter 13**, the narrative shifts focus to the significance of snow cover in shaping hydrological dynamics. Here, the innovative use of remote sensing to monitor snow cover is brought to the forefront. The authors elucidate the intricacies of snow cover health and its ties to atmospheric river flow, seasonal variability, and overall watershed health. This chapter emphasizes the profound influence of snowmelt on riverine systems and the communities that depend on them. **Chapter 14** delves deep beneath the surface to explore the hidden world of groundwater in the Godavari Sub-Basin. Groundwater, the silent provider for countless communities, is facing increasing stress due to overexploitation. This chapter employs sophisticated GIS techniques and an analytical hierarchy process to map and evaluate groundwater potential zones. This critical information is indispensable for sustainable management, ensuring that this life-sustaining resource is available for generations to come.

Chapter 15 presents an exhaustive study of the LULC changes in Dhanbad district of Jharkhand, India, over a period of five decades. The district, renowned for its Jharia Coalfield, has witnessed drastic LULC alterations chiefly due to coal mining activities, leading to changes not only in the physical landscape but also impacting air and water quality. Leveraging advanced geospatial techniques and satellite imagery, the authors meticulously chart the evolution of this district, shedding light on the anthropogenic forces at play. Moving to the state of Uttar Pradesh in **Chapter 16**, we find a comprehensive assessment of land susceptibility to degradation in Siddharthnagar district. Recognizing the inherent relationship between the quality of land and agricultural productivity, this chapter probes the reasons and degrees of this susceptibility based on various indicators. The findings of this chapter are crucial for policymakers as they offer insights into the areas requiring immediate attention, ensuring the sustainability and preservation of the region's land resources. **Chapter 17** shifts focus to the urban milieu, emphasizing the critical role played by small and medium cities in the grand tapestry of urbanization. Concentrating on Mirzapur city in Uttar Pradesh, the research paints a vivid picture of the city's urban restructuring and development patterns over recent decades. By exploring facets like urban housing, transport, civic infrastructure, and more, this chapter offers a holistic view of the challenges and opportunities presented by urban development in medium-sized Indian cities. Finally, **Chapter 18** delves into cultural tourism in Kushinagar, Uttar Pradesh, India, conducting a SWOT analysis to assess its strengths, weaknesses, opportunities, and threats. It reveals tourism trends showing growth until 2018–19, a decline due to the COVID-19 pandemic, and a recovery phase in 2021–22. The chapter offers recommendations like developing theme-based tourism circuits and improving infrastructure to enhance Kushinagar's tourism sector. These strategies aim to enrich visitors' cultural experiences, promote intercultural dialogue, and boost the local economy.

1

Hydro-Chemical Characterization and Geospatial Analysis of Groundwater for Drinking and Agriculture Usage in Bhandara District, Central India

Nanabhau Kudnar[1*], Nand Lal Kushwaha[2], Madiga Rajashekhar[3], Varun Narayan Mishra[4], Rongali Mahesh[2] and Malkhan Singh Jatav[2]

[1]*C. J. Patel College Tirora, Gondia, Maharashtra, India*
[2]*Division of Agricultural Engineering, ICAR-Indian Agricultural Research Institute, New Delhi, India*
[3]*UGC SRF- Research Scholar, Dept. of Geology, Yogi Vemana University, Kadapa, Andhra Pradesh, India*
[4]*Amity Institute of Geoinformatics & Remote Sensing, Noida, India*

Abstract

In the present study, the effect of groundwater on agriculture and drinking water in the Bhandara district and the correlation between rainfall and groundwater were studied. Parametric and non-parametric tests were used in the study of rainfall to show the rainfall distribution trend. Rainfall data were collected for 43 years from 1971 to 2013. The rainfall trend here is negative at the rainfall stations Tumsar, Lakhandur, Bhandara, Pavani, and Mohadi, whereas the positive trend is at the rainfall station Sakoli 0.84 mm/year increase. While studying the groundwater, the water quality index was calculated using several parameters. It was found that 62.5% of the water quality was good, while the remaining 37.5% water was not potable and suitable for agriculture. While studying whether groundwater is potable, a total of 16 parameters were considered, out of which two samples were found to have excessive hardness and ten samples were found to have low nitrate content. Population growth, urbanization, industrialization, and agricultural activities have resulted in significant pressure on groundwater resources. The degradation of groundwater quality due to natural and anthropogenic activities

**Corresponding author*: nanabhaukudnar@gmail.com

Shruti Kanga, Gowhar Meraj, Suraj Kumar Singh, Majid Farooq and M. S. Nathawat (eds.) Geospatial Technology for Natural Resource Management, (1–24) © 2024 Scrivener Publishing LLC

is a matter of increasing concern. This study can be used by many researchers for regional planning.

Keywords: Hydro-chemical characterization, geospatial analysis, groundwater, Bhandara district

1.1 Introduction

The origin of all cultures worldwide is linked to the presence of water. The stability that came to the nomadic man was due to water. Coming to the banks of the river, as most of his needs had to be met, he settled down on the banks of the river, and from there, his true development began. He and his animals had uninterrupted access to water, and his desire to settle in the river grew stronger (Singh *et al.*, 2023; Rani *et al.*, 2023; Ahmad *et al.*, 2023). After starting farming, he realized the importance of water. As the use of water for personal hygiene, animals, agriculture, and communication began to increase, the water culture developed and stabilized. He bestowed divinity on those things that gave him stability (Sharma *et al.*, 2023).

A different question has become important before that of India and the world. Because of the increasing population, the per capita availability of water is decreasing, and because of increasing water pollution, even the water that is available is not suitable for use. Surface rivers and lakes become sewage pools (Choudhury *et al.*, 2023; Debnath *et al.*, 2023). There is already less fresh water on the earth; if it is available in such a large amount in polluted form, then the question arises as to how humans can survive. This question does not end. Due to the pollution of surface water, this polluted water has started seeping into the ground, and because of this, the groundwater has also started to become polluted (Zolekar *et al.*, 2021). The visible effects of this pollution are beginning to be observed. There are various types of epidemics on a daily basis. Epidemics such as cholera, jaundice, malaria, typhoid, dengue, various types of influenza, gastrointestinal disease, and diarrhea are spreading on a large scale (Kushwaha, 2022). Domestic waste and effluents from industrial units that contain phosphorus and nitrogenous fertilizers are discharged into rivers. Due to the presence of elements such as phosphorus and nitrogen in water, its fertility increases, but such water is not suitable for drinking.

When absorbed by fishes, industrial chemicals and pesticides have a very negative effect on human health because humans eat these fishes and humans are affected by eating these fishes. When the use of pesticides is high, the water inside the ground becomes polluted, and this pollution reaches the drinking water very easily (Kumar *et al.*, 2023; Singh *et al.*, 2023). In ancient times, most civilizations have emerged on the banks of rivers, which indicates that water has not only been essential, but also an important resource to meet all the needs of life. In the past several decades, rapid urbanization, a continuous increase in population, an increase in demand for drinking water supply and irrigation, and the expansion of industrial activities have increased the pressure on water resources. On one hand, due to the uncontrolled exploitation of surface and underground water to supply the increasing demand for water, the groundwater level is declining. However, the quality and usefulness of water is decreasing due to the increasing number of pollutants. Disasters, such as erratic rainfall, drought, and floods have greatly affected underground water recharge. Today, in the blind race for development, the expansion of industrial activities and rapid urbanization have polluted the country's major rivers, especially Ganga, Yamuna, Godavari, Kaveri, Narmada, and Krishna. One of the main reasons for water quality deterioration is increasing water pollution (Rajasekhar *et al.*, 2020; Kudnar, 2022).

1.2 Study Area

Bhandara district latitudes range from 20°38' to 21°36' north, while its longitudes range from 79°27' to 80°06' east (Figure 1.1). Bhandara is a district in Maharashtra, India. The population of Bhandaria is 11,98,810. The district covers an area of 3,716 km^2. This district is famous for its tala and fragrant tandulachya castes. Yala is called the Talavacha district. There are 3,648 Lahan Lahan Talis in Bhandariat. This district is rich in forests and minerals. Bhandara is the city or district headquarters. Iron ore is abundant in the district, and the main crops in this area are rice, wheat, and millet (a cereal grain). The Tandalachya Chinnor, Dubhraj, Kali Kamod, and Suvasik castes are famous. Bhandara is famous for Maharashtra's largest rice-growing district. Wainganga (Kudnar 2020 a, b; Kudnar and Rajshekhar, 2020; Kudnar *et al.*, 2022) is the main river in the district. Alternatively, the flow of the river in the district does not corrode even in summer. In districts Bawanthadi, Chulband, Kanhan, Bagh, Gose Dharan, Kavalevda Dharan, there are only Dharnas.

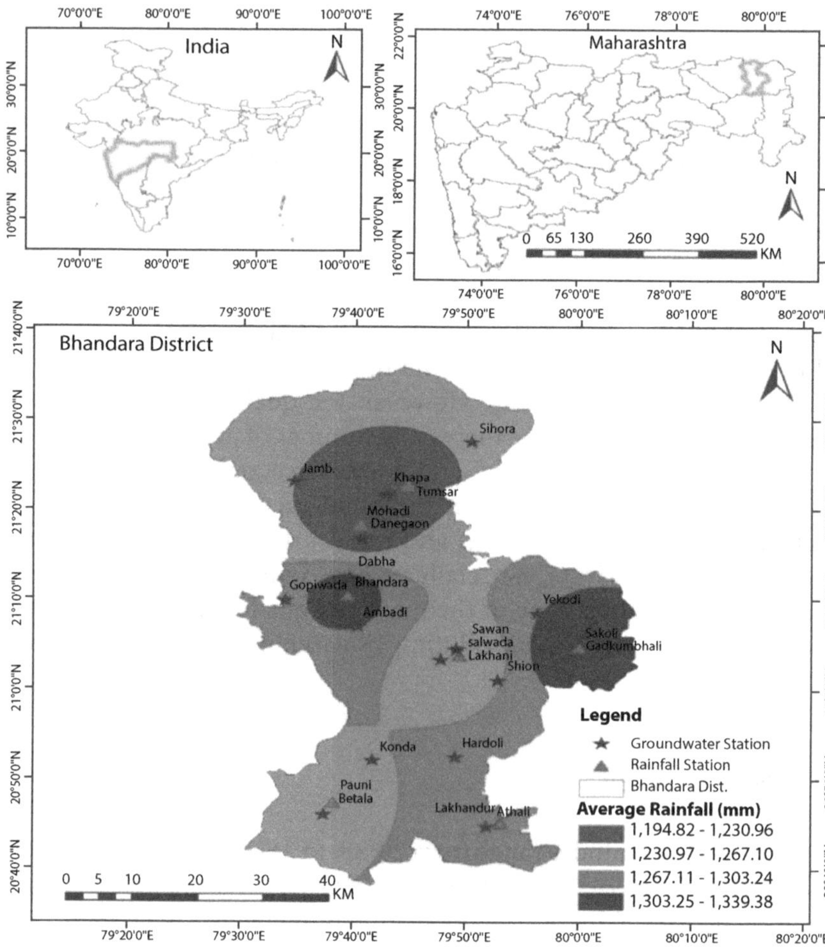

Figure 1.1 Study area.

1.3 Methodology

Rainwater remains pure in the beginning, but with the flow of land, organic and inorganic minerals dissolve. In this way, the minerals and other nutrients present in natural water are not only essential but also important from the point of view of health for all living beings, including humans. From the perspective of quality, drinking water should have the following characteristics.

Physical quality: Water should be completely colorless, odorless, tasteful, and cool.
Chemical quality: Water soluble oxygen, pH value, and mineral content should be within acceptable limits.
Biological Quality: Should be completely free of water-borne pathogenic impurities.

Water is polluted by mixing harmful substances such as microorganisms, chemicals, and contaminated water generated from industrial, domestic, or commercial establishments. This is known as water pollution.

The physical, chemical, and biological properties of water are affected by the presence of harmful substances. Due to the harmful side effects of pollutants on water quality, polluted water is unsuitable for domestic, commercial, industrial agriculture, or any other general use.

In addition to drinking, there is heavy consumption of water in domestic works, irrigation, agricultural work, use of cattle, industrial and commercial activities, etc., and this water, which is used, becomes contaminated water after use. In this contaminated water, traces of substances or chemicals that come in contact with the water during commercial activities remain in the form of residues. Their presence renders water unsuitable for use. When contaminated water is found in a clean water source, it can also contaminate it. Contaminated water contains viruses, bacteria and other harmful microorganisms, along with organic and inorganic compounds and chemicals that pollute water sources according to their nature.

1.3.1 Data Used

While studying the Bhandara district, the data of seven rainfall stations in seven tehsils from 1971 to 2013 were taken from IMD, Pune, and worked. It is divided into pre-monsoon, monsoon, post-monsoon, and winter monsoon periods (Figures 1.2 and 1.3).

1.3.2 Descriptive Statistics Analysis

$$\mathrm{Sd} = \sqrt{\frac{\sum_{i=1}^{n}(xi - x)2}{n-1}} \tag{1.1}$$

$$\mathrm{Sk} = \frac{n}{(n-1)(n-2)} \sum \frac{(xj - x)3}{s} \tag{1.2}$$

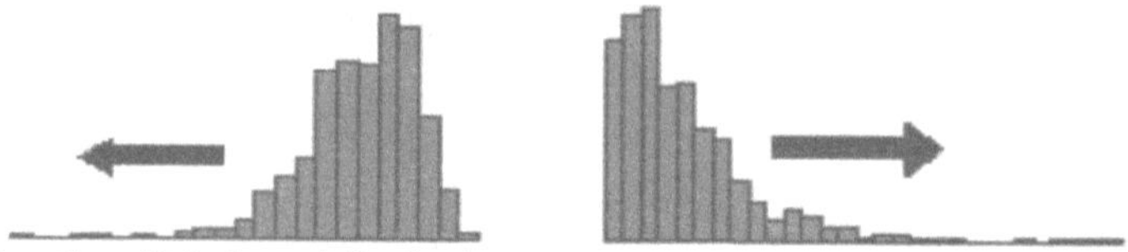

Figure 1.2 Two types of distribution patters with different skewness.

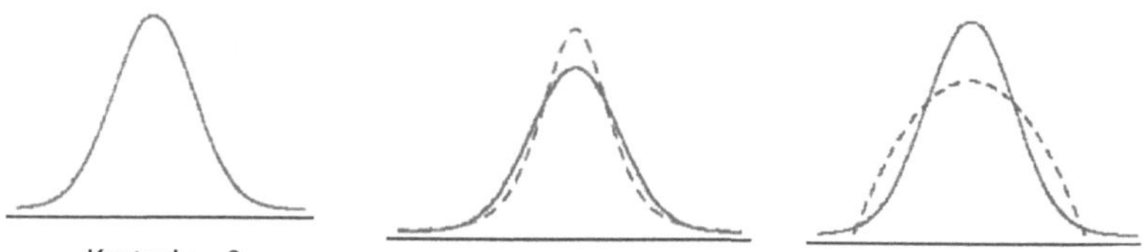

Figure 1.3 Three types of distribution patterns with different kurtosis values.

Kurtosis value is calculated by the following formula (Figure 1.3):

$$\mathrm{Kr} = \{\frac{n(n+1)}{(n-1)(n-2)(n-3)} \Sigma \frac{(xj-x)4}{s}\} - \frac{3(n-1)2}{(n-2)(n-3)} \tag{1.3}$$

1.3.3 Mann–Kendall Test (Non-Parametric Test)

The non-parametric test (Mann 1945; Kendall 1975; Fathian *et al.* 2014; Hamed and Rao 1998) was used in the study area, as shown in the following equations:

$$S = \sum_{i=1}^{n-1} \sum_{j=i+1}^{n} sgn(X_j - X_i) \tag{1.4}$$

$$Sgn(Y) = \left\{ \begin{Bmatrix} +1 & if & \mathrm{Y} > 0 \\ 0 & if & \mathrm{Y} = 0 \\ -1 & if & \mathrm{Y} < 0 \end{Bmatrix} \right\} \tag{1.5}$$

$$Var(S) = \frac{[n(n-1)(2n+1) - \Sigma_t t(t-1)(2t+5)]}{18} \tag{1.6}$$

$$Z_{mk} = \begin{cases} \frac{S-1}{\sqrt{V(S)}} if \quad S > 0 \\ 0 \quad if \quad S = 0 \\ \frac{S+1}{\sqrt{V(S)}} if \quad S < 0 \end{cases} \tag{1.7}$$

1.3.4 Regression Model (Parametric Test)

The regression model applies a geographical rainfall pattern (Rupa *et al.* 1992; Shakeel *et al.* 2020).

$$y = mx + c \tag{1.8}$$

1.3.5 Spearman's Rank Correlation Coefficient

Spearman's rank correlation coefficient was applied to two stations (Misra and Tripathi 2021).

$$\rho = 1 - 6\sum_{n-1}^{n} d^2 / n(n^2 - 1) \tag{1.9}$$

1.3.6 Inverse Distance Weighing

Define IDW as,

$$z = \frac{\sum_{i=1}^{n} w_i z_i}{\sum_{i=1}^{n} w_i}$$

$$w_i = \frac{1}{d_i^p} \tag{1.10}$$

1.3.7 Groundwater Quality

A total of 16 groundwater wells were studied in this study area and these data were collected during the pre-monsoon period in 2011 show in Figure 1.4.

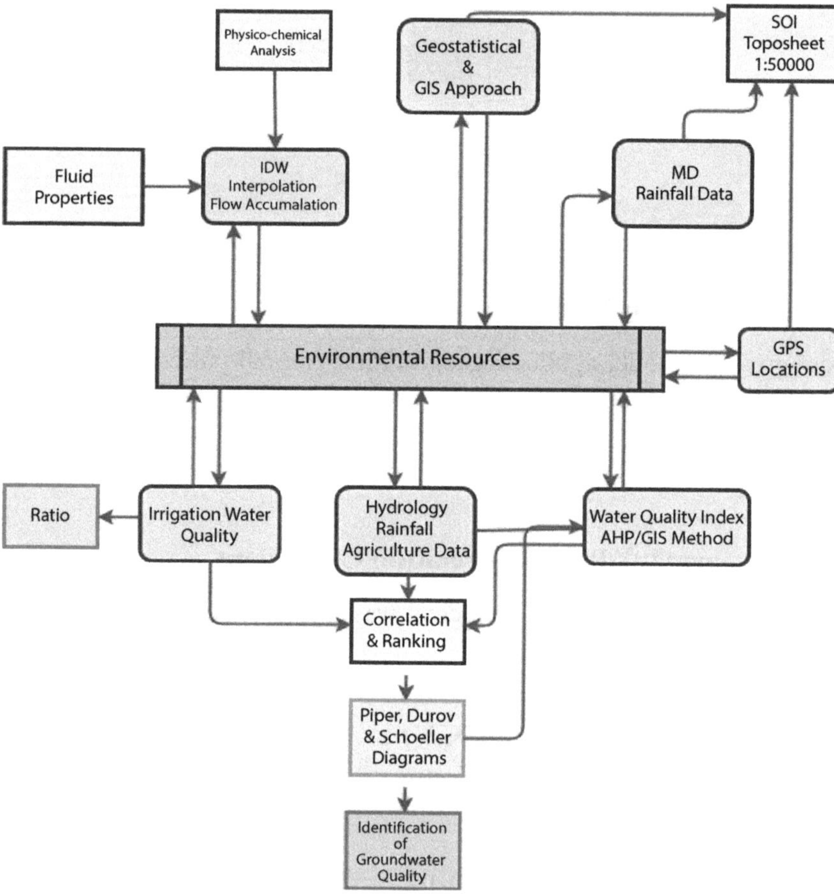

Figure 1.4 Flow chart of methodology.

1.3.8 Piper Diagram

This diagram explains the chemical properties of the water samples using a graphical method. It shows the information of sulfate, chloride, carbonate, hydrogen carbonate by creating different plots of calcium, magnesium, sodium, potassium, and all this information through a matrix (Figure 1.5) used by Hilly (1940) and modified by Piper (1944).

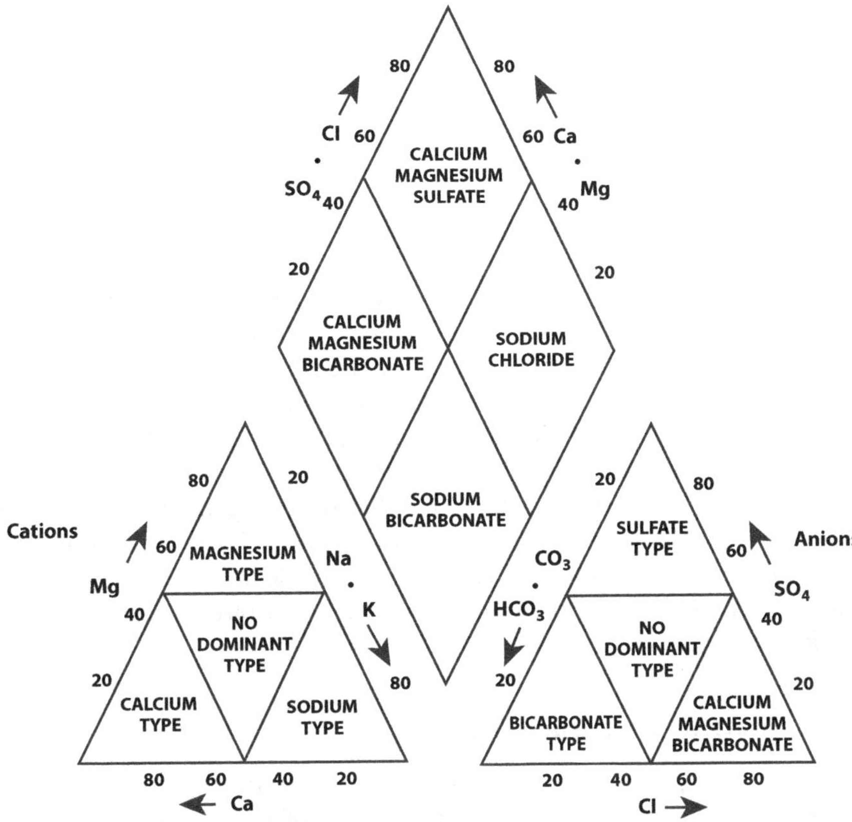

Figure 1.5 Piper diagram.

1.3.9 Groundwater for Drinking and Irrigation Purposes

Whether groundwater is potable is defined by its properties or water quality and has been studied by many scientists. While irrigating a farm, whether the water is suitable for agriculture or not is clear through the parameter of electrical conductivity in that water, and the quality of the water is determined from it, and this quality is taken.

1.4 Results and Discussion

The Bhandara district is located in the eastern part of the state, or the district is situated in the northern Madhya Pradesh state, eastern Gondia district, southern Chandrapur district, and western Nagpur district. Wainganga is the biggest river in Bhandara district, or in the district Wagh, Chulband, Bawanthadi, Ambagad, Bodalkasa, Gadhvi, Sur, Maru, etc. rivers flow. Bhandara District is also known as Talavancha District. For 400 years, the people of Kohli Jamati have been using the traditional method of collecting water, draining, or building ponds, which is why this district is known as the district of ponds. Rice is the main crop of Bhandara district, or in the district, Oats, Tur, Soyabean, Jowari, Moong, Wheat, Harbhara, Barley are ripened in the district. In some places, only flying drinks are consumed. Bhandara District is a rich district in the state in terms of mineral wealth. In the northern part of the district, manganese was found in abundance. Pandhari soil, Mubalak, was found in Tumsar and Kavadsi. Apart from this, only other minerals were available in the district to some extent. There are mines containing manganese, chromite, quartzite, sillimanite, and kyanite.

1.4.1 Statistical Characteristics

In winter, high-pressure belt created in morthwest India gradually moves to the south, since the sun is near the Tropic of Cancer. In April, there was more pressure in southern India. Most of the territory is Sagaravarhi. Northwest India is a pressure region. But I am still weak. In the pressure area, more pressure originates from the region. In April and May, they flow from the region of Maharashtra and rain. That is why, or monsoon, Embassy, it is said like this. In general, the arrival of the monsoon is a complex process. Terrain in the summer and the body of water are the mechanisms of the monsoon. These are characteristics of the Indian subdivision. Normally, it rains in Maharashtra. Approximately 85% of these rains come from the southwest monsoon winds. In the eastern part of Maharashtra, evidence of more rain continues to increasing. This is because the Bhandara region is a branch of the monsoon winds over the Bay of Bengal, and receives rain between August and September. The study area receives 110–160 m of rainfall. Evidence of rain has continued to increase in the mountains of Gaimukha. It rains 160 cm and a half in the southern part or a cold haven. This month, there was strong sunlight. Since there is much evidence of vapors in the air, there is a feeling of nausea (Table 1.1).

Table 1.1 Rainfall statistics (1971-2013).

Tehsil	Maximum	Minimum	Mean/ Average	Std. deviation	Kurtosis	Skewness
Mohadi	1,845.1	636.5	1,194.82	247.89	0.20	0.37
Bhandara	2,057.5	787.7	1,327.5	306.35	0.21	0.7
Pauni	2,117.3	748.9	1,257.54	334.72	0.07	0.86
Lakhandur	2,174.9	659	1,247.63	339.68	0.16	0.68
Tumsar	2,208.7	433.1	1,223.5	323.66	1.12	0.51
Lakhani	1,994.4	630	1,239.68	291.14	0.58	0.42
Sakoli	2,041.3	836	1,339.39	287.03	0.10	0.52

1.4.2 Trend Analysis

Rainfall data from 1971 to 2013 are available. Rainfall has decreased in the north, east and northeast study areas, and rains have increased only in rain-deficient regions such as Mohadi. However, considering the entire area, there was no significant momsoon trend. He also mentioned that the monsoon is variable, and there is wide variation in it. Due to the increase in atmospheric instability caused by climate change, there has been an increase in incidents such as thunder and lightning accompanied by torrential rains. The intensity of cyclones in the Arabian Sea is increasing, and due to the increase in adversity, it is challenging to change the weather forecast. Climate change affects the ability of meteorological organizations to predict heavy rain. With the help of radar, satellites, automatic weather stations, and rain gauges, the monitoring of weather accounts has increased. Due to improvements in computing systems, monitoring, and modeling, in the last five years, the forecasting of extreme rainfall, heat and cold waves, and smoke has increased by 30–40%.

1.4.3 Linear Regression Analysis

The linear regression equation R^2 was predicted for each station. Bhandara ($R^2 = 0.009$) in Figure 1.6(a), Lakhandur ($R^2 = 1.914E{-}4$) in Figure 1.6(b), Mohadi ($R^2 = 0.007$) in Figure 1.6(d), Pauni ($R^2 = 0.001$) in Figure 1.6(e), and Tumsar ($R^2 = 0.001$)in Figure 1.6(g), increasing trend in Lakhani (R^2= 7.225E–4) in Figure 1.6(b) and Sakoli ($R^2 = 0.001$) in Figure 1.6(f) along with their R^2 values presented in above.

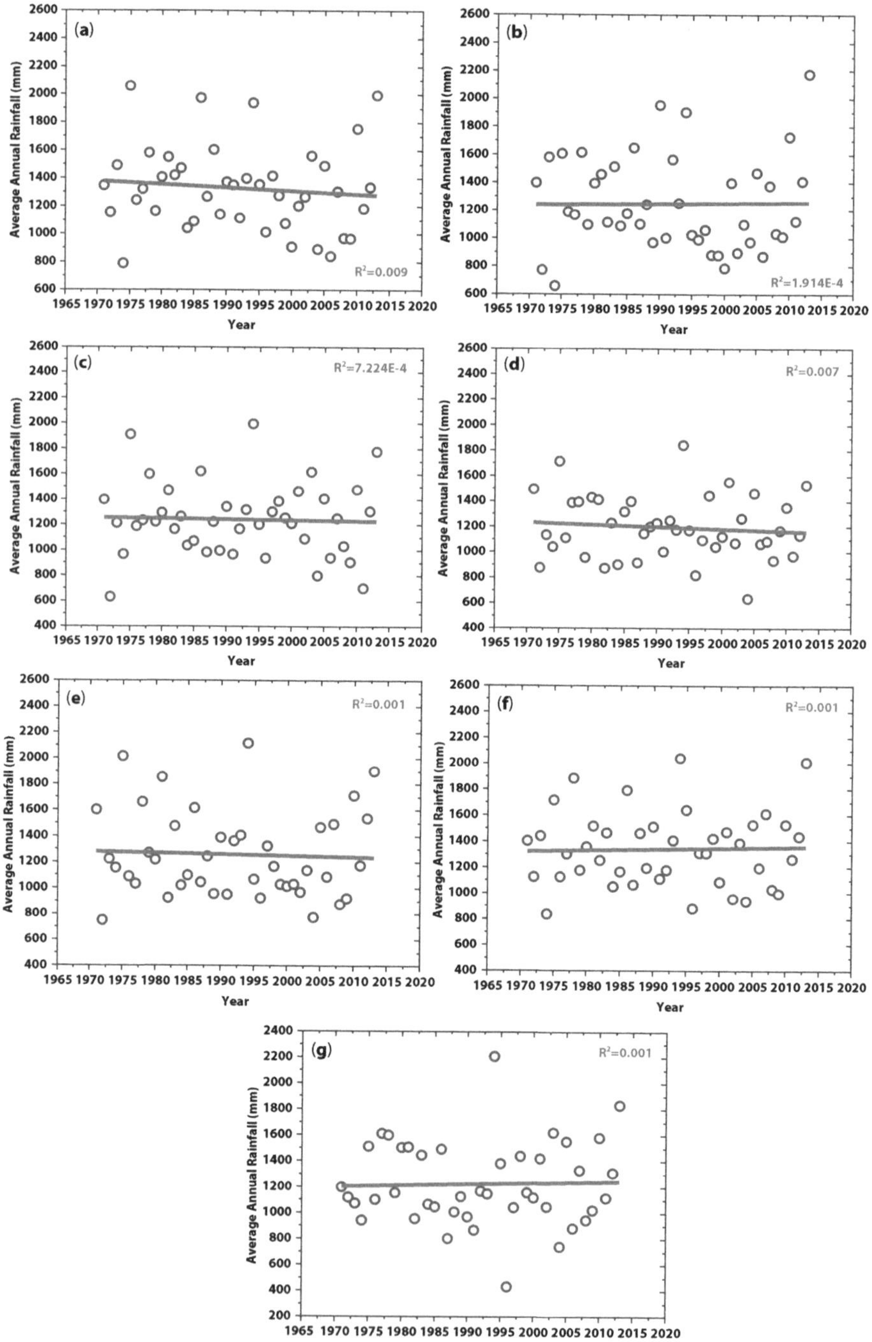

Figure 1.6 Linear regression analysis (a) Bhandara, (b) Lakhandur, (c) Lakhani, (d) Mohadi, (e) Pauni, (f) Sakoli, and (g) Tumsar stations over the period of 2071-2013.

1.4.4 Spatial–Temporal Rainfall Distribution

When the water vapor content increases and reaches a higher value, it suddenly becomes a cold region. This results in a reduction in evaporation, and the water droplet tends to form thanks to hygroscopic condensation nuclei. These may be dust or suspended particles found in the environment of the condensation center. When they harden, they lose weight, because they stop the action of gravity. Convection showers: These are the types of rain in which warm air comes in contact with the surface of the earth and is heated by the action of the sun. Once upon a time, drops of water fall, and because of the lack of drops of water, it rains. When a mountain rests in dust, it is are ready. The wind is used to move up and down and immerse all the moisture on the other side of the mountain until it becomes completely dry. Due to the collision of the two dust particles at different temperatures, they are formed. Usually, there is one quality and the other is cold. These rain events are generally windy or cyclonic. They are created by the intensity of the temperature along the Earth's surface. These usually occur at the end of summer and are accompanied by severe storms that cause severe damage, especially in agriculture (Figure 1.7).

Lack of water for drinking, lack of shelter for animals and fodder, lack of employment in rural areas, and due to this, movement of villagers to nearby cities or annual drought are characteristic features. Drought was placed in the years 1972, 1974, 1979, 1984, 1996, 2002, and 2009, but there would have been a big drought. If I compare both negative aspects, it would have been tolerable to consider 1972 as a drought or drought. Therefore, the condition of drinking water during drought would have been acceptable today (Figure 1.8) with less than 50% of the average rainfall. However, the rainfall distribution was different in the different parts of the study area. Similarly, climate change would have played an important role in covering the whole of Maharashtra on a large scale. Which includes the completed parts? The local meteorological conditions indicated the presence of a low-pressure system in the Bay of Bengal and westerly winds of the monsoon blowing from the Arabian Sea at Nangarun. In addition, the excessive amount of moisture brought in from the western warm Arabian Sea caused heavy rain in 1975, 1990, 1994, 2003, and 2013 (Figure 1.9).

1.4.5 Effects of Water Pollution

By drinking polluted water, cholera, dysentery, decay, and abdominal diseases can arise in humans. Consuming polluted water causes hormonal changes in the body of pregnant women, which can cause disorders or

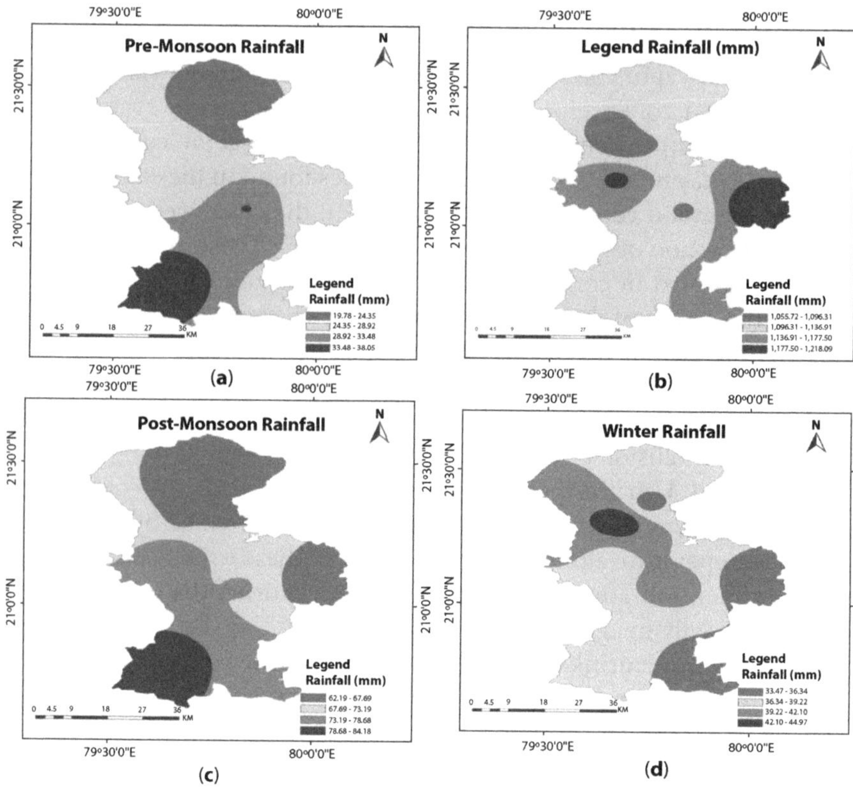

Figure 1.7 Rainfall distribution: (a) pre-monsoon rainfall, (b) monsoon rainfall, (c) post-monsoon rainfall, and (d) winter rainfall.

death of the unborn child. It is clear from some studies that one of the major causes of skin cancer is the consumption of polluted water. The consumption of polluted water is also a major cause of disability caused by polio.

1.4.5.1 Solution Steps

The Water Pollution Prevention and Control Act was enacted in 1974 to control water pollution in the country and to maintain its quality. There are various sources of water pollution. In such situations, for effective control, it is necessary to stop the sources generated and purify them through proper management and purification treatment. The following measures can be used to solve this problem: water is a valuable resource, keeping in mind its importance, and all the residents of the country should make a voluntary contribution in protecting it from being polluted and protected.

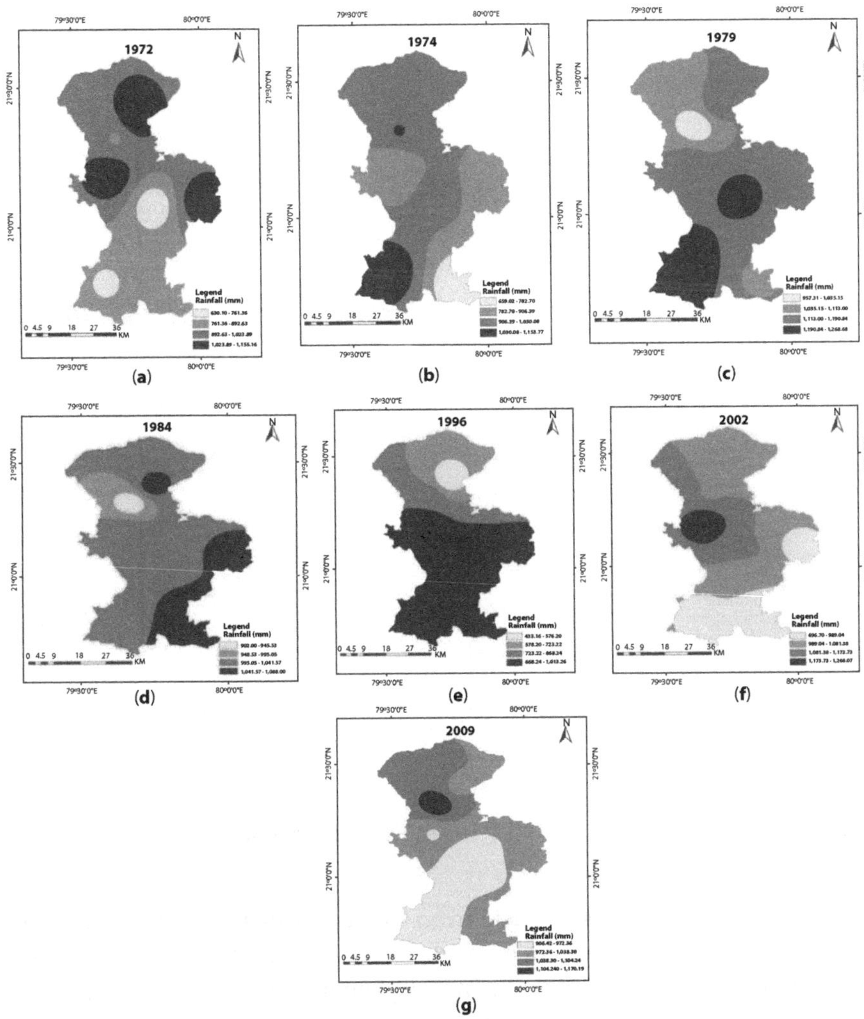

Figure 1.8 Drought rainfall (a) 1972, (b) 1974, (c) 1979, (d) 1984, (e) 1996, (f) 2002, and (g) 2009.

Recycling and reusing water is one of the best ways to control pollution; it should be included in your habits as far as possible. Industrial groups should take joint steps in setting up common effluent treatment plants in areas where there are clusters of industrial units so that the quality of water is not affected. New alternatives should be sought for the production of other useful materials by recycling wastes such as ferrous and non-ferrous

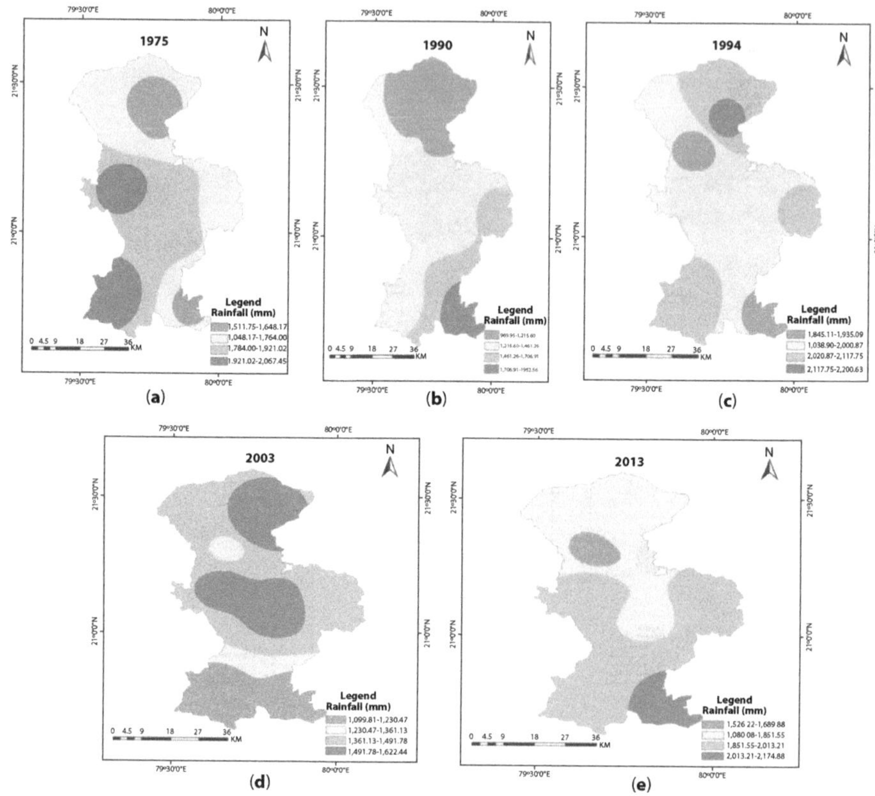

Figure 1.9 Heavy rainfall (a) 1975, (b) 1990, (c) 1994, (d) 2003, and (e) 2013.

materials, paper, and plastic waste from treatment plants so that the number of pollutants can be reduced at the source site. Through soil testing, pollution can be reduced by reducing seepage in the reservoir using fertilizers in appropriate quantities according to the requirements of the crops. Farmers should use nutrient management and commercial fertilizers in appropriate quantities. In the future, a comprehensive plan should be prepared and implemented in a time-bound manner to control of pollution caused by industrial effluents. Polluting industrial units discharging untreated industrial effluents should cancel their licenses by restricting the water supply. The use of eco-friendly household products and toiletries has few harmful effects on the environment; their use can play a helpful and active role in preventing pollution.

1.4.5.2 *Chemical Constituents of Groundwater*

Sodium (Na): Sodium is widely found in soil, plants, water, and food. Sodium chloride is present in most sodium-containing minerals. Sodium is usually naturally found in groundwater. The sodium contained in groundwater does not smell, but if its concentration is 200 Mg/liter or more, its taste is known. Its compounds are often found in most rocks and soils, where sodium dissolves easily. Some of the common sources of excess sodium in groundwater are the erosion of sodium-bearing rocks and minerals, and brackish water is naturally found naturally in some aquifers. Salt-contaminated surface water seepage and groundwater pollution by sewage effluents. Leachate seepage from landfills or industrial sites

Potassium (K): It is commonly found in soil and rocks. Potassium has no odor or color in water. However, water may taste salty. Some of the main sources of potassium are weathering and erosion of potassium-based minerals (such as feldspar) from fertilizer spills, and from seawater in areas vulnerable to saltwater encroachment.

Calcium (Ca): It enters fresh water through the weathering of rocks, especially limestone (marble, calcite, dolomite, gypsum, fluoride, and apatite), soil seepage, and leaching. The amount of calcium in surface water was less than that in groundwater. Calcium is a factor affecting water hardness because it is found in water as a (Ca++) ion.

Magnesium (Mg): As a result, the taste of drinking water is undesirable. If its concentration is 100 mg per liter (Mg/liter) or higher, the drinking water may taste unpleasant to sensitive people. The taste is unpleasant to the average person at concentrations above 500 mg per liter (Mg/L). Excess magnesium in drinking water can have a laxative effect, especially when the concentration of magnesium sulfate is greater than 700 Mg/L.

Carbonate and Bicarbonate (Co3 & HCo3): Carbon dioxide dissolved in rainwater and snow is the major source of carbonate and bicarbonate ions found in groundwater. Carbon dioxide (Co2) is released during the decay of organic matter. The pH (Table 1.2) of the water itself indicates the form of carbon dioxide present. Generally, the concentration of bicarbonate in groundwater ranges from 700 ppm to 800 ppm. The analyzed parameters, included pH, Electrical Conductivity (EC), TDS, Total Alkalinity (TA), Total Hardness (TH), Na+, DO,Nitrate (NO3), and fluoride (F).

The type of groundwater present in the study area was also analyzed using Piper chart. It is observed from the Piper plot (Figure 1.10). This diagram explains the chemical properties of the water sample through a

Table 1.2 Decision matrix of the selected water quality parameter.

Parameter	pH	EC	TDS	TA	TH	Na+	Do	K+	COD	Ca2+	Mg+2	HCO3	Cl–	SO2– 4	NO–; 3	F-
pH	1															
EC	0.5	1														
TDS	0.33	0.12	1													
TA	0.25	0.11	0.14	1												
TH	0.2	1	0.12	0.5	1											
Na+	0.17	0.5	0.11	0.33	0.2	1										
Do	0.14	0.25	1	0.25	0.17	0.14	1									
K+	0.12	0.2	0.5	0.2	0.14	0.12	0.12	1								
COD	0.11	0.14	0.33	0.17	0.12	0.11	0.11	0.12	1							
Ca2+	1	0.12	0.25	0.14	0.11	1	1	0.11	0.14	1						
Mg+2	0.5	1	0.2	0.12	1	0.5	0.5	1	0.12	0.2	1					
HCO3	0.33	0.5	0.17	0.11	0.5	0.33	0.33	0.5	0.11	0.17	0.5	1				
Cl–	0.25	0.33	0.14	1	0.33	0.25	0.25	0.33	1	0.14	0.33	0.14	1			
S 2– 4	0.2	0.25	0.12	0.5	0.25	0.2	0.2	0.25	0.5	0.12	0.25	0.12	0.5	1		
NO–; 3	0.17	0.2	0.11	0.33	0.2	0.17	0.17	0.2	0.33	0.11	0.2	0.11	0.33	0.2	1	
F–	0.14	0.17	1	0.25	0.17	0.14	0.14	0.17	0.25	1	0.17	1	0.25	0.17	0.14	1

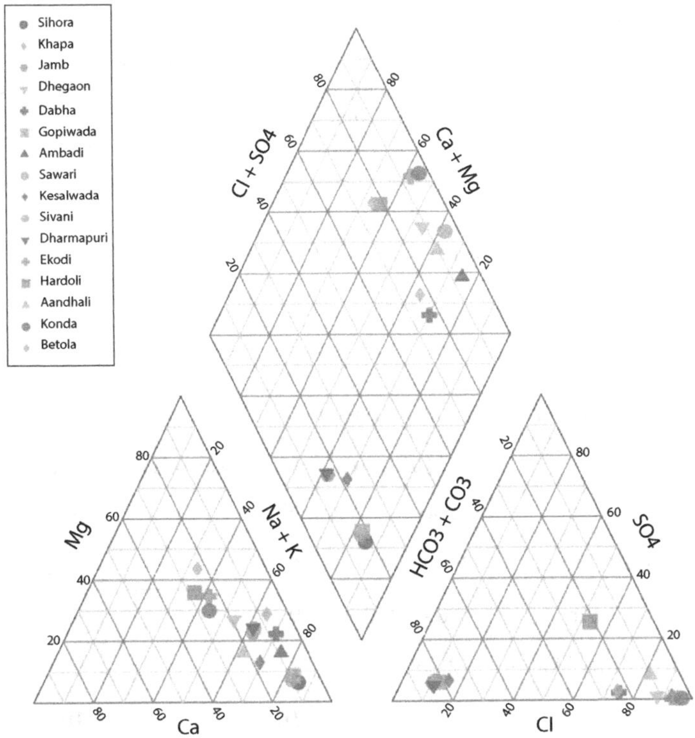

Figure 1.10 Piper graph.

graphical method. It shows information on sulfate, chloride, carbonate, and hydrogen carbonate by creating different plots of calcium, magnesium, sodium, potassium, and all this information through a matrix.

1.4.6 Drinking and Irrigation Purposes

While studying whether groundwater is potable, a total of 16 parameters were considered, out of which two samples were found to have excessive hardness and 10 samples were found to have low nitrate content, as shown in Table 1.3.

While determining whether the water is suitable for irrigation of the agricultural crop based on the parameter of electrical conductivity in the water, it was noticed that a total of 16 samples were taken and one place with the highest salinity was found, followed by nine samples with high salinity concentrations, six with moderate salinity, and none with the lowest salinity (Table 1.4). From this, it can be seen that the water here

Table 1.3 Groundwater samples for drinking based.

Parameters	DL	MPL	Samples with conc. <DL	Samples with conc. in DL-MPL	Samples with conc. >MPL
TA(mg/L)	200	600	15	4	–
TH(mg/L)	300	600	8	9	2
NO_3(mg/L)	45	No relaxation	8	–	10
F(mg/L)	1.0	1.5	18	–	–

Table 1.4 Groundwater for irrigation based on EC.

Type	EC (μS/cm)	No. of samples	% of samples
Low Salinity Water	<250	Nil	Nil
Medium SalinityWater	250–750	6	37.5
High SalinityWater	750–2,250	9	56.25
Very High Salinity Water	>2,250	1	6.25
Total		**16**	**100.0**

is suitable for agriculture, and 6.25% of the total parameter was found to be very high salinity water, followed by high salinity water at 56.25%, and medium salinity water at 37.5%.

1.5 Conclusion

The study in Bhandara District explains how the pattern of rainfall changed from 1971 to 2013 and the trend of rainfall. In this study, seven rainfall stations were studied, and the distribution of rainfall was found to be negative at these stations; only in the rainfall station Sakoli, the rainfall trend was positive. In this study, 16 groundwater samples were studied to determine whether groundwater is suitable for agriculture and potable. Based on the total sample, 62.5% of the water was potable, whereas the remaining water was contaminated by 37.5%. The main reason for this is that the chemical fertilizers and pesticides used in these fields affect the water quality. The Piper diagram explains the chemical properties of the water samples using a graphical method. It shows the information of sulfate, chloride, carbonate, hydrogen carbonate by creating different plots of calcium, magnesium, sodium, potassium, and all this information through a matrix. This study can be used by many researchers for regional planning.

References

Singh SK, Kanga S, Gulati B, Raič M, Sajan B, Đurin B, Singh S (2023). Spatial and Temporal Analysis of Hydrological Modelling in the Beas Basin Using SWAT+ Model. *Water* 15, 3338. https://doi.org/10.3390/w1519333

Rani A, Gupta SK, Singh SK, Meraj G, Kumar P, Kanga S, Đurin B, Dogančić D (2023). Predicting Future Land Use Utilizing Economic and Land Surface Parameters with ANN and Markov Chain Models. *Earth* 4(3):728-751.

Ahmad T, Gupta SK, Singh SK, Meraj G, Kumar P, Kanga S (2023). Unveiling Nature's Resilience: Exploring Vegetation Dynamics during the COVID-19 Era in Jharkhand, India, with the Google Earth Engine. *Climate* 11(9):187.

Sharma M, Upadhyay RK, Tripathi G, Kishore N, Shakya A, Meraj G, Kanga S, Singh SK, Kumar P, Johnson BA, Thakur SN (2023). Assessing Landslide Susceptibility along India's National Highway 58: A Comprehensive Approach Integrating Remote Sensing, GIS, and Logistic Regression Analysis. *Conservation* 3(3):444-459.

Choudhury U, Singh SK, Kumar A, Meraj G, Kumar P, Kanga S (2023) Assessing land use/land cover changes and urban heat island intensification: A case study of Kamrup Metropolitan District, Northeast India (2000–2032). *Earth* 4(3):503-521.

Rai NK, Mathur S, Singh SK, Tiwari M, Singh VK, Haque R, Tiwari S, Sharma LK (2023) Differential regulation of mitochondrial complex I and oxidative stress based on metastatic potential of colorectal cancer cells. *Oncol. Lett.* 26(1):1-1.

Debnath J, Sahariah D, Lahon D, Nath N, Chand K, Meraj G, Kumar P, Singh SK, Kanga S, Farooq M (2023) Assessing the impacts of current and future changes of the planforms of river Brahmaputra on its land use-land cover. *Geosci. Front.* 14(4):101557. Elsevier.

Kudnar NS, Rajasekhar M (2020) A study of the morphometric analysis and cycle of erosion in Waingangā Basin, India. Model. Earth Syst. Environ. 6: 311–327. https://doi.org/10.1007/s40808-019-00680-1.

Kudnar NS, Diwate P, Mishra VN et al. (2022) Spatio-temporal variability and trend analysis of rainfall in Wainganga river basin, Central India, and forecasting using state-space models. Theor. Appl. Climatol. 150: 469–488. https://doi.org/10.1007/s00704-022-04168-4.

Kudnar NS (2022) Geospatial modeling in the assessment of environmental resources for sustainable water resource management in a semiarid region:

A GIS approach, in: Current Directions in Water Scarcity Research, 7, pp. 133–151. https://doi.org/10.1016/B978-0-323-91910-4.00009-1.

Kumar P, Singh SK, Gupta SK, Kanga S, Meraj G (2023) Uncovering the hydro-meteorological drivers responsible for forest fires utilizing geospatial techniques. *Theor. Appl. Climatol.* https://doi.org/10.1007/s00704-023-04497. Springer Nature.

Kushwaha, N., Elbeltagi, A., Mehan, S. et al., Comparative study on morphometric analysis and RUSLE-based approaches for micro-watershed prioritization using remote sensing and GIS. Arab J. Geosci., 15, 564, 2022, https://doi.org/10.1007/s12517-022-09837-2.

Mann HB (1945) Nonparametric tests against trend. Econometrica, 13(3): 245–259, 1945. https://doi.org/10.2307/1907187.

Misra AK, Tripathi A (2021) Mathematical Models of how the Transpiration Affects Rainfall through Agriculture Crops. *Environ. Model. Assess.* 26, 837–848. https://doi.org/10.1007/s10666-021-09769-3

Mishra VN, Rai PK (2016) A remote sensing aided multi-layer perceptron-Markov chain analysis for land use and land cover change prediction in Patna district (Bihar), India. *Arab. J. Geosci.* 9, 249, https://doi.org/10.1007/s12517-015-2138-3

Mishra VN, Rai PK, Prasad R, Punia M, Nistor MM (2018) Prediction of spatiotemporal land use/land cover dynamics in rapidly developing Varanasi district of Uttar Pradesh, India using Geospatial approach: a comparison of hybrid models. *Appl. Geomat.* 10 (3), 257–276.

Nair SC, Mirajkar AB (2021)Spatio–temporal rainfall trend anomalies in Vidarbha region using historic and predicted data: a case study. Model. *Earth Syst. Environ.* 7, 503–510. https://doi.org/10.1007/s40808-020-00928-1

Rajasekhar M, Gadhiraju SR, Kadam A *et al.* (2020) Identification of groundwater recharge-based potential rainwater harvesting sites for sustainable development of a semiarid region of southern India using geospatial, AHP, and SCS-CN approach. *Arab. J. Geosci.*, pp 13-24. https://doi.org/10.1007/s12517-019-4996-6

Rajasekhar M, SudarsanaRaju et al (2021) Multi-criteria Land Suitability Analysis ForAgriculture in SemiArid Region of Kadapa District, Southern India: *Geospatial Approaches, Remote Sensing of Land,* 5(2), 59-72. https://doi.org/10.21523/gcj1.2021050201

Rajankar PN, Tambekar DH, Wate SR (2011) Groundwater quality and water quality index at Bhandara District. *Environ. Monit. Assess.* 179, 619–625. https://doi.org/10.1007/s10661-010-1767-y

Reddy GPO, Mouli K, Srivastav SK *et al.* (2000) Evaluation of Ground Water Potential Zones Using Remote Sensing Data -A Case Study of Gaimukh Watershed, Bhandara District, Maharashtra. *J. Indian Soc. Remote Sens.* 28, 19. https://doi.org/10.1007/BF02991858

Rupa Kumar K, Pant GB, Parthasarathy B, Sontakke NA (1992) Spatial and sub-seasonal patterns of the long term trends of Indian summer monsoon rainfall. *Int. J. Climatol.* 12(3):257–268. https://doi.org/10. 1002/joc.3370120303

Salunke VS, Bhagat RS *et al.* (2020) *Geography of Maharashtra*, Prashant Publication, Jalgaon, pp- 1-229.

Shakeel Ahmad, Muhammad Israr *et al.* (2020). Spatio-temporal trends in snow extent and their linkage to hydro-climatological and topographical factors in the Chitral River Basin (Hindukush, Pakistan), *Geocarto Int.*, 35:7, 711-734, DOI: 10.1080/10106049.2018.1524517

Shortridge J (2019). Observed trends in daily rainfall variability result in more severe climate change impacts to agriculture. *Clim. Change* 157, 429–444. https://doi.org/10.1007/s10584-019-02555-x

Sinha S, Chakraborty S, Shome D (2019). Mining footprint: a spatial indicator of environmental quality—a case study of a manganese mine in Bhandara district, Maharashtra. *Arab. J. Geosci.* 12, 96. https://doi.org/10.1007/s12517-019-4260-0

Singh SK, Kotecha MJ, Kanga S, Pankhaniya SK, Agrawal S, Meraj G (2023) Urban Heat Island (UHI) Resilience Plan in Varying Climatic Conditions Using Geospatial Approach: A Case Study of Rajkot City. In: *Climate Change, Agriculture and Society*, vol 1. pp 307-338.

Singh PK, Kumar S, Singh UC (2011). Groundwater resource evaluation in the Gwalior area, India, using satellite data: an integrated geomorphological and geophysical approach. *Hydrogeol. J.* 19, 1421–1429. https://doi.org/10.1007/s10040-011-0758-6

Sivakumar MVK, Das HP, Brunini O (2005). Impacts of Present and Future Climate Variability and Change on Agriculture and Forestry in the Arid and Semi-Arid Tropics. *Clim. Change* 70, 31–72. https://doi.org/10.1007/s10584-005-5937-9

Subash N, Sikka AK (2014) Trend analysis of rainfall and temperature and its relationship over India. *Theor. Appl. Climatol.* 117(3–4):449–462. https://doi.org/10.1007/s00704-013-1015-9

Thakur JK, Singh SK, Ekanthalu VS (2016). Integrating remote sensing, geographic information systems and global positioning system techniques with hydrological modeling. *Appl. Water. Sci.* 7(4):1595–1608

Yue S, Pilon P, Phinney B, Cavadias G (2002). The influence of autocor- relation on the ability to detect trend in hydrological series. *Hydrol. Process.* 16(9):1807–1829. https://doi.org/10.1002/hyp.1095

Zolekar RB, Todmal RS, Bhagat VS *et al.* (2021). Hydro-chemical characterization and geospatial analysis of groundwater for drinking and agricultural usage in Nashik district in Maharashtra, India. *Environ. Dev. Sustain.* 23, 4433–4452. https://doi.org/10.1007/s10668-020-00782-2

2

Technology-Driven Approaches to Enhance Disaster Response and Recovery

Chandni Kirpalani

Department of Chemistry & Environmental Science, Poornima University, Jaipur, India

Abstract

The increasing frequency and intensity of natural and man-made disasters create significant challenges for disaster response and recovery efforts. In recent years, technology has emerged as a vital tool to improve the effectiveness and efficiency of these processes. This chapter describes various technology-driven approaches that have been utilized to supplement disaster response and recovery strategies.

The chapter begins by examining the role of geospatial technologies such as geographic information systems (GIS), remote sensing, and satellite imagery in disaster management. These technologies enable rapid data collection, visualization, and analysis, aiding early warning systems, situational awareness, and resource allocation during response operations.

Furthermore, this chapter describes the use of unmanned aerial vehicles (UAVs) or drones in disaster scenarios. UAVs facilitate the collection of real-time aerial data, including damage assessment, search and rescue operations, and delivery of critical supplies to inaccessible areas. Additionally, emerging technologies, such as autonomous robots and artificial intelligence (AI), have been explored for their potential in disaster response and recovery, including tasks such as debris removal, infrastructure inspection, and decision support systems.

Moreover, this chapter discusses the role of communication technologies, including social media platforms, mobile applications, and Internet-based platforms, in enhancing coordination and information dissemination among responders, affected communities, and volunteers. These technologies enable real-time reporting, crowd-sourced data collection, and efficient resource mobilization.

Finally, this chapter highlights the importance of data analytics and predictive modeling in disaster management. By leveraging big data analytics, machine

Email: chandni@poornima.edu.in

Shruti Kanga, Gowhar Meraj, Suraj Kumar Singh, Majid Farooq and M. S. Nathawat (eds.) Geospatial Technology for Natural Resource Management, (25–82) © 2024 Scrivener Publishing LLC

learning, and predictive algorithms, decision makers can gain insights into disaster patterns, assess vulnerabilities, and proactively plan response and recovery strategies. Ultimately, harnessing the power of technology can contribute to more efficient, resilient, and sustainable disaster management practices, thereby minimizing human suffering and fostering effective recovery in the face of adversity.

Keywords: Natural and man-made disasters, geospatial technologies, remote sensing, satellite imagery, unmanned aerial vehicles, data analytics, predictive modeling, sustainable disaster management

Introduction

a) The importance of disaster management and the challenges faced during emergencies.
b) The critical role of technology in improving disaster response, mitigation, and recovery efforts.
c) Literature review related to various technology-driven approaches to supplement disaster response and recovery strategies

Section 1: Early Warning Systems

1.1 Remote Sensing and Satellite Technology:

a) Use of remote sensing and satellite technology for monitoring natural disasters such as hurricanes, floods, and wildfires.
b) Role of satellite imagery in helping early detection, tracking and prediction of disaster events.
c) Examples of successful early warning systems implemented using remote sensing data.

1.2 Sensor Networks and Internet of Things (IoT):

a) The role of sensor networks and IoT devices in disaster management.
b) Use of sensors to monitor various parameters like temperature, humidity, seismic activity, and water levels to provide early warnings.
c) Importance of data integration and real-time communication in ensuring timely responses.

Section 2: Emergency Communication and Information Management
2.1 Mobile Technologies:

a) Use of mobile phones, SMS alerts and mobile apps in disseminating emergency alerts and information.
b) How mobile technologies aid in coordinating rescue operations, locating survivors and providing critical updates to affected populations.
c) Case studies where mobile technologies played a pivotal role in disaster management.

2.2 Social Media and Crowdsourcing:

a) The significance of social media platforms in disaster management.
b) Role of social media to gather real-time information, identify affected areas and mobilize volunteers.
c) The concept of crowdsourcing and how it can assist in data collection, damage assessment and resource allocation.

Section 3: Geospatial Technologies for Situational Awareness
3.1 Geographic Information Systems (GIS):

a) GIS and its applications in disaster management.
b) Role of GIS in creating spatial databases, mapping affected areas, and analysing vulnerability and risk.
c) Integration of GIS with other technologies for better situational awareness.

3.2 Unmanned Aerial Vehicles (UAVs) and Drones:

a) Role of UAVs and drones in disaster response and recovery.
b) Applications of UAVs in aerial surveys, damage assessment and search and rescue operations.
c) Challenges and ethical considerations associated with the use of drones in disaster management.

Section 4: Data Analytics and Decision Support Systems
4.1 Big Data and Predictive Analytics:

a) Big data analytics application in disaster risk assessment, resource allocation and decision-making.
b) Potential of predictive analytics in forecasting disaster impacts, identifying vulnerable populations and optimizing response strategies.

4.2 Artificial Intelligence (AI) and Machine Learning (ML):

a) The use of AI and ML techniques in disaster management.
b) Applications of AI-powered algorithms for analyzing large datasets, automate damage assessment and support evacuation planning.
c) Ethical implications and challenges of using AI and ML in disaster response.

Conclusion:

a) Key advancements in technology for disaster management are discussed in the chapter.
b) The need for continued research, innovation, and collaboration to harness technology's full potential in mitigating the impacts of disasters.
c) Vision for the future, where technology plays a central role in building resilient communities and enhancing disaster preparedness.

2.1 Introduction

2.1.1 The Importance of Disaster Management and the Challenges Faced During Emergencies

Disaster management is a crucial process aimed at minimizing the impact of natural or man-made disasters on human life, infrastructure, and the environment. It encompasses a range of activities including preparedness, response, recovery, and mitigation. The importance of effective disaster management can be understood through the following points:

Saving Lives: The primary objective of disaster management is to save lives. By implementing appropriate preparedness measures, such as early warning systems and evacuation plans, and ensuring a coordinated response during emergencies, disaster management efforts can significantly reduce the loss of human life.

Minimizing Damage: Disasters can cause extensive damage to infrastructure, property, and the environment. Through proactive planning and mitigation strategies, disaster management aims to minimize the impact and destruction caused by disasters. This includes measures, such as reinforcing buildings, constructing flood control systems, and implementing wildfire prevention initiatives.

Ensuring Public Safety: Disaster management focuses on the safety and well-being of the public. This involves providing timely and accurate information to the affected population, organizing emergency shelters, and coordinating search and rescue operations. By prioritizing public safety, disaster management helps maintain confidence and trust in response efforts.

Enhancing Resilience: Disaster management aims to build resilient communities that can withstand and recover from adverse effects of disasters. This involves long-term planning, community engagement, and capacity-building. By promoting resilience, disaster management helps communities bounce back quickly and adapt to future hazards.

Despite its importance, disaster management faces several challenges during emergencies:

Limited Time for Response: Disasters often strike unexpectedly, leaving little time for preparation and response. The urgency of the situation requires swift decision making and mobilization of resources, which can be challenging without adequate preparedness measures in place.

Uncertainty and Complexity: Disasters can be highly complex and dynamic with rapidly changing conditions. Uncertainty regarding the scale, intensity, and trajectory of a disaster makes it challenging to develop effective response strategies. Additionally, the interdependencies of various sectors and cascading effects of disasters further complicate the management process.

Resource Constraints: Adequate resources, including personnel, equipment, and supplies, are essential for effective disaster management. However, emergencies often strain the available resources, leading to shortages and logistical difficulties. Limited resources can hinder response and recovery efforts, especially during large-scale disasters.

Communication and Coordination: Effective communication and coordination among various stakeholders, including government agencies, emergency responders, and the affected population, is vital during emergencies. However, challenges, such as disrupted infrastructure, overwhelmed communication networks, and information overload, can impede timely and efficient coordination.

Vulnerable Populations: Disasters often disproportionately affect vulnerable populations, including the elderly, children, people with disabilities, and those living in poverty. Ensuring the safety and well-being of these groups during emergencies presents additional challenges due to their specific needs and limited access to resources.

Overcoming these challenges requires a comprehensive and integrated approach to disaster management that incorporates advanced technologies, robust planning, community involvement, and international cooperation. By addressing these challenges, disaster management can effectively mitigate the impact of emergencies and contribute to a safer and more resilient future.

2.1.2 The Critical Role of Technology in Improving Disaster Response, Mitigation, and Recovery Efforts

The critical role of technology in improving disaster response, mitigation, and recovery efforts is defined by its ability to enhance various aspects of the disaster management cycle. The key areas in which technology plays a crucial role are as follows:

Early Warning Systems: Technology, such as remote sensing and satellite technology, enables early detection, monitoring, and prediction of disasters. It provides essential data on weather patterns, seismic activities, and other indicators, thereby enabling authorities to issue timely warnings and alerts. This early warning system helps evacuate vulnerable populations, prepare response teams, and minimize the impact of disasters.

Communication and Information Management: During emergencies, technology facilitates efficient communication and information sharing. Mobile technologies, including SMS alerts, mobile apps, and social media platforms, allow authorities to disseminate emergency notifications, provide real-time updates to the affected communities, and coordinate response efforts. This improves situational awareness, enables better resource allocation, and enhances stakeholder coordination.
Geospatial Technologies for Situational Awareness: Geospatial technologies such as Geographic Information Systems (GIS), unmanned aerial vehicles (UAVs), and drones play a vital role in disaster management. GIS helps map affected areas, analyze vulnerability and risk, and make informed decisions. UAVs provide high-resolution imagery, aid in damage assessment, search and rescue operations, and assess the extent of disasters. These technologies improve situational awareness, facilitate efficient resource deployment, and support the decision-making processes.
Data Analytics and Predictive Modeling: Technology enables the collection, analysis, and interpretation of large volumes of data during disasters. By leveraging big data analytics and predictive modeling techniques, authorities can identify trends, patterns, and potential risks. This information supports proactive decision-making, resource allocation, and the formulation of effective disaster response strategies. Predictive models assist in forecasting disaster impacts, identifying vulnerable areas, and optimizing resource distributions.
Remote Monitoring and Sensing: Technology allows for remote monitoring and sensing of critical infrastructure, such as bridges, dams, and power grids. IoT devices and sensor networks provide real-time data on the structural integrity, water levels, and other parameters. This enables the early detection of potential failures or vulnerabilities, helps authorities take preventive measures, and reduces the risk of infrastructure damage during disasters.
Collaboration and Resource Management: Technology facilitates collaboration among various stakeholders involved in disaster management, including government agencies, non-governmental organizations (NGOs), and volunteers. Digital platforms and cloud-based systems enable real-time information-sharing, resource coordination, and task

management. This improves the efficiency of response efforts, streamlines resource allocation, and enables effective collaboration between different organizations and agencies.

Overall, technology empowers disaster management by enhancing early warning systems, enabling efficient communication, providing situational awareness, facilitating data-driven decision making, and improving collaboration among stakeholders. By leveraging technology effectively, disaster response, mitigation, and recovery efforts can be made more efficient, targeted, and resilient, ultimately saving lives and reducing the impact of disasters.

Table 2.1 Outlining the different technologies and their applications in disaster management.

Technology	**Application**
Remote Sensing and Satellites	Monitoring natural disasters (hurricanes, floods, and wildfires)
Sensor Networks and IoT	Real-time monitoring of parameters (temperature and humidity)
Mobile Technologies	Dissemination of emergency alerts and information
Social Media Platforms	Real-time information sharing and mobilization
Geographic Information Systems (GIS)	Creating spatial databases, mapping affected areas
Unmanned Aerial Vehicles (UAVs) and Drones	Aerial surveys, damage assessment, and search and rescue
Big Data Analytics	Disaster risk assessment and resource allocation
Artificial Intelligence (AI) and Machine Learning (ML)	Data analysis, damage assessment, and decision support
Early Warning Systems	Detection and prediction of disasters
Communication Technologies	Real-time communication and coordination
Virtual and Augmented Reality	Training and simulation of disaster scenarios

Table 2.1 is showing different technologies and their application in disaster management.

2.1.3 Literature Review Related to Various Technology-Driven Approaches to Supplement Disaster Response and Recovery Strategies

Disasters require innovative methods from local and national governments to mitigate detrimental effects on society and the economy. The involvement of E-government in delivering vital services is especially important during COVID-19. A thorough literature review was conducted from SAGE, Science Direct, Scopus, Springer, Taylor & Francis, and Wiley Online Library for relevant materials. A total of 36 publications from 2004 to 2022 were analyzed. This study guides future research and assists disaster managers in effectively using e-government [1].

Disasters are sudden and unexpected catastrophes that can destroy property and life. Thus, disasters are unpredictable and destructive, making it difficult for academics, scientists, and governments to manage the situation and information flow adequately. The Internet of Things (IoT) uses networked mobile devices to move data and information more efficiently. IoT has numerous uses. Pre- and post-disaster IoT-aided systems meet such needs. IoT-aided systems employ strategically placed sensors to forecast disasters for predisaster management. In conclusion, IoT-enabled solutions improve evacuation, relief, and victim tracking after disasters [2].

Studies on the Mount Krakatau earthquake and tsunami are important because they occur regularly. Mount Krakatau is on the southern Lampung and Banten seabeds. This qualitative case study examines Mount Krakatau's local government leadership in earthquake and tsunami catastrophe management. Observations, documentation, literature analysis, and in-depth interviews were conducted to gather data and plan remedial measures [3].

Over the past decade, drones have been used in agriculture, commerce, and disaster management. Their use in large disasters remains uncertain and rare. The study was conducted using the Arksey and O'Malley framework and Joanna Briggs Institute Framework for Scoping Reviews technique to summarize 52 research publications from 2009 to 2020 on drone use in disasters. The use of drones in catastrophes can be divided into four categories: mapping or disaster management, search and rescue, transportation, and training. The number of papers on drone usage in disasters has increased significantly in the past five years; however, there has been little discussion on post-disaster healthcare, particularly disaster victim identification. Drone applications should rely more on human support, notably, victim identification. Drones have the potential to aid in disaster management with appropriate development [4].

The use of the spatial and temporal aspects of textual data from Twitter during Hurricane Harvey in 2017 was studied. This study presents a Hazard Risk Awareness (HRA) index that incorporates factors such as tweet volume, population size, Internet usage rate, and natural hazard characteristics. The index is cartographically represented throughout the southeastern region of Texas using Natural Language Processing (NLP) and statistical methodologies. These findings can help disaster managers and responders mitigate the impact of natural hazards and enhance preparedness for affected residents. The HRA index relies on meteorological observations and demographic data, and the availability of these datasets and hazard types determines its feasibility [5].

Natural disasters resulting from climate change, such as droughts and floods, exhibit inherent unpredictability for human beings and engender significant levels of devastation. The utilization of Information and Communication Technologies (ICTs) has demonstrated efficacy in global disaster management. Africa, widely regarded as the most vulnerable continent, continues to exhibit substantial evidence to support this assertion. A study was conducted on the effects of Information and Communication Technologies (ICTs) on the management of natural disasters in the African context. The econometric methodology relies on estimating a panel data model using various statistical techniques, including ordinary least squares (OLSs), generalized method-of-moments (GMMs), and fixed-effect Poisson estimators (FEs). The findings indicate a strong association between the prevalence of cell phones and Internet usage, and a decrease in adverse outcomes resulting from natural disasters. Therefore, it is recommended that the governments of these nations enhance the development and dissemination of information and communication technology (ICT) capabilities [6].

Urban environments are characterized by their ability to accommodate diverse human activities and functions, leading to challenges such as water resource depletion, deforestation, land use changes, environmental degradation, health issues, and increased epidemic risks. Resilience is a prominent concept in disaster management that prioritizes recovery and rebounding after disturbances [7].

Third-generation geostationary meteorological satellites (GEOs) offer advanced imagery and atmospheric measurements of the Earth's weather, oceans, and terrestrial environments. They improve the environmental change detection capabilities and contribute to disaster monitoring and risk mitigation. The 2019 Typhoon No. 19 Hagibis case study showcases the collaboration between GEOs and Low Earth orbit satellites for fine spatial-resolution observations [8].

Knowledge management (KM) practices are frequently observed in disaster contexts. Nevertheless, the field of disaster management comprises numerous interconnected elements, and the precise applicability of knowledge management (KM) practices across all the significant stages of the disaster management cycle remains uncertain. This review examines the application of knowledge management (KM) throughout all stages of disaster management to validate the benefits of utilizing KM in mitigating the impacts of disasters and bolstering disaster resilience [9].

Natural hazards have the inherent capacity to induce extensive devastation and substantial socioeconomic detriment. Empirical evidence from recent decades has demonstrated a notable upward trajectory in the observed damage and loss. Consequently, disaster managers must assume an increasingly significant role in proactively safeguarding their communities by formulating and implementing effective management strategies. Several studies have utilized artificial intelligence (AI) methodologies to analyze data related to disasters to facilitate well-informed disaster management. The implementation of artificial intelligence (AI) in the field of disaster management across four distinct stages—mitigation, preparedness, response, and recovery—is crucial. A significant proportion of artificial intelligence (AI) applications primarily concentrate on the phase of disaster response [10].

Due to technology and online sources, huge volumes of data are created daily in many industries. Despite breakthroughs and promises of greater understanding, big data management and analysis remain a problem, requiring more rigorous and extensive study, and the creation of methodologies and approaches to confront and exploit large data. Big data have emerged as an amalgam of storage, statistics, technology, and research fields with interlinked aspects, such as data hosting and computing, data management, data refining, data patterns, and machine learning. The results also suggest that the seven Vs of big data—variety, volume, variability, value, visualization, veracity, and velocity—can be summarized. The proposed frameworks show that big data can reduce customer regrets related to poor quality or a lack of information in smart real estate by using an intermediate organization to process and monitor the data provided to customers by sellers and real estate managers. Social media, drones, multimedia, and search engines may also be used to manage disasters, such as floods, bushfires, and earthquakes, and prepare emergency responses. A merger framework for smart real estate and disaster risk management shows that big data from occupant data, facility management, and building integration and maintenance can be shared with disaster risk management and emergency response teams to help prevent, prepare for, respond to, and recover from disasters [11].

The significant increase in the magnitude and frequency of disasters worldwide has predominantly been ascribed to global climate change. Similar to other continents, Africa has also encountered a fair share of disasters, albeit not to the same extent as other countries in Asia. Although the scale of disasters in Africa may be small, their impact on the population is devastating. This study employed an action research methodology to investigate the National Disaster Management Agency (NDMA) and its institutional structure. By conducting an online survey and semi-structured interviews with disaster management leadership, this study reveals the need for improvements in NDMA's current institutional framework to foster enhanced efficiency and effectiveness in disaster management practices. Consequently, a novel framework for national institutions is suggested, aiming to mitigate political and administrative inefficiencies while concurrently advancing the nation's newly established developmental objectives and vision [12].

Collaborative disaster management initiatives based on trust and networks between governmental entities and community actors help address constraints in government-centric human and material resources. The involvement of public officials in establishing, maintaining, and advancing cooperative disaster management is crucial. This study examined the correlation between the expertise of disaster management officials and their ability to collaborate effectively in disaster management. The experts were categorized into four sub-components: knowledge, experience, skills, and attitudes toward public service. The analysis found that knowledge positively impacts intergovernmental and public–private collaboration. Skill acquisition positively impacts both intergovernmental and public–private collaborations [13].

The application of information technology in various phases of disaster management, including disaster response, recovery, preparedness, and risk reduction, has been studied. Information systems are of the utmost importance in documenting, transmitting, and manipulating information. The integration of various roles contributes to the optimization of the system performance. Additionally, it emphasizes the need for standardized data protocols to facilitate information exchange among diverse systems and stakeholders [14].

Pakistan is more vulnerable to flood disasters due to rising urbanization and environmental imbalance, which has hampered its rescue efforts. A study used systems thinking methods such as the Causal Loop Diagram

(CLD) and Driver-Pressures-States-Impacts-Responses (DPSIR) framework to identify disaster risk reduction stakeholders and analyze social, technical, institutional, cultural, infrastructural, and environmental factors that cause flooding in Pakistan. Comprehensive flood management and mitigation recommendations would raise community awareness and strengthen institutional capacities at the federal, provincial, and district government levels in Pakistan and other developing nations facing catastrophic floods [15].

Big data analytics (BDA) and the Internet of Things (IoT) offer disaster management systems and disaster-related authorities (emergency responders, police, public health, and fire departments) cutting-edge support and better insights for accurate and timely decision-making. Conventional disaster management systems lack support for many new data sources and real-time big data processing technologies that can help decision makers obtain fast and accurate answers. A conceptual reference model for BDA- and IoT-based disaster management and its integrated parameters may help gather, transmit, manage, and analyze disaster data from diverse data sources to provide current and meaningful disaster management information [16].

Satellite remote sensing is a fundamental tool for disaster management. The establishment of the Sentinel Asia (SA) initiative in 2006 marked a collaborative effort between regional space and disaster management agencies. This initiative aimed to leverage space technology, specifically satellite remote sensing, and Web GIS technology to provide support in disaster management within the Asia-Pacific region. The establishment of the SA framework and determination of the implementation plan were undertaken by the Japan Aerospace Exploration Agency (JAXA) as the SA secretariat [17].

The main goals of disaster management are to lessen potential damage from disasters, deliver prompt and appropriate aid to those in need, and achieve an effective and quick recovery. Achieving these objectives necessitates the implementation of a well-organized and efficient rescue operation following such catastrophic events. Planning an efficient and timely relief operation requires acquisition of diverse forms of information about the consequences of a disaster. The current state of IoT technology exhibits a significant level of maturity, thereby presenting substantial potential for application in disaster scenarios [18].

Technology plays a crucial role in facilitating agile decision-making processes during catastrophic events in the age of social media, data analysis, and Industry 4.0. A resilient and innovative disaster decision support system is essential for modern well-equipped communities that rely on real-time big data. This study proposes a novel tool that enhances disaster management readiness, response, recovery, and mitigation with the aim of building resilient societies. This tool integrates virtual experiments, machine learning, blockchain, and database management technologies, employing innovative techniques for information registration, distribution, and exploration [19].

2.2 Early Warning Systems

2.2.1 Remote Sensing and Satellite Technology

2.2.1.1 Use of Remote Sensing and Satellite Technology for Monitoring Natural Disasters such as Hurricanes, Floods, and Wildfires

Remote sensing and satellite technology play crucial roles in monitoring and analyzing natural disasters such as hurricanes, floods, and wildfires. The following is an elaboration of their use in disaster monitoring.

Hurricanes: Satellite imagery: Satellites equipped with sensors capture high-resolution images of hurricanes in space. These images provide valuable information regarding the size, structure, and intensity of storms. Tracking and prediction: By continuously monitoring hurricanes using satellite data, meteorological agencies can track their movement, speed, and potential landfall. This information aids in issuing timely warnings, evacuating vulnerable areas, and preparing response teams.
Floods: Rainfall and river monitoring: Remote sensing satellites can estimate rainfall patterns over large areas, aiding flood forecasting. They can also monitor river levels and assess the extent of flooding by analyzing changes in water bodies. Mapping flood extent: Satellite imagery can capture the extent of flooded areas, allowing emergency management agencies to assess affected regions, plan rescue operations, and allocate resources [20].
Wildfires: Fire detection: Satellites equipped with infrared sensors can detect and monitor wildfires in space. They can identify areas with high-heat signatures, allowing authorities to detect fires early, track their spread, and estimate the size of the affected area.

> Hotspot mapping: Satellite imagery helps map active fire hotspots and identify at-risk areas. This information assists firefighting teams in effectively deploying resources and prioritizing containment efforts.

Benefits and Advantages: Wide coverage: Satellites provide a comprehensive view of large geographical areas, allowing for monitoring and assessment at a regional or global scale.
Timeliness: Satellite data are available in near real-time, enabling quick response and timely decision-making during disasters.
Remote and inaccessible areas: Remote sensing and satellite technology can be used to monitor disaster-prone regions that are difficult to access on the ground, providing critical information for emergency management [21].
Historical data analysis: Satellite archives contain historical data that can be used to identify long-term trends, patterns, and changes in disaster occurrences. This aids in better understanding and forecasting of future events.

Challenges:

Cloud cover: Persistent cloud cover can obstruct satellite imagery, limiting its effectiveness in monitoring disaster events.
Resolution limitations: Satellite imagery may have limitations in terms of spatial resolution, which can affect its ability to identify fine-scale features or details during disaster monitoring.
Cost and accessibility: Access to high-resolution satellite data and advanced remote sensing technology may be costly or limited, posing challenges for organizations and agencies with resource constraints.
Despite these challenges, remote sensing and satellite technologies continue to be invaluable tools for monitoring natural disasters. They enhance early warning capabilities, support decision-making processes, and contribute to effective disaster response and mitigation. Ongoing advancements in satellite technology and data processing techniques have the potential to improve disaster monitoring and management in the future.
In addition, the use of remote sensing and satellite technology for monitoring natural disasters offers several other advantages and applications.
Damage assessment: After a natural disaster, satellite imagery can be used to assess the extent of the damage. By comparing pre- and post-disaster images, emergency management

agencies can identify affected areas, evaluate infrastructure damage, and prioritize response efforts [22].

Evacuation planning: Satellite data combined with geographic information systems (GIS) can assist in identifying high-risk areas prone to natural disasters. This information helps develop evacuation plans and determine safe zones for affected populations, minimizing the potential loss of life.

Resource allocation: Remote sensing technology enables the identification of areas where resources such as food, water, and medical supplies are most needed during a disaster. This information assists in optimizing resource allocation and distribution to ensure that the affected communities receive adequate support.

Environmental monitoring: Satellites equipped with specialized sensors can monitor environmental factors relevant to disaster management. For example, they can track changes in sea surface temperatures to detect the formation and intensity of hurricanes, or monitor vegetation health to identify areas prone to wildfires.

Climate change analysis: Remote sensing and satellite data contribute to the study of long-term climate change patterns and their impacts on natural disasters. By monitoring changes in sea level, land cover, and atmospheric conditions, scientists can gain insights into the broader trends of climate-related events and their potential future implications.

International collaboration: Satellite technology enables international collaboration and data-sharing in disaster management. Satellites operated by different countries can provide valuable data and imagery to support global disaster response efforts, allowing for improved coordination and a more comprehensive understanding of transboundary disasters.

It is important to note that, while remote sensing and satellite technology offer significant benefits, they should be used in conjunction with other data sources and ground-based observations for a comprehensive understanding of the situation. Integration with ground-based monitoring systems, weather radar, and other technologies enhances the accuracy and effectiveness of disaster monitoring and response.

Finally, the use of remote sensing and satellite technology for monitoring natural disasters such as hurricanes, floods, and wildfires enhances early warning capabilities, facilitates damage assessment, aids in evacuation planning, optimizes resource allocation, contributes to environmental monitoring, and supports global collaboration. These technologies continue to evolve, providing increasingly valuable data and insights to improve disaster management strategies and help build more resilient communities [23].

2.2.1.2 Role of Satellite Imagery in Helping Early Detection, Tracking, and Prediction of Disaster Events

Satellite imagery plays a crucial role in enabling early detection, tracking, and prediction of disaster events. Here is a breakdown of its role in each of these aspects.

Early Detection: Satellite imagery provides a bird's-eye view of the Earth's surface, enabling the detection of potential disaster events. It helps to identify early signs and indicators that can signify the onset of a disaster. For example:
Weather Monitoring: Satellites equipped with weather sensors monitor atmospheric conditions, including cloud formation, temperature gradients, and wind patterns. By analyzing these data, meteorologists can identify the development of severe weather phenomena such as tropical storms, hurricanes, and cyclones.
Fire Detection: Satellites with infrared sensors can detect the thermal anomalies associated with wildfires. By capturing the heat signatures emitted by fires, satellite imagery enables early detection of fire outbreaks in remote or inaccessible areas.
Tracking: Once a disaster event is detected, satellite imagery assists in tracking its progression, movement, and intensity. This tracking helps assess potential impact areas and guide response efforts. Examples include:

> Hurricane Tracking: Satellites provide continuous monitoring of hurricanes, capturing high-resolution images of their structure, cloud patterns, and wind speeds. By tracking hurricane movement, authorities can issue timely warnings, plan evacuations, and allocate resources to areas in the storm's predicted path.
> Flood Monitoring: Satellite imagery enables the tracking of flood events by observing changes in water levels, river discharge, and extent of flooded areas. This information helps emergency management agencies to monitor the spatial extent and progression of flooding, assisting in evacuation planning and resource allocation.

Prediction: Satellite imagery combined with advanced data analysis techniques supports the prediction of disaster events by providing valuable insights and patterns. This predictive capability aids in preparedness measures and decision-making. Examples include:
Cyclone Intensity Prediction: By analyzing satellite imagery, scientists can assess the structure and characteristics of cyclones, helping predict their intensity and potential impact on coastal regions. This information assists in issuing warnings, implementing evacuation plans, and allocating resources.
Landslide Susceptibility Mapping: Satellite imagery can be used to identify areas prone to landslides by analyzing factors such as slope, vegetation cover, and land surface changes. By mapping landslide susceptibility, authorities can implement mitigation measures and provide early warning to communities living in high-risk areas.

Overall, satellite imagery provides a valuable advantage for the early detection, tracking, and prediction of disaster events. By leveraging these data, emergency management agencies can enhance preparedness, issue timely warnings, allocate resources effectively, and minimize the impact of disasters on communities and infrastructure.

2.2.1.3 Examples of Successful Early Warning Systems Implemented Using Remote Sensing Data

Several examples of successful early warning systems have been implemented using remote sensing data. Below are a few notable examples:

Global Disaster Alert and Coordination System (GDACS): The GDACS, a collaboration between the United Nations and the European Commission, utilizes remote sensing data and other sources to provide early warnings and information on various natural hazards. It integrates data from satellites, including weather satellites, to globally detect and monitor cyclones, floods, earthquakes, and other disasters. The system combines this data with other sources, such as seismic networks and meteorological stations, to issue alerts and provide impact assessments for the affected regions.
Indian Ocean Tsunami Warning System: The Indian Ocean Tsunami Warning System was established after the devastating Indian Ocean tsunami in 2004. The system employs remote-sensing technologies, including

satellite-based sea-level monitoring and ocean buoy networks, to detect and monitor tsunamigenic earthquakes and sea-level anomalies. It uses these data to issue timely tsunami warnings to countries in the Indian Ocean region, helping to save lives and minimize the impact of tsunamis.

Drought Early Warning Systems: Drought early warning systems often rely on remote sensing data to monitor various indicators of drought such as vegetation health, soil moisture, and precipitation patterns. For example, the Famine Early Warning Systems Network (FEWS NET) uses satellite imagery to monitor vegetation conditions, which serve as indicators of drought and food security risks in vulnerable regions. These data can help governments, humanitarian organizations, and communities anticipate and respond to drought-related challenges, such as food shortages and water scarcity.

Wildfire Early Warning Systems: Remote sensing data, particularly satellite imagery and thermal sensors, are used in wildfire early warning systems. For instance, the European Forest Fire Information System (EFFIS) incorporates satellite data to detect active fires, map burned areas, and assess fire risk across Europe. This information assists firefighting agencies in allocating resources, planning containment strategies, and issuing timely warnings to communities that are at risk of wildfires. These examples highlight the effective integration of remote sensing data into early warning systems for different types of natural hazards. By leveraging satellite imagery and other remote sensing technologies, these systems can detect, monitor, and provide timely alerts regarding potential disasters, allowing authorities and communities to take proactive measures and mitigate the impact of these events.

Cyclone Early Warning Systems: Countries prone to cyclones have implemented early warning systems that rely on remote sensing data. For instance, the Regional Specialized Meteorological Centre (RSMC) in India uses satellite data to track cyclones in the Indian Ocean region. The system combines remote sensing information, including cloud patterns, sea surface temperatures, and wind speeds, to monitor cyclones and issue warnings to coastal areas along the storm path.

Landslide Early Warning Systems: Landslide early warning systems often incorporate remote sensing data to monitor slope, land cover, and precipitation patterns. For example, in the Philippines, the Nationwide Operational Assessment of Hazards (NOAH) program uses remote sensing data, including satellite imagery and weather radar, to monitor areas susceptible to landslides. The system provides real-time alerts and warnings to vulnerable communities, helping them evacuate and take necessary precautions.

Volcano Monitoring Systems: Remote sensing plays a vital role in volcano monitoring and early warning systems. Satellites equipped with specialized

sensors can detect volcanic activity such as changes in gas emissions, thermal anomalies, and ground deformation. For example, the Global Volcano Monitoring System (GVMS) uses remote sensing data to track volcanic eruptions worldwide. This information will enable scientists and authorities to issue warnings, monitor volcanic hazards, and evacuate nearby populations.
Coastal Flood Early Warning Systems: Coastal areas prone to flooding use remote sensing data for early warning systems. For instance, in the Netherlands, Remote Sensing-based Information and Decision Support System for Coastal Zones (DISCovery) integrates satellite data with hydrodynamic models to monitor sea level, wave height, and coastal erosion. The system provides timely alerts and forecasts to coastal authorities, helping them implement flood protection measures and evacuate at-risk areas.
Heatwave Early Warning Systems: Heatwave early warning systems leverage remote sensing data such as land surface temperature and vegetation health to monitor heat stress and forecast extreme heat events. For example, the European Heat-Med Health Early Warning System utilizes satellite data to monitor thermal conditions, assess heat stress levels, and issue heat wave warnings to vulnerable populations. This information can assist public health agencies in implementing heatwave mitigation measures and protecting public health.

These examples demonstrate the diverse applications of remote sensing data in early warning systems for a range of natural hazards. By integrating remote sensing information with other data sources, authorities can provide timely warnings, monitor hazardous conditions, and implement proactive measures to protect lives and property.

2.2.2 Sensor Networks and Internet of Things (IoT)

2.2.2.1 The Role of Sensor Networks and IoT Devices in Disaster Management

Sensor networks and Internet of Things (IoT) devices play a significant role in disaster management by providing real-time data, enhancing situational awareness, and enabling efficient response and recovery. Here is a closer look at their roles:

Data Collection and Monitoring: Sensor networks and IoT devices are deployed in disaster-prone areas to collect and transmit data related to various parameters. These devices can monitor factors, such as temperature, humidity, air quality, water levels, and seismic activity. The collected data help to understand the evolving conditions and potential risks associated with a disaster event.

Early Warning Systems: Sensor networks and IoT devices contribute to the development of early warning systems by detecting and transmitting critical information in real-time. For example, seismic sensors can detect earthquakes, and weather sensors can track atmospheric conditions. When these devices detect anomalies or changes that indicate an impending disaster, they can trigger automatic alerts and warnings, thereby allowing authorities to take immediate action.

Situational Awareness: Sensor networks and IoT devices provide real-time information about disaster-affected areas, enabling decision-makers to have an accurate understanding of the situation. For instance, sensor networks can monitor the movement of floodwaters or the spread of wildfires, provide timely updates on affected areas, and help emergency management teams plan their responses accordingly.

Infrastructure Monitoring: IoT devices equipped with sensors can monitor the condition and integrity of critical infrastructure, such as bridges, buildings, and utility systems. By continuously monitoring structural parameters, such as vibrations, stress, and deformation, these devices can detect early signs of damage or potential failures, allowing for proactive maintenance and reducing the risk of infrastructure collapse during disasters.

Resource Management: IoT devices can play a vital role in resource management during disaster response and recovery operations. For example, they can track and manage the inventory of emergency supplies, monitor the usage of medical equipment, and even control and optimize the distribution of resources based on real-time needs and demands.

Resilient Communication: IoT devices and sensor networks can establish resilient communication networks, particularly in areas where the traditional communication infrastructure may be disrupted during a disaster. These devices can form ad-hoc networks, enabling data transmission and communication between emergency responders and facilitating coordination and decision making in challenging environments.

Post-Disaster Assessment and Recovery: Sensor networks and IoT devices aid in post-disaster assessment and recovery efforts. They can provide valuable data on the extent of damage, monitor environmental conditions, and support the evaluation of the recovery progress. For example, sensor networks can monitor air quality and water contamination levels in the aftermath of a chemical spill, or assess structural damage to buildings.

By leveraging sensor networks and IoT devices, disaster-management authorities can access real-time data, improve situational awareness, and make informed decisions. These technologies enable faster responses, more efficient resource allocation, and better coordination among response teams, ultimately contributing to more effective disaster management and enhanced resilience.

2.2.2.2 Use of Sensors to Monitor Various Parameters Like Temperature, Humidity, Seismic Activity, and Water Levels to Provide Early Warnings

The use of sensors to monitor various parameters such as temperature, humidity, seismic activity, and water levels plays a crucial role in providing early warnings for potential disasters. Here is a breakdown of how sensors are used to monitor these parameters and facilitate early warning systems.

Temperature and Humidity Sensors: Temperature and humidity sensors are deployed in areas prone to heat waves, wildfires, and extreme weather events. These sensors continuously measure the temperature and humidity levels in the surrounding environment. When the readings exceed predefined thresholds, indicating potentially hazardous conditions, the sensors can trigger early warning alerts. These warnings help authorities and communities to take preventive measures, such as implementing cooling centers during heatwaves or increasing fire safety precautions during periods of high fire risk.

Seismic Sensors: Seismic sensors are used to monitor ground vibrations and detect earthquakes. These sensors can even detect minor tremors and transmit data in real time. By analyzing seismic data, experts can assess the magnitude and location of an earthquake, issuing early warnings to the affected regions. This enables timely evacuation, the activation of emergency response plans, and the potential to save lives.

Water Level Sensors: Water level sensors are employed to monitor river levels, coastal tides, and water bodies, such as lakes and reservoirs. By continuously measuring the water levels, these sensors can detect rapid increases or significant deviations from normal levels, indicating a risk of flooding. Early warning systems integrated with water-level sensors can trigger alerts when water levels reach critical thresholds, allowing authorities to issue evacuation orders and implement flood preparedness measures.

Rainfall and Precipitation Sensors: Rainfall and precipitation sensors are used to measure the intensity and volume of the rainfall in an area. By monitoring precipitation levels, these sensors contribute to flood early warning systems. When rainfall exceeds certain thresholds or when heavy rainfall is forecasted, the sensors can trigger warnings, enabling communities to prepare for potential flooding and take necessary precautions.

Environmental Monitoring Sensors: Various environmental monitoring sensors can be deployed to detect parameters, such as air quality, gas emissions, and chemical pollutants. These sensors help identify hazardous conditions resulting from industrial accidents, chemical spills, or air pollution events. Real-time data from environmental sensors enable the early detection of harmful substances, allowing authorities to issue public health advisories, evacuate affected areas, and implement measures to mitigate the impact of the incident.

By utilizing sensors to monitor these parameters, early warning systems can be developed to detect and predict potential disasters. The real-time data collected by these sensors can help authorities and emergency management agencies make informed decisions, issue timely warnings, and implement appropriate measures to protect lives and minimize the impact of disasters on communities and infrastructure.

2.2.2.3 The Importance of Data Integration and Real-Time Communication in Ensuring Timely Responses

Data integration and real-time communication are of utmost importance in ensuring timely responses during disaster situations. Here, we provide a closer look at the significance of these aspects.

Comprehensive Situational Awareness: Data integration involves bringing together information from various sources such as sensors, satellite imagery, weather forecasts, and social media feeds. By integrating these diverse datasets in real time, emergency management agencies can gain comprehensive situational awareness of disaster events. This holistic understanding enables better decision making and allows for a more effective and coordinated response.

Early Detection and Rapid Response: Real-time data integration facilitates the early detection of disaster events. By continuously monitoring multiple data streams, authorities can identify abnormal patterns and anomalies, or trigger thresholds that are surpassed, indicating the onset of a disaster. Real-time communication allows the immediate dissemination of this information to relevant stakeholders, enabling a rapid response. Timely alerts and warnings can be issued, evacuation plans can be activated, and emergency services can be mobilized promptly, increasing the chances of saving lives and reducing the impact of the disaster.

Resource Allocation and Mobilization: Effective disaster response requires the efficient allocation and mobilization of resources. Real-time data integration provides up-to-date information on the evolving situation, including affected areas, population density, infrastructure status, and resource availability. By integrating these data, emergency management agencies can optimize the allocation of resources such as personnel, equipment, medical supplies, and relief materials. Real-time communication enables swift coordination among response teams and the seamless deployment of resources to areas where they are most needed.

Timely Decision-Making: During a disaster, decision making needs to be swift and well-informed. Real-time data integration ensures that decision makers have access to the latest information, allowing them to make timely and accurate decisions. By aggregating and analyzing real-time data, authorities can assess the severity of a situation, identify emerging trends, and adjust response strategies accordingly. Real-time communication enables decision makers to share critical information, collaborate, and coordinate their efforts in real time, facilitating a more effective response.

Adaptive Planning and Continual Improvement: Data integration and real-time communication also support adaptive planning and the continual improvement of disaster response efforts. By analyzing real-time data and feedback from the field, authorities can evaluate the effectiveness of their response strategies, identify areas for improvement, and make necessary adjustments in real-time. This iterative approach enhances the efficiency and effectiveness of response efforts, ensuring that the lessons learned from one disaster can be applied to future events.

In summary, data integration and real-time communication are vital in disaster management, as they enable comprehensive situational awareness, early detection, rapid response, efficient resource allocation, informed decision-making, and continual improvement. By harnessing the power of real-time data and effective communication, emergency management agencies can enhance their capabilities to respond promptly and effectively to disasters, ultimately saving lives and minimizing their impact on communities.

2.3 Emergency Communication and Information Management

2.3.1 Mobile Technologies

2.3.1.1 Use of Mobile Phones, SMS Alerts and Mobile Apps in Disseminating Emergency Alerts and Information

The use of mobile phones, SMS alerts, and mobile apps has significantly transformed the dissemination of emergency alerts and information during disasters. Here, we explore their importance, and how they facilitate effective communication.

Wide Reach and Accessibility: Mobile phones are widely accessible, making them an effective communication tool during emergencies. Most populations, including those in remote areas or vulnerable communities, have access to mobile phones. This widespread reach ensures that emergency alerts and information can be disseminated to many people, enabling authorities to reach a broad audience quickly.
Timely and Direct Communication: Mobile phones allow immediate and direct communication with individuals. SMS alerts can be sent to mobile phone users, to directly deliver important information and emergency warnings to their devices. This direct communication ensures that people receive critical alerts promptly, even if they are not connected to the Internet or are engaged in other media channels.
Personalized and Location-specific Information: Mobile apps and SMS alerts provide personalized and location-specific information during disasters. By leveraging location-based services and user profiles, emergency management agencies can send alerts and notifications tailored to specific regions or users. Personalized information ensures that people receive relevant updates and instructions based on their specific locations and circumstances.
Multilingual Communication: Mobile phones and SMS alerts offer flexibility to disseminate emergency information in multiple languages. In diverse communities or areas where different languages are spoken, SMS alerts can be sent in different languages to ensure that the message is understood by a broader audience. This helps overcome language barriers and ensures that critical information reaches all segments of the population.
Two-way Communication and Feedback: Mobile applications and SMS alerts facilitate two-way communication during emergencies. Users can provide feedback, report incidents, request assistance, or ask questions through mobile apps or dedicated SMS response systems. This interactive communication enables emergency management agencies to gather real-time information from the affected individuals, assess their needs, and respond accordingly.
Integration with Other Technologies: Mobile applications can be integrated with other technologies such as GPS, mapping systems, and real-time data feeds to provide comprehensive information during emergencies. This integration allows users to access live updates regardibng evacuation routes, shelter locations, road closures, and real-time hazards.

By combining multiple sources of information in a mobile app, users can have a holistic view of the situation and make informed decisions.

Information Dissemination to Vulnerable Populations: Mobile phones and SMS alerts are particularly valuable in reaching vulnerable populations, including the elderly, individuals with disabilities, or those with limited access to traditional media channels. These individuals may rely heavily on their mobile phones for communication, making it a crucial tool for delivering emergency alerts, evacuation notices, and supporting information to those who may need it the most.

In summary, mobile phones, SMS alerts, and mobile apps have revolutionized the dissemination of emergency alerts and information. They offer wide-reach, timely communication, personalized messaging, two-way communication, and integration with other technologies. By leveraging these tools, emergency management agencies can ensure that vital information reaches the public quickly and efficiently, thereby facilitating a more informed and responsive approach to disaster management.

2.3.1.2 Benefit of Using Mobile Technologies in Coordinating Rescue Operations, Locating Survivors and Providing Critical Updates to Affected Populations

The use of mobile technologies to coordinate rescue operations, locate survivors, and provide critical updates to affected populations offers several benefits. Here are some of the key advantages:

Rapid and Real-time Communication: Mobile technologies enable rescue teams and emergency responders to communicate in real time, facilitating the swift coordination of rescue operations. Rescuers can exchange information, update their statuses, share critical updates instantly, enhance collaboration, and ensure an efficient response.

Geolocation and Mapping: Mobile technologies, including GPS and mapping applications, play a vital role in locating survivors and guiding rescue teams to their precise locations. Rescuers can use mobile devices to track their positions, navigate through the affected areas, and identify specific coordinates shared by survivors in distress. This capability helps expedite search and rescue efforts, increasing the chances of promptly finding and rescuing survivors.

Multimedia Capabilities: Mobile devices allow for the capture and sharing of multimedia content such as photos, videos, and audio recordings. Rescuers can document the situation on the ground and capture critical information that aids in decision making and resource allocation. Survivors can also use their mobile devices to share visual or audio cues, helping rescuers understand their immediate needs and prioritize assistance.

Crowd-sourced Information: Mobile technologies empower affected populations to contribute to rescue and relief efforts by providing crowd-sourced information. Through mobile apps or social media platforms, individuals can report their location, share updates on the situation in their area, and communicate their urgent needs. This user-generated data supplements official information and helps emergency management agencies gain a comprehensive understanding of the crisis.

Emergency Alert Systems: Mobile technologies enable the deployment of emergency alert systems through SMS, push notifications, or dedicated mobile apps. Authorities can send mass alerts and updates to affected populations, providing critical information on evacuation routes, safety guidelines, and available resources. These alerts rapidly reach a wide audience, ensuring that people receive timely and potentially life-saving information.

Language and Accessibility Considerations: Mobile technologies offer language options and accessibility features that cater to diverse populations. Emergency information can be disseminated in multiple languages, ensuring that nonnative speakers receive critical updates in the language they understand. Additionally, mobile devices can accommodate accessibility features such as text-to-speech or visual aids, making information accessible to individuals with disabilities.

Psychological Support: Mobile technologies can provide psychological support to affected populations during and after a disaster. Mobile apps or messaging platforms can offer counseling services, helplines, or mental health resources. Digital support helps alleviate distress, provides guidance, and connects individuals with professional help when needed. By harnessing the power of mobile technologies, rescue operations become more efficient, survivors can be located swiftly, and affected populations can receive critical updates in real time. These benefits significantly contribute to the effectiveness of disaster response efforts, saving lives and minimizing the impact of the disaster on affected communities.

2.3.1.3 Case Studies Where Mobile Technologies Played a Pivotal Role in Disaster Management

The following are a few case studies in which mobile technologies play a pivotal role in disaster management:

The Philippines—Typhoon Haiyan (2013): During Typhoon Haiyan, one of the strongest tropical cyclones on record, mobile technologies played a crucial role in coordinating response efforts. In the aftermath of the disaster, mobile networks were quickly restored, enabling the affected communities to send SMS messages and make phone calls to communicate their needs and locations. This information helped rescue teams to locate survivors, distribute aid, and prioritize rescue operations effectively.

Nepal—Earthquake (2015): After the devastating earthquake in Nepal, mobile technologies have proved invaluable in facilitating response and recovery efforts. Mobile apps such as the "Nepal Quake" app provide real-time information on emergency services, shelter locations, and medical assistance. SMS-based systems are used to send alerts and updates to affected populations, providing critical information on safety measures and aid distribution. Mobile networks play a significant role in enabling communication and coordination among rescue teams, government agencies, and humanitarian organizations.

Haiti—Earthquake (2010): Following the earthquake in Haiti, mobile technologies played a pivotal role in organizing rescue operations and delivering aid. Mobile networks remain operational, allowing responders to communicate and coordinate effectively. Mobile payment platforms facilitate the distribution of financial assistance, enabling people to receive funds directly from their mobile devices. Social media platforms have been utilized to share real-time updates, emergency information, and coordinate volunteer efforts.

India—Cyclone Fani (2019): During Cyclone Fani, mobile technologies played a critical role in disaster management. Mobile apps, such as the "Disaster Alert" app, provide real-time updates on a cyclone's path, evacuation routes, and emergency contact information. SMS alerts were sent to the population in the affected areas, informing them about the cyclone's intensity and necessary precautions. Mobile networks enable individuals to communicate their distress, seek help, aid rescue operations, and ensure timely assistance.

United States—Hurricane Harvey (2017): During Hurricane Harvey, mobile technologies played a vital role in disaster response and recovery efforts. Mobile apps, like the "Zello" app, provided a platform for citizens to communicate and request assistance. Social media platforms serve as channels for sharing information, connecting volunteers with those in need, and coordinating rescue efforts. Mobile payment platforms facilitate donations and provide financial assistance to affected individuals and communities.

These case studies highlight how mobile technologies are instrumental in disaster management, enabling communication, coordination, information sharing, and assistance delivery. By leveraging the power of mobile technologies, response efforts have become more efficient, effective, and inclusive, ultimately saving lives and mitigating the impacts of disasters.

2.3.2 Social Media and Crowdsourcing

2.3.2.1 The Significance of Social Media Platforms in Disaster Management

Social media platforms have emerged as powerful tools for disaster management, offering significant advantages in communication, information sharing, and community engagement. The key significances of social media platforms in disaster management are as follows:

Rapid Information Dissemination: Social media platforms enable rapid dissemination of information during disasters. Emergency management agencies, government authorities, and humanitarian organizations can use social media to share real-time updates, emergency alerts, evacuation orders, and safety instructions. This immediate access to information helps keep the affected population informed and allows them to make timely decisions to protect themselves.

Crowd-sourced Data and Situational Awareness: Social media platforms allow individuals to share their experiences, observations, and photos/videos of a disaster. This user-generated content provides valuable insights and complements official data, thereby contributing to comprehensive situational awareness. Emergency responders can monitor social media platforms to gather real-time information about evolving situations, identify emerging needs, and prioritize response efforts accordingly.

Two-way Communication and Engagement: Social media facilitate two-way communication between emergency management agencies and the affected population. Individuals can ask questions, seek assistance, or provide feedback through comments, direct messages, or dedicated hashtags. This direct engagement helps authorities address concerns, clarify information, and offer support, fostering a sense of community and trust during crises.

Community Support and Volunteer Coordination: Social media platforms serve as hubs for community support and volunteer coordination. People can offer help, share resources, and organize relief efforts through social media channels. Volunteer groups can form online communities, making it easier to coordinate efforts, mobilize resources, and provide aid to the affected areas. Social media also enables the dissemination of information about donation drives, fundraisers, and support services.

Crisis Mapping and Real-time Updates: Social media data can be utilized for crisis mapping, visualizing the extent of a disaster, and identifying areas of immediate concern. By analyzing social media posts, hashtags, and geolocation data, emergency management agencies can gain insights into affected areas, resource needs, and response gaps. Real-time updates on social media platforms keep both responders and the public informed of evolving situations, road closures, shelter availability, and other critical information.

Public Sentiment Analysis and Psychological Support: Social media platforms allow authorities to analyze public sentiment during disasters. By monitoring social media conversations, emergency management agencies can gauge the emotional well-being of affected populations, identify areas of distress, and provide psychological support. Social media can serve as a platform for disseminating mental health resources, counseling services, and helpline information.

Remote Support and Virtual Coordination: Social media platforms enable remote support and virtual coordination among response teams, government agencies, and humanitarian organizations. Professionals from various locations can collaborate, exchange information, and coordinate relief efforts via online platforms. This virtual coordination helps streamline response activities, share best practices, and optimize resource allocation.

Social media platforms have revolutionized disaster management by providing real-time information dissemination, crowd-sourced data, two-way communication, community support, crisis mapping, public sentiment analysis, and virtual coordination. Leveraging the power of social media platforms allows for more effective and inclusive disaster management, fostering community resilience and enabling timely responses to crises.

2.3.2.2 Role of Social Media to Gather Real-Time Information, Identify Affected Areas, and Mobilize Volunteers

The role of social media in gathering real-time information, identifying affected areas, and mobilizing volunteers during disasters is significant. Here, is an elaboration of each aspect.

Gathering Real-Time Information: Social media platforms serve as real-time information sources during disasters. Users can share updates, photos, videos, and eyewitness accounts on the ground. This user-generated content provides immediate and first-hand information regarding the extent of the disaster, impacted areas, and emerging needs. Emergency management agencies and responders can monitor social media feeds to gather crucial real-time information, assess situations, and make informed decisions regarding resource allocation and response strategies.

Identifying Affected Areas: Social media platforms enable the identification of affected areas through geolocation data, hashtags, and user-generated content. By monitoring social media posts, emergency management agencies can pinpoint locations where people report emergencies, request assistance, and share information about hazards. This information helps in mapping the affected areas, prioritizing rescue and relief operations, and directing resources to where they are most needed.

Mobilizing Volunteers: Social media platforms are powerful tools for mobilizing volunteers and for coordinating relief efforts. During a disaster, individuals can use social media to express their willingness to help and offer services. Volunteer organizations and community groups can create dedicated social media pages or groups to recruit volunteers, disseminate information about volunteer opportunities, and coordinate volunteer activities. Social media allows for the rapid dissemination of calls for assistance, sharing of volunteer needs, and coordination of volunteer efforts, thereby facilitating the efficient mobilization of resources and manpower.

Crowdsourcing and Community Engagement: Social media platforms enable crowdsourcing of information and engage communities in disaster management efforts. Hashtags related to a disaster can be used to aggregate relevant information and create a searchable update repository. Communities affected by a disaster can participate in discussions, share experiences, and provide mutual support through social media platforms. This engagement fosters a sense of solidarity and empowers individuals to actively contribute to their response and recovery efforts.

Rapid Communication and Alerts: Social media platforms facilitate rapid communication and dissemination of alerts to a wide audience. Emergency management agencies and organizations can use social media to send official announcements, emergency alerts, evacuation notices, and safety instructions. The shareability of social media posts allows information to reach a broader network of users, thereby increasing the likelihood of it being seen and shared by others. This helps to ensure that critical information reaches as many people as possible, including those who may not directly follow traditional media channels.

Overall, social media play a crucial role in disaster management by gathering real-time information, identifying affected areas, mobilizing volunteers, enabling crowdsourcing, and facilitating rapid communication. By harnessing the power of social media platforms, emergency management agencies, responders, and communities can collaborate effectively, respond promptly to disasters, and provide timely assistance to those in need.

2.3.2.3 The Concept of Crowdsourcing and How it Can Assist in Data Collection, Damage Assessment, and Resource Allocation

Crowdsourcing involves harnessing the collective intelligence, skills, and contributions of a large group of people to accomplish a specific task or to solve a problem. In the context of disaster management, crowdsourcing plays a significant role in data collection, damage assessment, and resource allocation. This explains how crowdsourcing can assist in these areas.

Data Collection: Crowdsourcing allows the collection of vast amounts of data from a wide range of sources and perspectives. During a disaster, affected individuals, eyewitnesses, and volunteers can contribute valuable information through various channels including social media, dedicated platforms, and mobile apps. These data include real-time updates, situational reports, photos, videos, and geolocation information. Using crowdsourcing data, emergency management agencies can obtain a more comprehensive and dynamic understanding of the situation on the ground.
Damage Assessment: Crowdsourcing can aid in assessing the extent of damage caused by a disaster. Individuals within affected communities can report damage to their property, infrastructure, and environment. By sharing photos, videos, and descriptions of the damage, they contribute to a collective database that helps emergency responders prioritize their efforts and effectively allocate resources. Crowdsourced data can provide a broader and more detailed perspective on the impact of disasters across different areas, enabling more accurate damage assessment.
Resource Allocation: Crowdsourcing can assist in identifying needs and efficiently allocating resources. By actively engaging affected populations and volunteers through crowdsourcing platforms, emergency management agencies can gather information on specific needs such as food, water, shelter, medical supplies, and transportation. This information helps in determining resource requirements and directing assistance to areas and individuals who need it the most. By involving the crowd in

resource-allocation decisions, the process becomes more inclusive and responsive to the actual needs of the affected communities.

Expertise and Specialized Skills: Crowdsourcing allows for tapping into the expertise and specialized skills of individuals within the crowd. During a disaster, professionals such as engineers, medical personnel, and translators can remotely contribute their knowledge and skills. Through expert advice and guidance, emergency management agencies can benefit from a diverse pool of expertise and make informed decisions related to infrastructure assessments, medical support, language translation, and other specialized needs.

Validation and Verification: Crowdsourcing can also assist in the validation and verification of information. Given the abundance of data generated during a disaster, it is necessary to ensure the accuracy and reliability of the information collected. By involving multiple contributors, cross-referencing data, and leveraging the collective wisdom of the crowd, emergency management agencies can verify and validate the authenticity of information before using it for decision-making.

Overall, crowdsourcing in disaster management enables the collection of timely and diverse data, facilitates damage assessment, helps allocate resources effectively, leverages specialized expertise, and enhances the validation and verification processes. By harnessing the power of the crowd, emergency management agencies can leverage collective intelligence and make informed and targeted decisions to mitigate the impact of disasters and provide efficient support to affected communities.

2.4 Geospatial Technologies for Situational Awareness

2.4.1 Geographic Information Systems (GISs)

2.4.1.1 GIS and Its Applications in Disaster Management

GIS, which is Geographic Information System, is a powerful tool used in disaster management for capturing, analyzing, and visualizing geospatial data. It allows the integration of various types of data, including maps, satellite imagery, and attribute data, to provide valuable insights and support decision-making processes. Some key applications of GIS in disaster management are as follows:

Risk Assessment and Vulnerability Analysis: GIS is used to assess and map areas prone to natural hazards such as floods, earthquakes, or wildfires. By overlaying different layers of data, including elevation, land use, population density, and infrastructure, GIS helps to identify high-risk areas and vulnerable populations. This information aids in planning and preparedness efforts, enabling authorities to allocate resources and implement appropriate mitigation strategies.

Emergency Planning and Response: GIS plays a crucial role in emergency planning and response. It allows emergency management agencies to create dynamic incident maps that display real-time information about the disaster, including affected areas, evacuation routes, emergency shelters, and critical infrastructure. GIS helps responders make informed decisions by visualizing data related to the location of emergency resources such as hospitals, fire stations, and transportation networks, optimizing response efforts, and ensuring effective resource allocation.

Damage Assessment and Recovery: Following a disaster, a GIS was used for damage assessment and recovery planning. By integrating data from various sources, including aerial imagery, field surveys, and satellite data, GIS helps create detailed damage assessment maps. These maps provide an overview of the affected areas, infrastructure damage, and status of critical facilities. Such information aids in prioritizing recovery efforts, estimating resource requirements, and facilitating the allocation of financial aid and reconstruction projects.

Evacuation Planning and Route Optimization: GIS supports evacuation planning by identifying evacuation zones, determining optimal evacuation routes, and estimating evacuation time. By considering factors such as road networks, population density, traffic patterns and shelter locations, GIS can help authorities develop evacuation plans that minimize congestion, ensure the safety of evacuees, and maximize the efficiency of the evacuation process.

Resource Management and Logistics: GIS is used for resource management and logistics during disaster response. It enables the tracking and visualization of available resources such as emergency supplies, personnel, and equipment in real time. GIS helps to identify gaps in resource distribution, optimizes supply chain logistics, and facilitates coordination among different response teams and agencies involved in the disaster management process.

Public Awareness and Communication: GIS supports public awareness and communication efforts by providing interactive maps, web-based applications, and mobile platforms. These tools allow the public to access vital information about disasters, including evacuation orders, shelter locations, road closures, and hazard zones. GIS-based applications facilitate two-way communication, enabling users to report incidents, request assistance, provide feedback, and enhance public engagement and community resilience.

Overall, GIS applications in disaster management provide geospatial insights and decision-support tools for emergency management agencies. By leveraging GIS technology, authorities can enhance risk assessment, improve emergency planning, optimize resource allocation, and facilitate efficient response and recovery efforts, ultimately minimizing the impact of disasters and ensuring the safety and well-being of the affected communities.

2.4.1.2 Role of GIS in Creating Spatial Databases, Mapping Affected Areas, and Analyzing Vulnerability and Risk

Geographic Information System (GIS) plays a vital role in creating spatial databases, mapping affected areas, and analyzing vulnerability and risk in the context of disaster management. The following is an explanation of how GIS serves these functions.

Creating Spatial Databases: GIS allows the creation of spatial databases that store geographically referenced data. These databases integrate various types of information including maps, satellite imagery, sensor data, and attribute data related to infrastructure, demographics, and environmental factors. By organizing and structuring data in a spatial context, GIS enables the efficient storage, retrieval, and analysis of information relevant to disaster management.

Mapping Affected Areas: GIS is instrumental in mapping and visualizing affected areas during and after a disaster. By overlaying geospatial data, such as satellite imagery, aerial photographs, and topographic maps, with real-time updates from field surveys or remote sensing technologies, GIS provides a comprehensive view of impacted regions. Mapping affected areas helps emergency management agencies and responders to understand the extent and severity of the disaster, identify hotspots, plan response efforts, and allocate resources effectively.

Analyzing Vulnerability and Risk: GIS supports the analysis of vulnerability and risk by integrating multiple layers of spatial data. By overlaying data related to hazards (e.g., flood zones and fault lines), exposure (e.g., population density and critical infrastructure) and vulnerability (e.g., socioeconomic factors and building codes), GIS enables the assessment of areas susceptible to disasters. Through spatial analysis techniques, such as buffering, interpolation, and spatial modeling, GIS helps quantify and visualize vulnerability and risk levels, aiding in decision-making processes related to mitigation strategies, land-use planning, and emergency response.

Hazard Modeling and Scenario Analysis: GIS allows for hazard modeling and scenario analysis to understand the potential impact of different disaster scenarios. GIS helps forecast potential paths, intensities and affected areas by simulating and modeling hazards, such as hurricanes, earthquakes, or wildfires. This information assists emergency preparedness by enabling authorities to develop contingency plans, identify evacuation routes, and evaluate the effectiveness of various response strategies.

Spatial Statistical Analysis: GIS provides spatial statistical analysis capabilities to assess patterns, relationships, and trends in geospatial data. This analysis helps to identify areas at higher risk, understand the spatial distribution of vulnerabilities, and prioritize interventions based on evidence-based decision-making. GIS tools can perform statistical operations such as hotspot analysis, clustering, and density mapping to uncover patterns that aid in risk assessment and resource allocation.

Decision Support and Visualization: GIS serves as a decision support system by providing visualizations, maps, and interactive tools for data exploration. Through GIS, decision makers can gain insights into complex spatial data, interpret the relationships between various factors, and evaluate different scenarios. GIS visualizations, such as heat maps, thematic maps, and 3D representations, enhance understanding, facilitate effective communication with stakeholders, and support informed decision-making in disaster management.

In summary, GIS plays a critical role in creating spatial databases, mapping affected areas, and analyzing vulnerability and risk in disaster management. By leveraging GIS technology, emergency management agencies can make informed decisions, plan mitigation strategies, allocate resources efficiently, and enhance disaster resilience in communities.

2.4.1.3 Integration of GIS with Other Technologies for Better Situational Awareness

The integration of Geographic Information System (GIS) with other technologies enhances situational awareness in disaster management. By combining GIS with various technologies, emergency management agencies can gather and analyze geospatial data more effectively, thereby improving their understanding of the situation on the ground. Below are some examples of how GIS can be integrated with other technologies for better situational awareness.

Remote Sensing and Satellite Imagery: GIS can be integrated with remote sensing technologies, such as satellites and aerial platforms, to acquire high-resolution imagery and data. This integration allows for the real-time monitoring of disaster-affected areas, tracking changes over time, and providing up-to-date visual information. By overlaying satellite imagery onto GIS platforms, emergency management agencies can analyze the extent of damage, identify areas of concern, and make informed decisions based on current conditions.

Internet of Things (IoT) Devices and Sensor Networks: Integrating GIS with IoT devices and sensor networks provides real-time data on environmental conditions, infrastructure status, and human activity. Sensors embedded in various devices can collect data on parameters such as temperature, humidity, air quality, water level, and seismic activity. GIS can then analyze and visualize these data, allowing emergency management agencies to identify potential risks, monitor critical infrastructure, and make informed decisions based on real-time information.

Mobile Technologies: The integration of GIS with mobile technologies such as smartphones and tablets enables field data collection and real-time updates. Field personnel can collect geospatial data, capture images, and input information directly into a mobile GIS application. These data include damage assessments, resource needs, and situational updates. By integrating mobile technologies with GIS, emergency management agencies can access and analyze real-time data from the field, enabling quicker decision-making and response coordination.

Communication and Collaboration Tools: GIS can be integrated with communication and collaboration tools, such as web-based platforms and social media, to enhance situational awareness and facilitate information sharing. Real-time communication channels allow emergency management agencies, responders, and the public to share updates, coordinate their response efforts, and disseminate critical information. By integrating GIS with these tools, agencies can visualize and share geospatial data, collaborate with stakeholders, and provide timely information to the public.

Data Analytics and Artificial Intelligence: Integrating GIS with data analytics and artificial intelligence (AI) technologies enhances the analysis and interpretation of geospatial data. Advanced analytical techniques, such as spatial modeling, machine learning, and predictive analytics, can be applied to GIS data to identify patterns and forecast trends and to make data-driven predictions. This integration enables emergency management agencies to gain deeper insights into the situation, anticipate future risks, and optimize resource allocation strategies.

By integrating GIS with these technologies, emergency management agencies can gather, analyze, and visualize geospatial data in a comprehensive and timely manner. This integration enhances situational awareness, supports informed decision making, and enables more effective response and recovery efforts in disaster management.

2.4.2 Unmanned Aerial Vehicles (UAVs) and Drones

2.4.2.1 Role of UAVs and Drones in Disaster Response and Recovery

Unmanned Aerial Vehicles (UAVs) and drones play a crucial role in disaster response and recovery efforts. They offer unique capabilities for data collection, situational assessment, and support at various stages of the disaster management cycle. The key roles of UAVs in disaster response and recovery are as follows:

Rapid Damage Assessment: UAVs can quickly survey disaster-affected areas and provide high-resolution aerial imagery or video footage. This allows emergency management agencies to assess the extent of damage to the infrastructure, buildings, and critical facilities. By capturing real-time data, UAVs enable rapid damage assessment, helping prioritize response efforts and resource allocation.

Search and Rescue Operations: UAVs equipped with thermal imaging cameras and other sensors can assist with search and rescue operations. They can cover large areas in a short time, locate survivors, and identify hazards or inaccessible areas. UAVs provide aerial views, enhance situational awareness, and guide search and rescue teams to areas where survivors may be trapped or need assistance.

Hazard Mapping and Risk Assessment: UAVs can be used to map and monitor hazardous areas or potential risks. By capturing aerial imagery, UAVs help identify and map flood-prone zones, landslide areas, and areas at risk of wildfires. This information aids risk assessment, evacuation planning, and proactive mitigation measures.

Infrastructure Inspection: UAVs equipped with cameras or sensors can inspect critical infrastructure such as bridges, power lines, or telecommunications towers after a disaster. They can quickly identify structural damage, monitor the stability of buildings, and assess the conditions of utility networks. UAV-based inspections provide valuable data for making informed decisions regarding repairs, reconstructions, and temporary fixes.

Delivery of Supplies and Aid: UAVs have the potential to deliver essential supplies, such as medical kits, food, or water, to inaccessible or remote areas affected by disasters. In situations in which traditional transportation routes are disrupted or unsafe, UAVs can be deployed to reach stranded populations or provide immediate relief until ground logistics are established.

Communication and Mapping Support: UAVs equipped with communication devices can serve as temporary communication relays in areas where the infrastructure has been damaged or destroyed. They can establish communication networks, enabling emergency responders to communicate and coordinate effectively. In addition, UAVs can generate detailed 3D maps and topographic models of disaster-affected areas, assisting in urban planning, infrastructure reconstruction, and resource allocation.

Environmental Monitoring: UAVs can monitor environmental conditions such as air quality, water quality, and the spread of pollutants following a disaster. They can collect samples, capture imagery, or use specialized sensors to assess environmental impacts and guide appropriate response measures.

The use of UAVs in disaster response and recovery enhances situational awareness, accelerates data collection, and provides valuable insights for decision making. By leveraging their capabilities, emergency management agencies can respond more efficiently, allocate resources effectively, and aid in the recovery process, ultimately saving lives and minimizing the impact of disasters.

2.4.2.2 Applications of UAVs in Aerial Surveys, Damage Assessment, and Search and Rescue Operations

Unmanned Aerial Vehicles (UAVs) have a wide range of applications in aerial surveys, damage assessments, and search and rescue operations during disaster management. Let us explore specific UAV applications in these areas.

Aerial Surveys and Mapping: UAVs equipped with high-resolution cameras or sensors can capture aerial imagery and generate detailed maps of disaster-affected areas. These data can be used for rapid assessment, situational analysis, and the identification of critical infrastructure, hazards, and resources. UAVs enable the creation of up-to-date high-resolution maps that aid decision-making, resource allocation, and planning during the response and recovery phases.

Damage Assessment: UAVs play a vital role in assessing the extent of damage to infrastructure and buildings after a disaster. Equipped with cameras or sensors, UAVs can capture detailed images of damaged structures, allowing emergency management agencies to evaluate the severity of the destruction, identify collapsed or unstable buildings, and assess potential risks. This information helps prioritize rescue and recovery operations as well as determine the need for structural assessments and repairs.

Search and Rescue Operations: UAVs are valuable tools in search and rescue operations, particularly in hard-to-reach or hazardous areas. Equipped with thermal cameras, UAVs can detect heat signatures and locate survivors even under low-light or obscured conditions. They provide aerial views and real-time data to aid in identifying distressed people, directing ground teams to their locations, and guiding evacuation routes. UAVs enhance the speed and effectiveness of search and rescue efforts, potentially saving lives.

Monitoring Hazardous Environments: UAVs equipped with gas sensors or other environmental monitoring devices can assess and monitor hazardous environments after a disaster. They can be used to measure air quality, detect chemical leaks, or identify the spread of pollutants. This information is critical for identifying risks to responders and affected populations, and for guiding appropriate safety measures.

Communication and Mapping Support: UAVs equipped with communication devices can establish temporary communication networks in disaster-affected areas, where infrastructure has been damaged or disrupted. They can act as communication relays, enabling emergency responders to coordinate and exchange vital information in real-time. In addition, UAVs can create detailed 3D maps and topographic models of affected areas, assisting in planning and resource allocation during the recovery phase.

Delivery of Supplies and Aid: UAVs have the potential to deliver essential supplies, such as medical kits, food, or water, to areas that are inaccessible or have damaged transportation routes. They can provide immediate relief and support until the ground logistics are established. UAVs can transport lightweight payloads to stranded populations, thereby improving the access to critical resources during emergencies.

These applications highlight the versatility and effectiveness of UAVs for disaster management. By leveraging their capabilities in aerial surveys, damage assessment, and search and rescue operations, UAVs significantly contribute to enhancing situational awareness, accelerating response efforts, and ultimately, saving lives in the face of disasters.

2.4.2.3 *Challenges and Ethical Considerations Associated with the Use of Drones in Disaster Management*

The use of drones in disaster management has numerous benefits, but also poses certain challenges and ethical considerations. Some of the key challenges and ethical considerations associated with the use of drones in this context are as follows:

Regulatory Compliance: Operating drones in disaster zones require compliance with relevant regulations and airspace restrictions. Obtaining necessary permits and authorizations, adhering to flight restrictions, and ensuring the safety of airspaces can be challenging, especially in dynamic and rapidly changing emergencies. Strict compliance with regulations is essential to prevent interference with other aerial operations and to ensure the safety of both drone operators and the public.
Privacy Concerns: Drones equipped with cameras or sensors can capture images and videos of the affected areas and individuals. This raises privacy concerns because the collection of personal data and footage in disaster-stricken communities must be handled with care. Respecting privacy rights, obtaining informed consent when applicable, and implementing safeguards to protect sensitive data are essential ethical considerations when deploying drones for data collection.
Security and Safety Risks: Drones themselves can become potential security risks if they are misused or fall into the wrong hands. Unauthorized individuals may exploit drones for malicious purposes, such as the reconnaissance of critical infrastructure or interference with emergency operations. Ensuring proper security measures, including secure communication channels, encryption, and restricted access to drone operations, is crucial for mitigating these risks.
Interference with Emergency Operations: In disaster situations, coordination between various emergency response teams is essential. However, the presence of drones in airspace can potentially interfere with the operations of manned aircraft, such as helicopters or airplanes. Proper communication protocols, coordination with relevant authorities, and adherence to airspace management guidelines are necessary to prevent disruptions and maintain safety during emergencies.
Technical Limitations and Environmental Conditions: Drones may have limitations in terms of flight endurance, payload capacity, and maneuverability. Adverse weather conditions, such as strong winds, heavy rain, or low visibility, can affect the ability of drones to operate effectively. Understanding these technical limitations and assessing the environmental

conditions before deploying drones is crucial to ensure their safe and efficient use.

Public Perception and Acceptance: Drones are relatively new technologies, and public perceptions can vary. Some individuals may view drones as intrusive or perceive them as threats to privacy and safety. Building public trust, ensuring transparency in drone operations, and engaging with affected communities to address concerns and provide information about the purpose and benefits of drone use are important for fostering acceptance and minimizing potential conflicts.

Ethical Use and Data Management: Responsible and ethical use of drone-collected data is essential. Clear guidelines for data storage, handling, and sharing should be in place to protect the privacy and confidentiality of individuals and communities affected by disasters. Data should be used for legitimate purposes, and efforts should be made to minimize potential biases or discriminatory practices in data analysis and decision making.

Addressing these challenges and ethical considerations requires a comprehensive approach involving regulatory frameworks, stakeholder engagement, community outreach, and ongoing dialogue among drone operators, emergency management agencies, and the public. By promoting responsible and ethical drone use, we can maximize the benefits of these technologies in disaster management, while safeguarding privacy, ensuring safety, and respecting the rights of affected communities.

2.5 Data Analytics and Decision Support Systems

2.5.1 Big Data and Predictive Analytics

2.5.1.1 Big Data Analytics Application in Disaster Risk Assessment, Resource Allocation, and Decision Making

Big data analytics plays a significant role in disaster management, specifically in disaster risk assessment, resource allocation, and decision-making. Some applications of big data analytics in these areas are as follows:

Disaster Risk Assessment: Big data analytics can analyze large volumes of data from diverse sources, including historical disaster data, weather data, population demographics, infrastructure data, and social media feeds. By applying advanced analytics techniques such as machine learning and predictive modeling, big data analytics can identify patterns, assess vulnerabilities, and predict potential risks. This helps to understand and quantify disaster risks more accurately, enabling proactive measures to mitigate and prepare for disasters.

Resource Allocation: Efficient resource allocation is crucial during disaster response. Big data analytics can analyze real-time data from various sources, such as emergency calls, sensor networks, and social media, to identify the areas and populations most affected by a disaster. By analyzing the data on infrastructure damage, population density, and resource availability, big data analytics can optimize resource allocation, ensuring that critical supplies, personnel, and equipment are directed to areas with the highest need.

Decision-Making: Big data analytics provides valuable insights for decision-making during disasters. By analyzing large datasets, including real-time data streams, historical data, and geospatial information, decision makers can gain a comprehensive understanding of the situation on the ground. Big data analytics can help identify evacuation routes, prioritize response efforts, assess the effectiveness of interventions, and support strategic decision making in resource allocation and recovery planning. Real-time analytics can also enable decision makers to respond quickly to changing conditions and make data-driven decisions for effective disaster management.

Early Warning Systems: Big data analytics can enhance early warning systems by processing large amounts of data from multiple sources. By analyzing weather data, sensor data, satellite imagery, and historical patterns, big data analytics can identify early indicators of potential disasters such as hurricanes, floods, and wildfires. This allows for early detection, timely warnings, and proactive evacuation measures, reducing the impact on vulnerable populations and enabling more effective disaster responses.

Social Media Analysis: Social media platforms generate a vast amount of data during disasters, including real-time updates, situational reports, and requests for assistance. Big data analytics can analyze social media feeds, extract relevant information, and derive valuable insights into the evolving situation, public sentiment, and emerging needs. This information can inform decision makers, aid in resource allocation, and support targeted response efforts.

Predictive Analytics: Big data analytics can utilize predictive modeling to forecast the future behavior of disasters and their potential impacts. By analyzing historical data, weather patterns, geographical features, and other relevant factors, predictive analytics can be used to generate models that estimate the likelihood of certain events and their potential consequences. These insights can help in planning and preparedness efforts, enabling decision makers to allocate resources and implement measures to minimize the impact of disasters.

Overall, big data analytics provides a powerful tool for understanding, assessing, and responding to disasters. By leveraging large datasets and advanced analytics techniques, it supports evidence-based decision making, enhances risk assessment, optimizes resource allocation, and improves the overall effectiveness of disaster management efforts.

2.5.1.2 Potential of Predictive Analytics in Forecasting Disaster Impacts, Identifying Vulnerable Populations, and Optimizing Response Strategies

Predictive analytics holds immense potential in disaster management, particularly in forecasting disaster impacts, identifying vulnerable populations, and optimizing response strategies. Here is a closer look at how predictive analytics can be leveraged in these areas:

Forecasting Disaster Impacts: Predictive analytics uses historical data, real-time data, and advanced modeling techniques to forecast the potential impacts of disasters. By analyzing factors, such as weather patterns, geographical features, infrastructure data, and historical disaster data, predictive analytics can generate models that estimate the severity and extent of a disaster. These forecasts provide valuable insights to emergency management agencies, enabling them to anticipate the scale of a disaster, plan response efforts, and allocate resources accordingly.

Identifying Vulnerable Populations: Predictive analytics can help to identify vulnerable populations that are most likely to be affected by a disaster. By analyzing demographic data, socioeconomic indicators, and historical patterns, predictive models can identify areas with higher vulnerability and predict which communities are at greater risk. This information allows emergency responders to prioritize their efforts, allocate resources, and tailor interventions to protect and assist those who are most vulnerable.

Optimizing Response Strategies: Predictive analytics can optimize response strategies by analyzing various factors and generating actionable insights. By integrating data from multiple sources, including weather forecasts, infrastructure data, historical response data, and social media feeds, predictive models can determine the most effective response strategy. For example, predictive analytics can help determine optimal evacuation routes, identify suitable shelter locations, and allocate resources based on the predicted needs. These insights enable emergency management agencies to make informed decisions and respond more efficiently to disasters.

Resource Allocation and Planning: Predictive analytics aids in optimizing resource allocation and planning by forecasting the demand for resources based on predicted disaster impacts. By analyzing historical data, population density, infrastructure data and response time metrics, predictive models can estimate the resources needed, such as personnel, medical supplies, equipment, and transportation. This information enables emergency management agencies to allocate resources effectively, pre-position supplies in strategic locations and optimize logistical planning to ensure a timely and targeted response.

Situational Awareness: Predictive analytics enhances situational awareness by continuously analyzing real-time data and updating predictive models accordingly. By integrating data from various sources, including sensors, satellite imagery, social media and emergency calls, predictive analytics can provide real-time insights into an evolving situation. This helps emergency responders to stay informed about changing conditions, anticipate potential challenges, and make data-driven decisions for effective disaster response.

Risk Mitigation and Preparedness: Predictive analytics supports risk mitigation and preparedness efforts by identifying potential hazards and vulnerabilities before a disaster strikes. By analyzing historical data, geographic information, and other relevant factors, predictive models can identify areas prone to specific types of disasters and assess their vulnerability. This information allows for targeted mitigation measures such as reinforcing infrastructure, implementing early warning systems, and conducting community education programs.

The potential of predictive analytics lies in its ability to leverage vast amounts of data and advanced modeling techniques to generate insights that inform decision making, optimize resource allocation, and enhance overall disaster response. By harnessing the power of predictive analytics, emergency management agencies can improve their ability to forecast impact, identify vulnerable populations, and implement effective strategies to mitigate the effects of disasters and save lives.

2.5.2 Artificial Intelligence (AI) and Machine Learning (ML)

2.5.2.1 The Use of AI and ML Techniques in Disaster Management

The use of Artificial Intelligence (AI) and Machine Learning (ML) techniques in disaster management has the potential to revolutionize the way we prepare for, respond to, and recover from disasters. The following are some key areas where AI and ML techniques are applied.

Predictive Modeling: AI and ML algorithms can analyze historical data, weather patterns, geographical features, and other relevant factors to build predictive models for forecasting disaster events. These models can help anticipate the occurrence, intensity, and impact of disasters, thereby enabling early warning systems and proactive preparedness measures.
Image and Sensor Analysis: AI and ML techniques can analyze satellite imagery, aerial photographs, and sensor data to assess the extent of damage caused by disasters. Computer vision algorithms can automatically identify damaged infrastructure, collapsed buildings, and other critical features, aiding rapid damage assessment and resource allocation.
Risk Assessment and Vulnerability Analysis: AI and ML algorithms can analyze large datasets containing demographic information, socioeconomic indicators, and infrastructure data to assess vulnerability and identify high-risk areas. By identifying vulnerable populations and areas with a higher likelihood of damage, decision makers can prioritize resources and interventions for risk reduction and resilience-building.
Real-time Data Analysis: AI and ML algorithms can process real-time data from various sources, such as social media feeds, sensor networks, and emergency calls, to extract relevant information and generate situational awareness. This allows timely decision-making, resource allocation, and response coordination during a disaster.
Natural Language Processing (NLP): NLP techniques enable the analysis of text data such as social media posts, news articles, and emergency messages to extract valuable insights. Sentiment analysis and topic modeling can help identify critical needs, public sentiments, and emerging issues, thereby facilitating effective communication and response coordination.
Resource Allocation and Optimization: AI and ML algorithms can optimize the allocation of limited resources such as emergency personnel, equipment, and supplies based on real-time demand, resource availability, and logistical constraints. These techniques help to streamline resource distribution, minimize response time, and improve the overall effectiveness of resource management.
Decision Support Systems: AI and ML techniques can power decision support systems to assist emergency managers and responders in making informed decisions during a disaster. These systems integrate data from multiple sources, perform real-time analyses, and provide actionable insights for decision-making in resource allocation, evacuation planning, and response coordination.

Robotics and Autonomous Systems: AI-powered robotics and autonomous systems can be deployed in disaster scenarios to perform tasks that are dangerous or inaccessible. For example, drones equipped with AI algorithms can conduct search and rescue operations, deliver supplies to affected areas, and assess hazardous environments.

The integration of AI and ML techniques into disaster management systems enhances the speed, accuracy, and efficiency of the decision-making processes, enabling more effective preparedness, response, and recovery efforts. However, it is important to address ethical considerations, privacy concerns, and the need for human oversight to ensure responsible and beneficial use of these technologies in disaster management.

2.5.2.2 Applications of AI-Powered Algorithms for Analyzing Large Datasets, Automating Damage Assessment, and Supporting Evacuation Planning

AI-powered algorithms have numerous applications in disaster management for analyzing large datasets, automating damage assessments, and supporting evacuation planning. Some specific examples are as follows:

Large Dataset Analysis: AI-powered algorithms can process vast amounts of data collected from various sources, such as satellite imagery, social media feeds, sensor networks, and historical records. Using techniques such as data mining and machine learning, these algorithms can identify patterns, trends, and anomalies within the data. This analysis helps to understand the current situation, predict potential risks, and inform decision-making processes.

Automated Damage Assessment: AI algorithms can analyze satellite imagery, aerial photographs, and sensor data to automatically assess the extent of damage caused by a disaster. Computer-vision techniques enable algorithms to identify damaged infrastructure, collapsed buildings, and other critical features. This automation accelerates the damage assessment process, allowing emergency responders to prioritize their efforts and allocate resources more efficiently.

Evacuation Planning: AI-powered algorithms can support the planning and coordination of evacuation. By analyzing the population density, traffic patterns, road networks, and real-time data, these algorithms can optimize evacuation routes and provide real-time updates on traffic congestion or road closures. This information aids in making informed decisions regarding evacuation timing, resource allocation, and ensuring the safety of the affected populations.

Risk Assessment and Early Warning Systems: AI algorithms can analyze historical data, weather patterns, geographical features, and other relevant factors to assess risks and provide early warnings. Using machine learning techniques, these algorithms can identify patterns and correlations between different variables, enabling the prediction of potential disaster events. This information will help in implementing early warning systems that provide timely alerts to vulnerable populations, allowing them to take appropriate precautions.

Decision Support Systems: AI-powered decision support systems assist emergency managers and responders in making informed decisions during a disaster. These systems integrate real-time data from multiple sources and apply AI algorithms to analyze and visualize information. They provided actionable insights and recommendations for resource allocation, response coordination, and evacuation planning. By leveraging AI, decision-makers can access up-to-date and data-driven guidance for effective decision making.

Simulation and Modeling: AI algorithms can simulate and model different disaster scenarios based on historical data and real-time information. These simulations helps to understand the potential impacts of disasters, test different response strategies, and evaluate their effectiveness. By running multiple simulations, AI algorithms can provide insights into the potential outcomes of different intervention options, aiding the selection of the most effective response strategies.

These applications demonstrate that AI-powered algorithms can enhance the efficiency, accuracy, and speed of data analysis, damage assessment, and decision-making processes in disaster management. By leveraging the power of AI, emergency management agencies can better understand the situation, allocate resources effectively, and improve their overall response and recovery efforts.

2.5.2.3 Ethical Implications and Challenges of Using AI and ML in Disaster Response

The use of AI and ML in disaster response has ethical implications and challenges that need to be addressed. Some key considerations are as follows:

Bias and Fairness: AI algorithms are only as good as the data they are trained on. If the training data are biased or incomplete, it can lead to biased outcomes and unfair treatment in certain populations. It is crucial to ensure that the data used for training AI models are diverse, representative, and free from bias. In addition, ongoing monitoring and evaluation of the algorithms should be conducted to detect and rectify any biases that may emerge during deployment.

Privacy and Data Protection: AI and ML systems rely on large amounts of data, including personal and sensitive information. It is essential to handle this data with strict adherence to privacy regulations and ethical standards. Data anonymization, encryption, and secure storage practices should be implemented to protect individuals' privacy and prevent unauthorized access to or misuse of data.

Transparency and Explainability: AI algorithms are often complex and difficult to interpret. In disaster response scenarios, transparency and explainability are crucial in the decision-making process. The algorithms should be designed to allow stakeholders to understand how decisions are being made. Providing explanations and justifications for the outcomes generated by AI systems builds trust and facilitates accountability.

Human Oversight and Responsibility: While AI can automate certain tasks and support decision-making, it is important to maintain human oversight and responsibility. Humans should be able to intervene, question, and override AI-generated decisions when necessary. The ultimate responsibility for the outcomes of AI systems lies with human operators and decision makers.

System Robustness and Resilience: AI and ML systems are susceptible to adversarial attacks, system failures, or data limitations. In disaster response scenarios where accurate and reliable information is crucial; it is essential to ensure the robustness and resilience of AI systems. Regular testing, validation, and backup mechanisms should be implemented to minimize the risk of system failure or malicious manipulation.

Community Engagement and Inclusivity: Engaging local communities and stakeholders in the development and deployment of AI systems is essential for ensuring inclusivity and fairness. Community inputs can provide valuable insights, identify potential biases, and address the specific needs and concerns of different population groups. Collaboration and participatory approaches help to create AI systems that are more responsive to the diverse needs of affected communities.

Capacity Building and Training: Adequate training and capacity-building initiatives should be in place to ensure that users and operators of AI systems have the necessary skills and knowledge to understand the limitations and potential biases of technology. Training programs can also emphasize ethical considerations and responsibilities associated with using AI in disaster response.

Addressing these ethical implications and challenges requires a multidisciplinary approach that involves technologists, ethicists, policymakers, and communities. By promoting transparency, fairness, privacy protection, and responsible decision-making, the ethical use of AI and ML in disaster response can enhance the effectiveness and efficiency of emergency management efforts, while safeguarding the rights and well-being of affected populations.

2.6 Conclusion

2.6.1 Key Advancements in Technology for Disaster Management Discussed in the Chapter

In this chapter, several key advancements in disaster management technology are discussed. Some of the major advancements highlighted are as follows:

Remote Sensing and Satellite Technology: The use of remote sensing and satellite technology has revolutionized disaster management. It enables the monitoring and detection of natural disasters, such as hurricanes, floods, and wildfires from space. Satellite imagery provides valuable information for the early detection, tracking, and prediction of disaster events, helping emergency management agencies plan and respond effectively.

Sensor Networks and IoT Devices: Sensor networks and Internet of Things (IoT) devices play critical roles in disaster management. These devices can monitor various parameters such as temperature, humidity, seismic activity, and water levels in real time. By providing continuous data, they enable early warning, facilitate situational awareness, and support decision-making processes during emergencies.

Mobile Technologies and Applications: Mobile technologies, including mobile phones, SMS alerts, and mobile apps, have become invaluable tools for disaster management. They enable the dissemination of emergency alerts, provide information to affected populations, and facilitate coordination among the responders. Mobile technologies also aid in locating survivors, coordinating rescue operations, and providing critical updates in real-time.

Social Media Platforms: Social media platforms have emerged as important tools for disaster management. They enable real-time information sharing, help in gathering situational updates from affected individuals, and facilitate communication and coordination between response agencies and the public. Social media platforms can be used to mobilize volunteers, organize relief efforts, and amplify emergency messages.

Geographic Information Systems (GIS): GIS technology plays a vital role in disaster management by creating spatial databases, mapping affected areas, and analyzing vulnerability and risk. GIS allows for the integration and visualization of various data layers, aiding decision-making processes, resource allocation, and evacuation planning. It provides valuable insights into the spatial aspects of disasters and helps identify high-risk areas.

Artificial Intelligence (AI) and Machine Learning (ML): AI and ML techniques are being increasingly applied in disaster management. They enable the analysis of large datasets, automate damage assessment, support evacuation planning, and optimize resource allocation. AI and ML algorithms can assist in predictive modeling, real-time data analysis, decision support, and risk assessment, thereby enhancing the overall efficiency and effectiveness of disaster response efforts.

Unmanned Aerial Vehicles (UAVs) and Drones: UAVs and drones are instrumental in disaster response and recovery. They can be deployed for aerial surveys, damage assessments, and search and rescue operations. Equipped with cameras and sensors, drones provide high-resolution imagery and real-time data, allowing rapid and efficient assessment of disaster-affected areas.

These technological advancements have significantly improved disaster management capabilities, enabling better preparedness, response, and recovery. They enhance situational awareness, enable timely and informed decision-making, facilitate communication and coordination, help save lives, and mitigate the impacts of disasters.

2.6.2 The Need for Continued Research, Innovation, and Collaboration to Harness Technology's Full Potential in Mitigating the Impacts of Disasters

Although technology has made significant advancements in disaster management, there is still a need for continued research, innovation, and collaboration to harness its full potential in mitigating the impacts of disasters. There are several reasons why the ongoing efforts in these areas are crucial.

Addressing Evolving Challenges: Disasters are dynamic and their nature and impacts can change over time. Continued research and innovation are essential for keeping pace with emerging challenges. By understanding evolving risks, vulnerabilities, and hazards, researchers and innovators can develop new technologies and approaches to effectively respond to and mitigate the impacts of disasters.

Enhancing Technology Integration: Effective disaster management requires the integration of various technologies into a cohesive system. Continued research and collaboration are needed to develop interoperable systems that seamlessly integrate different technologies such as remote sensing, IoT devices, AI algorithms, and mobile applications. This integration facilitates data-sharing, enhances situational awareness, and enables efficient resource allocation and coordination.

Improving Data Collection and Analysis: Disaster management relies heavily on accurate and timely data. Research and innovation efforts can focus on improving data collection methods, enhancing data quality, and developing advanced data-analysis techniques. This includes leveraging emerging technologies, such as big data analytics, AI, and ML, to extract valuable insights from large datasets, enabling better risk assessment, resource allocation, and decision-making.

Enhancing Resilience and Adaptation: Research and innovation play a critical role in developing technologies and strategies to enhance community resilience and adaptation to disasters. This includes exploring innovative approaches to infrastructure design, developing early warning systems, designing resilient communication networks, and promoting community engagement and participation. Continued efforts in these areas can contribute to the building of more resilient and adaptive societies.

Ensuring Ethical and Responsible Use: As technology evolves, it is important to address the ethical and responsible uses of these tools in disaster management. Research can focus on understanding and addressing ethical implications, and ensuring transparency, fairness, and accountability in the use of AI, ML, and other emerging technologies. Collaboration among researchers, policymakers, and communities is essential for developing guidelines, standards, and best practices that promote ethical and responsible technology use.

Bridging Gaps and Promoting Collaboration: Continued research and collaboration efforts can bridge the gaps between academia, industry, government agencies, and communities. By fostering collaboration and knowledge sharing, stakeholders can collaborate to address common challenges, share best practices, and develop innovative solutions. This multidisciplinary approach can lead to more effective and holistic disaster-management strategies.

Capacity Building and Knowledge Transfer: Research and innovation efforts should focus on capacity building and knowledge transfer to ensure that communities and emergency management professionals can effectively utilize and benefit from new technologies. This includes training programs, workshops, and knowledge dissemination initiatives that promote the adoption and effective use of technology in disaster management.

By emphasizing continued research, innovation, and collaboration, we can harness the full potential of technology to mitigate the impacts of disasters. These efforts contribute to more effective preparedness, response, and recovery, ultimately saving lives and reducing the devastating consequences of disasters.

2.6.3 Vision for the Future, Where Technology Plays a Central Role in Building Resilient Communities and Enhancing Disaster Preparedness

In the future, technology will play a central role in building resilient communities and in enhancing disaster preparedness. The key aspects of this vision are as follows.

Smart and Connected Infrastructure: Communities are equipped with a smart and connected infrastructure that leverages technology to enhance resilience. This includes sensor networks integrated into buildings, bridges, and critical infrastructure to monitor the structural integrity in real time. These systems can provide early warnings of potential failures, enabling proactive maintenance and reducing risks associated with infrastructure failures during disasters.

Advanced Early Warning Systems: Early warning systems have evolved to become more accurate, timely, and localized. They leveraged a combination of remote sensing, satellite technology, IoT devices, and advanced predictive analytics to detect and predict disasters with higher precision. Citizens receive personalized alerts and recommendations on their mobile devices, which allows them to take immediate action and evacuate to safer locations.

Real-time Situational Awareness: Advanced technologies, such as AI, ML, and GIS, enable real-time situational awareness during disasters. Emergency management agencies have access to integrated data from various sources including satellite imagery, sensor networks, social media feeds, and drone surveillance. This comprehensive view of the situation facilitates informed decision-making, resource allocation, and coordination of response efforts.

Collaborative Platforms: Technology enables collaborative platforms to connect various stakeholders, including government agencies, first responders, NGOs, and the public. These platforms facilitate seamless communication, information sharing, and coordination during all the phases of disaster management. They enable rapid mobilization of resources, volunteer engagement, and real-time exchange of critical information.

Virtual and Augmented Reality: Virtual and augmented reality technologies were used to simulate disaster scenarios and train emergency responders. These immersive training experiences allow responders to practice their skills in realistic environments, thereby improving their preparedness and decision-making abilities. Virtual reality is also used to educate and raise public awareness of disaster risks, preparedness, and response strategies.

Robotics and Automation: Robotics and automation play significant roles in disaster response and recovery. Autonomous drones and robots were deployed to assess damage, search for survivors in hazardous areas, and facilitate the delivery of essential supplies. These technologies help reduce risks to human responders and improve the speed and efficiency of response operations.

Citizen Engagement and Crowdsourcing: Technology platforms facilitate citizen engagement and crowdsourcing during disasters. Social media, mobile apps, and online platforms allow citizens to report incidents, share information, and offer assistance. Citizen-generated data and real-time information contribute to situational awareness and aid decision-making processes.

Data Analytics and Predictive Modeling: Advanced data analytics and predictive modeling techniques are extensively used to analyze large datasets and forecast disaster impact. These technologies identify vulnerable populations, assess potential damage to infrastructure, and optimize resource allocation strategies. Decision makers have access to data-driven insights, enabling evidence-based planning and effective allocation of resources.

By embracing this vision, technology can play a central role in building resilient communities and in enhancing disaster preparedness. These advancements will save lives, minimize the economic and social impacts of disasters, and foster a culture of proactive disaster management. However, it is important to ensure that technology is deployed ethically, inclusively, and with a focus on equity to ensure that the benefits are accessible to all communities, including the most vulnerable ones.

References

Singh SK, Kanga S, Gulati B, Raič M, Sajan B, Đurin B, Singh S (2023) Spatial and Temporal Analysis of Hydrological Modelling in the Beas Basin Using SWAT+ Model. *Water* 15, 3338. https://doi.org/10.3390/w1519333

Rani A, Gupta SK, Singh SK, Meraj G, Kumar P, Kanga S, Đurin B, Dogančić D (2023) Predicting Future Land Use Utilizing Economic and Land Surface Parameters with ANN and Markov Chain Models. *Earth* 4(3):728-751.

Ahmad T, Gupta SK, Singh SK, Meraj G, Kumar P, Kanga S (2023) Unveiling Nature's Resilience: Exploring Vegetation Dynamics during the COVID-19 Era in Jharkhand, India, with the Google Earth Engine. *Climate* 11(9):187.

Sharma M, Upadhyay RK, Tripathi G, Kishore N, Shakya A, Meraj G, Kanga S, Singh SK, Kumar P, Johnson BA, Thakur SN (2023) Assessing Landslide Susceptibility along India's National Highway 58: A Comprehensive Approach Integrating Remote Sensing, GIS, and Logistic Regression Analysis. *Conservation* 3(3):444-459.

Saqib N, Rai PK, Kanga S, Kumar D, Đurin B, Singh SK (2023) Assessment of Ground Water Quality of Lucknow City under GIS Framework Using Water Quality Index (WQI). *Water* 15(17):3048.

Choudhury U, Singh SK, Kumar A, Meraj G, Kumar P, Kanga S (2023) Assessing land use/land cover changes and urban heat island intensification: A case study of Kamrup Metropolitan District, Northeast India (2000–2032). *Earth* 4(3):503-521.

Rai NK, Mathur S, Singh SK, Tiwari M, Singh VK, Haque R, Tiwari S, Sharma LK (2023) Differential regulation of mitochondrial complex I and oxidative stress based on metastatic potential of colorectal cancer cells. *Oncol. Let.* 26(1):1-1.

Debnath J, Sahariah D, Lahon D, Nath N, Chand K, Meraj G, Kumar P, Singh SK, Kanga S, Farooq M (2023) Assessing the impacts of current and future changes of the planforms of river Brahmaputra on its land use-land cover. *Geosci. Front.* 14(4):101557. Elsevier.

Kumar P, Singh SK, Gupta SK, Kanga S, Meraj G (2023) Uncovering the hydro-meteorological drivers responsible for forest fires utilizing geospatial techniques. *Theor. Appl. Climatol.* https://doi.org/10.1007/s00704-023-04497. Springer Nature.

Singh SK, Kotecha MJ, Kanga S, Pankhaniya SK, Agrawal S, Meraj G (2023) Urban Heat Island (UHI) Resilience Plan in Varying Climatic Conditions Using Geospatial Approach: A Case Study of Rajkot City. In: *Climate Change, Agriculture and Society*, vol 1. pp 307-338.

Singh S, Singh SK, Prajapat DK, Pandey V, Kanga S, Kumar P, Meraj G (2023) Assessing the Impact of the 2004 Indian Ocean Tsunami on South Andaman's Coastal Shoreline: A Geospatial Analysis of Erosion and Accretion Patterns. *J. Mar. Sci. Eng.* 11(6):1134.

Rathore SS, Mathur SC, Mathur S, Sharma S, Singh SK (2023) Late Cretaceous–Early Paleocene Geodiversity of Fatehgarh Formation of Petroliferous Barmer Basin, Western Rajasthan, India: A Potential Geopark Site. *Geoheritage* 15:54.

Singh S, Meraj G, Kumar P, Singh SK, Kanga S, Johnson BA, Prajapat DK, Debnath J, Sahariah D (2023) Decoding Chambal River Shoreline Transformations: A Comprehensive Analysis Using Remote Sensing, GIS, and DSAS. *Water* 15(9):1793.

Tripathi G, Shakya A, Upadhyay RK, Singh SK, Kanga S, Pandey SK (2023) Landslide Susceptibility Mapping of Tehri Reservoir Region Using Geospatial Approach. In: *Climate Change Adaptation, Risk Management and Sustainable Practices in the Himalaya*, pp 135-156.

Sud A, Kanga R, Singh SK, Meraj G, Kanga S, Kumar P, Ramanathan AL, Bhardwaj V (2023) Simulating Groundwater Potential Zones in Mountainous Indian Himalayas—A Case Study of Himachal Pradesh. *Hydrology* 10(3):65.

Upadhyay RK, Tripathi G, Đurin B, Šamanović S, Cetl V, Kishore N, Sharma M, Singh SK, Kanga S, Wasim Md, Rai PK, Bhardwaj V (2023) Groundwater Potential Zone Mapping in the Ghaggar River Basin, North-West India, Using Integrated Remote Sensing and GIS Techniques. *Water* 15(5):961.

Lavane K, Kumar P, Meraj G, Han TG, Ngan LHB, Lien BTB, Ty TV, Thanh NT, Downes NK, Nam NGD, Minh HVT, Singh SK, Kanga S (2023) Assessing the effects of drought on rice yields in the Mekong Delta. *Climate* 11(1):13.

Nath N, Sahariah D, Meraj G, Debnath J, Kumar P, Lahon D, Chand K, Farooq M, Chandan P, Singh SK, Kanga S (2023) Land use and land cover change monitoring and prediction of a UNESCO world heritage site: Kaziranga eco-sensitive zone using cellular automata-Markov model. *Land* 12(1):151.

Kanga S, Upadhyay RK, Sharma M, Sharma S, Tripathi G, Singh SK (2023) Geotechnical Stability Assessment of Road-cut Slopes: A Case Study of Srinagar, Garhwal Himalaya, India. *Biol. Forum* 15(5):1071-1080.

Gulati B, Sharma R, Kanga S, Singh SK, Sajan B, Meraj G, Kumar P, Ramanathan AL (2023) Unraveling the Relationship Between Stubble Burning and Air Quality Degradation in Punjab: A Temporal and Spatial Analysis (2019-2022). *J. Clim. Change* 9(2):43-53, IOS Press.

Tomar P, Singh SK, Kanga S, Pattanaik A (2023) Water Bodies mapping and monitoring using high-resolution satellite images. *Sustainability, Agri, Food and Environmental Research (SAFER)* 11.

Singh H, Meraj G, Singh S, Shrivastava V, Sharma V, Farooq M, Kanga S, Singh SK, Kumar P (2022) Status of Air Pollution during COVID-19-Induced Lockdown in Delhi, India. *Atmosphere* 13(12):2090.

Farooq M, Mushtaq F, Meraj G, Singh SK, Kanga S, Gupta A, Kumar P, Singh D, Avtar R (2022) Strategic Slum Upgrading and Redevelopment Action Plan for Jammu City. *Resources* 11(12):120.

Chand K, Jamwal AK, Meraj G, Thakur T, Farooq M, Kumar P, Singh SK, Kanga S and Debnath J (2014) Integrating geoenvironmental and socioenvironmental analyses for flood vulnerability assessment in the Kullu Valley, Himachal

Pradesh, India. *International Journal of Disaster Risk Reduction*, vol. 108, pp. 104494.

Debnath J, Sahariah D, Mazumdar M et al. (2023) Evaluating Flood Susceptibility in the Brahmaputra River Basin: An Insight into Asia's Eastern Himalayan Floodplains Using Machine Learning and Multi-Criteria Decision-Making. *Earth Syst. Environ.* 7, 733–760. https://doi.org/10.1007/s41748-023-00358-w

Tripathi G, Shakya A, Upadhyay RK, Singh SK, Kanga S, Pandey SK (2023) Landslide Susceptibility Mapping of Tehri Reservoir Region Using Geospatial Approach. In: *Climate Change Adaptation, Risk Management and Sustainable Practices in the Himalaya*, pp 135–156.

Singh S, Singh SK, Prajapat DK, Pandey V, Kanga S, Kumar P, Meraj G (2023) Assessing the Impact of the 2004 Indian Ocean Tsunami on South Andaman's Coastal Shoreline: A Geospatial Analysis of Erosion and Accretion Patterns. *J. Mar. Sci. Eng.*, 11(6): 1134.

3

Integrating Sustainable Development Goals with the Management of Natural and Technological Hazards and Disaster Risk Reduction

Prateek Srivastava[1] and Ambrina Sardar Khan[2]*

[1]Department of Botany, University of Allahabad, Prayagraj, Uttar Pradesh, India
[2]Department of Environmental Sciences, Integral University, Lucknow, Uttar Pradesh, India

Abstract

This chapter explores the critical need for integrating Sustainable Development Goals (SDGs) with the management of natural and technological hazards and disaster risk reduction (DRR). Natural and technological hazards pose significant threats to human life, infrastructure, and sustainable development. The SDGs provide a comprehensive framework for addressing various societal challenges including poverty, inequality, climate change, and sustainable development. This chapter highlights the importance of aligning DRR efforts with SDGs to enhance resilience, reduce vulnerability, and promote sustainable development in the face of hazards and disasters. It examines the intersections between specific SDGs, such as SDG 1 (No Poverty), SDG 2 (Zero Hunger), SDG 11 (Sustainable Cities and Communities), SDG 13 (Climate Action), and the management of hazards and DRR. The chapter also discusses the key principles, strategies, and policy recommendations for integrating SDGs into hazard management and DRR frameworks. By fostering synergies between SDGs and hazard management, this chapter aims to contribute to a holistic approach that enhances disaster resilience and advances sustainable development.

***Keywords*:** Sustainable development goals (sdgs), hazard management, disaster risk reduction (drr), resilience, poverty, inequality, climate change, policy recommendations

**Corresponding author*: ambrina@iul.ac.in; ambiikhan@gmail.com

Shruti Kanga, Gowhar Meraj, Suraj Kumar Singh, Majid Farooq and M. S. Nathawat (eds.) Geospatial Technology for Natural Resource Management, (83–132) © 2024 Scrivener Publishing LLC

3.1 Introduction

The management of natural and technological hazards and disaster risk reduction (DRR) are essential for building resilient societies and achieving sustainable development. Simultaneously, the United Nations' Sustainable Development Goals (SDGs) provide a comprehensive framework for addressing various global challenges including poverty, inequality, climate change, and environmental sustainability (Singh *et al.*, 2022; Rani *et al.*, 2023). Recognizing the interconnectedness of these goals, there is a growing need to integrate SDGs with the management of hazards and DRR to enhance resilience, reduce vulnerability, and promote sustainable development (Ahmad *et al.*, 2023).

Natural and technological hazards, such as earthquakes, floods, hurricanes, industrial accidents, and climate change, pose significant threats to human lives, ecosystems, infrastructure, and economic stability (Sharma *et al.*, 2023; Saqib *et al.*, 2023). These hazards can disrupt social systems, exacerbate poverty, and lead to progress towards sustainable development. Managing these risks requires a proactive and comprehensive approach that considers the interdependencies among SDGs and the unique challenges posed by hazards (Choudhury *et al.*, 2023; Rai *et al.*, 2023).

The SDGs encompass 17 interconnected goals that address a wide range of economic, social, and environmental issues. By integrating SDGs with hazard management and DRR, stakeholders can align their efforts to achieve multiple goals simultaneously. For example, integrating DRR with SDG 1 (No Poverty) can help protect vulnerable populations from the impact of hazards and prevent the exacerbation of poverty (Debnath *et al.*, 2023; Kumar *et al.*, 2023). Similarly, integrating DRR with SDG 13 (Climate Action) can foster climate resilience and mitigate the effects of climate change-induced hazards.

This integration requires a shift from a sectoral and fragmented approach towards a holistic and cross-cutting approach to sustainable development. This calls for collaboration among governments, international organizations, civil society, academia, and the private sector to jointly address the complex challenges posed by hazards and DRR while advancing the SDGs.

In this context, this chapter explores the importance of integrating SDGs with the management of natural and technological hazards and DRR (Singh *et al.*, 2023). It examines the intersection between specific SDGs and hazard management, highlighting the potential synergies and co-benefits that can be achieved. This chapter will discuss the principles, strategies, and policy recommendations for incorporating SDGs into hazard management and DRR frameworks (Singh *et al.*, 2023).

By integrating SDGs with hazard management and DRR, societies can enhance their resilience to hazards, reduce vulnerabilities, and promote sustainable development. This integration provides an opportunity to address the root causes of risks, build adaptive capacities, and foster sustainable and inclusive communities (Rathore *et al.*, 2023; Singh *et al.*, 2023). Ultimately, the integration of SDGs with hazard management and DRR can contribute to the realization of a more resilient and sustainable future, where hazards and disasters no longer impede progress but instead become opportunities for transformative change (Tripathi *et al.*, 2023; Sud *et al.*, 2023).

Disasters, risks, hazards, vulnerability, and resilience are crucial concepts for integrating Sustainable Development Goals (SDGs) with the management of natural and technological hazards and disaster risk reduction (DRR). Understanding these terms is essential for comprehending the complex dynamics involved in achieving sustainable development, while mitigating and managing hazards and risks (Upadhyay *et al.*, 2023).

- **Disaster:** A disaster is a sudden or prolonged event, such as a natural or technological hazard that causes significant damage, destruction, loss of life, and disruption to the functioning of a community or society. Disasters can result from hazards, such as earthquakes, floods, cyclones, industrial accidents, or epidemics. The integration of SDGs with hazard management and DRR aims to minimize the impacts and consequences of disasters on human well-being, infrastructure, and the environment (Lavane *et al.*, 2023).
- **Risk:** Risk refers to the potential for loss, damage, or harm caused by a hazard. It involves the interaction between the probability of hazard occurrence and the vulnerability of exposed elements (e.g., people, infrastructure, and ecosystems). Risk assessment and management are integral components of integrating the SDGs with hazard management and DRR. Understanding and evaluating risks associated with hazards help inform decision-making processes and the allocation of resources towards effective risk reduction and resilience-building measures (Nath *et al.*, 2023).
- **Hazard:** A hazard is a potential source of harm, threat, or danger that can cause adverse impacts on human lives, ecosystems, infrastructure, or the environment. Natural hazards include earthquakes, floods, storms, droughts, and wildfires,

whereas technological hazards include industrial accidents, chemical spills, and nuclear incidents. Integrating SDGs with hazard management and DRR involves identifying, assessing, and mitigating the impact of hazards through preventive measures, early warning systems, and preparedness efforts (Kanga *et al.*, 2023; Gulati *et al.*, 2023).

- **Vulnerability:** Vulnerability refers to the susceptibility of individuals, communities, or systems to harm or negative impacts from hazards or disasters. It is influenced by various factors, including social, economic, and environmental conditions; access to resources; governance structures; and exposure to hazards. Integrating SDGs with hazard management and DRR involves addressing and reducing vulnerabilities through measures that enhance social protection, promote inclusivity, strengthen infrastructure, and improve access to essential services (Tomar *et al.*, 2023).
- **Resilience:** The ability of individuals, communities, systems, or ecosystems to absorb, adapt, and recover from the impacts of hazards and disruptions. It encompasses the capacity to withstand shocks, maintain essential functions, and bounce back stronger. Integrating SDGs with hazard management and DRR aims to enhance resilience by building adaptive capacities, promoting sustainable development practices, strengthening early warning systems, and fostering community engagement. Resilience also involves addressing the underlying vulnerabilities and ensuring equitable access to resources and opportunities.

The idea of integrating SDGs with the management of natural and technological hazards and DRR involves understanding and addressing disasters, risks, hazards, vulnerability, and resilience. This integration requires a comprehensive risk assessment, preventive measures, inclusive approaches, and the pursuit of sustainable development goals to enhance resilience and promote the well-being of communities and the environment in the face of hazards and disasters.

3.1.1 Types of Disasters/Hazards

There are various types of natural and technological disasters and hazards. These events pose significant threats to human life, infrastructure, and the environment. Some common types of disasters and hazards are as follows:

3.1.1.1 Natural Disasters

- **Earthquakes:** Sudden shaking of the Earth's surface caused by the release of energy into the Earth's crust.
- **Floods:** Overflow of water onto normally dry land, often caused by heavy rainfall, river overflow, or coastal storm surges.
- **Hurricanes and Cyclones:** Intense tropical storms characterized by strong winds, heavy rainfall, and storm surges.
- **Tornadoes:** Violently rotating columns of air that extend from the thunderstorm to the ground, causing strong winds and potential destruction.
- **Wildfires:** Uncontrolled fires that spread rapidly through vegetation and are often fueled by dry conditions and high winds.
- **Landslides:** Movement of rock, soil, and debris down a slope, typically triggered by heavy rainfall, earthquakes, or human activity.
- **Droughts:** Prolonged periods of below-average precipitation leading to water scarcity, crop failure, and ecological stress.

3.1.1.2 Technological Hazards

- **Industrial Accidents:** Chemical spills, explosions, or release of hazardous materials from industrial facilities.
- **Nuclear Incidents:** Malfunctions or accidents at nuclear power plants result in radiation leaks or contamination.
- **Transportation Accidents:** Crashes or derailments involving trains, planes, ships, or vehicles carrying hazardous materials.
- **Structural Collapses:** Building or infrastructure failures, such as bridge or building collapses.

3.1.1.3 Biological Hazards

- **Epidemics and Pandemics:** Widespread outbreaks of infectious diseases that affect human populations, such as influenza, Ebola, and COVID-19.
- **Foodborne Illnesses:** Contamination of food products by pathogens, toxins, or chemicals leading to illness or outbreaks.

3.1.1.4 Climate-Related Hazards

- **Heatwaves:** Extended periods of excessively hot weather, often accompanied by high humidity, pose risks to human health.
- **Storm Surges:** Abnormal rise in sea level during storms, particularly in coastal areas, resulting in coastal flooding and erosion.
- **Melting Glaciers and Polar Ice:** Rapid melting of ice caps, glaciers, and polar ice leads to rising sea levels and ecological disruptions.
- **Ocean Acidification:** The increasing acidity of the ocean due to the absorption of carbon dioxide has adverse effects on marine ecosystems and biodiversity.

3.1.2 Disaster Management

Disaster management plays a crucial role in the integration of Sustainable Development Goals (SDGs) with the management of natural and technological hazards and disaster risk reduction (DRR). It involves a comprehensive approach to prevent, prepare for, respond to, and recover from disasters, while considering the principles of sustainable development (Singh *et al.*, 2022; Farooq *et al.*, 2022). Key aspects of disaster management in the context of integrating SDGs with hazard and DRR management:

1. **Risk Assessment and Mapping:** Effective disaster management begins with comprehensive risk assessment, including the identification and mapping of hazards, vulnerabilities, and exposed elements. This assessment helps to understand the potential impacts of disasters on human lives, infrastructure, ecosystems, and socioeconomic systems. Risk assessment forms the foundation for informed decision making and resource allocation.
2. **Prevention and Mitigation:** Disaster management integrates prevention and mitigation measures to reduce the likelihood and impact of a hazard. This involves implementing structural and nonstructural measures to strengthen infrastructure, improve land-use planning, and promote sustainable development practices. Integrating SDGs with hazard and DRR management emphasizes measures that

address the root causes of risks, such as poverty, inequality, and unsustainable practices.

3. **Preparedness and Early Warning Systems:** Disaster management includes measures to enhance the capacity of individuals, communities, and institutions to respond effectively to disasters. These include developing and testing emergency response plans, conducting drills and simulations, and establishing early warning systems to provide timely information and alerts to at-risk populations. Preparedness efforts focus on building resilience, promoting community engagement, and ensuring the availability of resources and capacity.
4. **Emergency Response and Coordination:** When disasters occur, disaster management involves an immediate response to save lives, provide essential services, and mitigate the impacts. Effective emergency response requires coordination among multiple stakeholders, including government agencies, civil society organizations, and the private sector. Integrating the SDGs emphasizes the importance of equity, inclusivity, and protection of vulnerable populations during emergency response efforts.
5. **Recovery and Reconstruction:** After disaster, disaster management focuses on recovery and reconstruction efforts to restore affected areas and communities. These efforts should align with the principles of sustainable development and SDGs, promoting social, economic, and environmental resilience. Recovery and reconstruction provide an opportunity to rebuild in a more resilient and sustainable manner, address underlying vulnerabilities, and incorporate risk reduction measures.
6. **Capacity Building and Knowledge Sharing:** Disaster management involves capacity building at all levels to enhance the skills, knowledge, and resources required for an effective risk reduction and response. Capacity-building initiatives focus on strengthening institutions, promoting community participation, and sharing best practices and lessons. Integrating SDGs with disaster management emphasizes knowledge sharing, the utilization of scientific research, and evidence-based approaches.
7. **Monitoring and Evaluation:** Disaster management integrates monitoring and evaluation mechanisms to assess the

effectiveness of policies, strategies, and interventions. This includes monitoring progress towards the SDGs, tracking disaster impacts, evaluating the implementation of risk reduction measures, and conducting post-disaster assessments. Monitoring and evaluation inform evidence-based decision making, adaptive management, and continuous improvements in disaster management practices.

Integrating SDGs with the management of natural and technological hazards and DRR recognizes the importance of holistic and sustainable approaches to disaster management. By aligning efforts towards achieving the SDGs with risk-reduction measures, disaster management can contribute to the creation of resilient, inclusive, and sustainable communities that are better prepared to withstand and recover from disasters.

3.1.3 Disaster Risk Reduction

Disaster risk reduction (DRR) plays a critical role in the integration of Sustainable Development Goals (SDGs) with the management of natural and technological hazards. DRR aims to minimize the impact of disasters by reducing the underlying risks, vulnerabilities, and exposure of communities and ecosystems. When integrated with the SDGs, DRR becomes a vital component of sustainable development strategies. Here are the key aspects of DRR in the context of integrating SDGs with hazard and disaster management:

1. **Risk Identification and Assessment:** DRR starts with the identification and assessment of risks associated with natural and technological hazards. This involves analyzing the potential impacts on human lives, infrastructure, ecosystems, and socioeconomic systems. By understanding the risks, including their causes, consequences, and spatial distribution, decision makers can prioritize interventions that address the most significant hazards and vulnerabilities.
2. **Risk Reduction Measures:** DRR integrates risk reduction measures to minimize the likelihood and impact of disasters. These measures include both structural and nonstructural approaches. Structural measures involve physical interventions such as building resilient infrastructure, constructing flood barriers, and retrofitting buildings to withstand hazards. Non-structural measures include land use planning,

early warning systems, education and awareness campaigns, and strengthening governance and policy frameworks.

3. **Community Engagement and Empowerment:** DRR emphasizes the importance of engaging and empowering communities in risk-reduction efforts. This involves involving local communities, including marginalized groups, decision-making processes, fostering participatory approaches, and promoting community-led initiatives. By integrating the SDGs, DRR strives for inclusive and equitable participation to address the underlying social, economic, and environmental factors that contribute to vulnerability.
4. **Climate Change Adaptation:** DRR intersects with climate change adaptation, recognizing the need to address the impacts of climate change on disaster risks. As climate change intensifies hazards and alters risk patterns, integrating DRR with SDGs calls for climate-resilient approaches. This includes promoting nature-based solutions, enhancing ecosystem services, and incorporating climate projections into risk assessment and planning processes.
5. **Knowledge and Information Sharing:** DRR integrates knowledge- and information-sharing as a core component. This involves sharing scientific research, best practices, and lessons learned among stakeholders to inform decision making and enhance resilience. Integrating SDGs emphasizes the utilization of scientific knowledge and evidence-based approaches, fostering collaboration between researchers, practitioners, and communities.
6. **Capacity Building and Institutional Strengthening:** DRR requires capacity building and institutional strengthening to implement effective risk-reduction measures. This includes enhancing technical skills, knowledge, and resources at all levels from local communities to national governments. Integrating the SDGs emphasizes the need for integrated and cross-sectoral approaches and encourages collaboration between different sectors and stakeholders.
7. **Monitoring and Evaluation:** DRR integrates monitoring and evaluation mechanisms to assess the effectiveness of risk-reduction measures, policies, and strategies. This includes monitoring disaster impacts, tracking progress towards risk reduction targets, and evaluating the implementation of DRR interventions. By integrating the SDGs,

monitoring and evaluation focus on measuring progress towards sustainable development objectives while assessing the resilience of communities and ecosystems.

Integrating SDGs with the management of natural and technological hazards and DRR emphasizes a proactive, sustainable, and inclusive approach to reducing disaster risks. By aligning risk-reduction efforts with the broader sustainable development agenda, DRR contributes to the creation of resilient communities, promotes social equity, and fosters sustainable development practices.

3.2 SD and SDGs

In the context of integrating SDGs with the management of natural and technological hazards and disaster risk reduction (DRR), Sustainable Development (SD), and Sustainable Development Goals (SDGs) play interconnected roles.

Sustainable Development (SD): Sustainable Development refers to an approach that seeks to meet the needs of the present generation without compromising the ability of future generations to meet their own needs. It recognizes the interlinkages between environmental, social, and economic dimensions, and aims to achieve a balance that ensures long-term well-being and resilience.

Integrating SD into hazard management and DRR involves considering the sustainability of strategies, policies, and interventions. This includes promoting environmentally sound practices, ensuring social equity and inclusiveness, and fostering economic viability in disaster risk reduction and management. By integrating SD principles, hazard and DRR management can contribute to sustainable and resilient development.

Sustainable Development Goals (SDGs): The SDGs are a set of 17 global goals adopted by the United Nations in 2015 as part of the 2030 Agenda for Sustainable Development. The SDGs provide a comprehensive framework for addressing various global challenges, including poverty, inequality, climate change, and sustainable development. They cover a wide range of interconnected issues from ending poverty and hunger to promoting sustainable cities and combating climate change.

Integrating the SDGs with the management of hazards and DRR involves aligning risk-reduction efforts and policies with the specific targets and indicators of the SDGs. For example, integrating SDG 1 (No Poverty) with DRR efforts can focus on addressing poverty as the root cause of vulnerability and promoting inclusive approaches to reduce disaster risks. Integrating SDG 13 (Climate Action) with hazard management can emphasize climate resilience and mitigation strategies to combat climate-related hazards.

By integrating SDGs into hazard and DRR management, stakeholders can work towards achieving multiple goals simultaneously. The SDGs provide a framework for guiding policy decisions, resource allocation, and collaboration between different sectors and stakeholders. Integrating SDGs with hazard and DRR management emphasizes the interconnectedness of social, economic, and environmental dimensions, and promotes sustainable and resilient development in the face of hazards and disasters.

In summary, integrating Sustainable Development (SD) and Sustainable Development Goals (SDGs) into the management of natural and technological hazards and disaster risk reduction involves incorporating sustainability principles, aligning efforts with the specific targets and indicators of SDGs, and promoting holistic and balanced approaches to achieve long-term well-being and resilience.

3.2.1 Timeline of SDGs

The timeline of the Sustainable Development Goals (SDGs) to integrate them with the management of natural and technological hazards and disaster risk reduction (DRR) is outlined as follows:

1. **September 2015: Adoption of the SDGs**—The United Nations General Assembly formally adopted the 2030 Agenda for Sustainable Development, including 17 SDGs. This marks the beginning of a global commitment to achieve sustainable development across various dimensions.
2. **Integration of SDGs and DRR Frameworks**—Efforts have started to align DRR frameworks, such as the Sendai Framework for Disaster Risk Reduction, with the SDGs. The Sendai Framework, adopted in 2015, highlights the importance of integrating disaster risk reduction into sustainable development efforts.
3. **Awareness and Policy Alignment**—Governments, international organizations, and stakeholders have begun to raise awareness about the interlinkages between the SDGs and disaster risk reduction. Discussions focus on integrating

disaster risk reduction and resilience-building measures into policies and strategies aimed at achieving the SDGs.

4. **Mainstreaming SDGs in DRR Plans**—Countries start integrating SDGs into their national and local disaster risk reduction plans. This involves considering the SDG targets and indicators while developing strategies to reduce disaster risks, enhance resilience, and promote sustainable development.
5. **Capacity Building and Knowledge Exchange**—Efforts are made to build capacity among stakeholders involved in hazard and DRR management regarding the integration of the SDGs. This includes knowledge exchange, sharing best practices, and promoting collaboration among different sectors and disciplines.
6. **Monitoring and Reporting**—Countries begin monitoring and reporting on progress towards the SDGs, including those related to disaster risk reduction. This involves tracking indicators, assessing risk-reduction measures, and evaluating the contributions of DRR efforts to the achievement of specific SDG targets.
7. **Review and Adaptation**—Periodic reviews and evaluations were conducted to assess the effectiveness of integrating SDGs with hazard and DRR management. The lessons learned and best practices inform adjustments to policies, strategies, and interventions to enhance the integration process.

3.2.2 Concept of Sustainability

The concept of sustainability is central to integrating Sustainable Development Goals (SDGs) with the management of natural and technological hazards and disaster risk reduction (DRR). Sustainability encompasses meeting the needs of the present generation without compromising the ability of future generations to meet their own needs. When applied to hazard and DRR management, sustainability guides efforts to reduce risks, enhance resilience, and promote long-term well-being. The concept of sustainability is relevant to integrating SDGs with hazard and DRR management.

1. **Environmental Sustainability:** Integrating the SDGs with hazard and DRR management involves promoting

environmental sustainability. This includes preserving and restoring ecosystems, reducing pollution, and mitigating the impact of climate change. By adopting sustainable practices in hazard and DRR strategies, such as nature-based solutions or low-carbon approaches, we can enhance resilience, reduce vulnerability, and protect the natural resources on which communities depend.

2. **Social Sustainability:** Social sustainability is integral to integrating SDGs with hazard and DRR management. It focuses on ensuring social equity, inclusivity, and wellbeing. By addressing underlying social vulnerabilities, promoting community engagement, and fostering inclusive decision-making processes, disparities can be reduced, marginalized groups empowered, and social cohesion enhanced. Social sustainability efforts aim to create resilient and supportive communities that can effectively respond to hazards and recover from them.
3. **Economic Sustainability:** Economic sustainability is another crucial aspect of integrating SDGs with hazard and DRR management. This involves promoting economic viability and resilience in the face of hazards. Sustainable economic practices, such as diversification, building local economies, and investing in resilient infrastructure, contribute to long-term economic stability and reduce vulnerability to future disasters. By aligning economic development with hazard and DRR objectives, we can ensure that risk-reduction efforts are economically feasible and contribute to sustainable growth.
4. **Integrated Approaches:** Sustainability requires integrated and cross-sectoral approaches. Integrating SDGs with hazard and DRR management encourages collaboration and coordination among stakeholders, sectors, and disciplines. This involves breaking down silos and fostering partnerships to address the complex challenges posed by hazards. Integrated approaches consider the interlinkages between environmental, social, and economic dimensions, ensuring that risk-reduction efforts are comprehensive and aligned with sustainable development goals.
5. **Resilient and Adaptive Strategies:** Sustainability in hazards and DRR management emphasizes building resilience and adaptability. Resilience refers to the ability of communities,

systems, and ecosystems to absorb, adapt, and recover from a hazard. By integrating the SDGs, hazard and DRR strategies can adopt adaptive management approaches, foster learning from past experiences, and prioritize measures that enhance long-term resilience. This includes investing in early warning systems, promoting capacity building, and incorporating climate change adaptation into risk-reduction plans.

Integration of SDGs with the management of natural and technological hazards and disaster risk reduction emphasizes sustainability across environmental, social, and economic dimensions. By adopting sustainable practices, promoting social equity, fostering economic resilience, and adopting integrated and adaptive approaches, we can work towards reducing risks, enhancing resilience, and achieving long-term sustainable development in the face of hazards.

3.2.3 Goal 1: No Poverty

The goal of "No Poverty" (SDG 1) in the context of integrating the Sustainable Development Goals (SDGs) with the management of natural and technological hazards and disaster risk reduction (DRR) is to address poverty as a root cause of vulnerability and enhance resilience in impoverished communities. Here is how the goal of No Poverty is relevant to integrating SDGs with hazard and DRR management:

1. **Poverty as a Vulnerability:** Poverty increases vulnerability to hazards and disasters. Impoverished communities often lack resources, infrastructure, and access to basic services, which makes them more susceptible to the impacts of hazards. Integrating the SDGs with hazard and DRR management involves addressing poverty to reduce the underlying vulnerabilities and enhance the resilience of impoverished communities.
2. **Inclusive Approaches:** Integrating the No Poverty goal with hazard and DRR management emphasizes inclusive approaches. This involves ensuring that vulnerable and marginalized populations, such as those living in poverty, are included in decision-making processes, risk assessments, and the development of risk reduction strategies. Their unique needs, perspectives, and priorities should be

considered to ensure that interventions are effective and to address the specific challenges faced by impoverished communities.

3. **Poverty Alleviation and Resilience Building:** Integrating the No Poverty goal with hazard and DRR management aims to alleviate poverty and build resilience simultaneously. Efforts to reduce poverty, such as providing livelihood opportunities, access to education, and social protection measures, have contributed to enhancing the adaptive capacities and coping mechanisms of impoverished communities. By promoting sustainable and inclusive development practices, poverty reduction efforts can contribute to reducing vulnerability and strengthening resilience to hazards.
4. **Social Safety Nets and Risk Reduction:** Integrating the No Poverty goal with hazard and DRR management involves implementing social safety nets and risk-reduction measures. Social safety nets, such as cash transfer programs or social insurance schemes, can provide support to impoverished communities during and after disasters, helping them recover and rebuild their lives. Combining poverty reduction efforts with risk-reduction measures ensures that the impact of hazards and disasters does not push communities back into poverty.
5. **Empowerment and Capacity Building:** Integrating the No Poverty goal with hazard and DRR management involves empowering impoverished communities and building their capacity to manage risk. This includes providing access to education, skills training, and resources to enhance their ability to prepare for, respond to, and recover from hazards. Empowered communities can actively participate in risk-reduction efforts, contribute local knowledge, and build resilience to future disasters.

The integration of the goal of No Poverty (SDG 1) with the management of natural and technological hazards and disaster risk reduction aims to address poverty as a root cause of vulnerability. By adopting inclusive approaches, promoting poverty alleviation measures, implementing social safety nets, empowering communities, and building their capacity to manage risks, poverty-related vulnerabilities can be reduced and resilience enhanced in impoverished communities. This integration ensures that hazard and DRR management efforts contribute to sustainable and inclusive development.

3.2.4 Goal 2: Zero Hunger

The goal of "Zero Hunger" (SDG 2) in the context of integrating the Sustainable Development Goals (SDGs) with the management of natural and technological hazards and disaster risk reduction (DRR) focuses on addressing food insecurity, malnutrition, and hunger, particularly in vulnerable communities affected by hazards and disasters. Here is how the goal of Zero Hunger is relevant to integrating SDGs with hazard and DRR management:

1. **Food Security in Hazard-Prone Areas:** Integrating the Zero Hunger goal with hazard and DRR management involves ensuring food security in hazard-prone areas. Hazards, such as floods, droughts, and storms, can disrupt agricultural production, damage crops and livestock, and disrupt food supply chains. By incorporating strategies to mitigate these risks, such as climate-resilient agriculture, crop diversification, and improved post-harvest management, food security can be enhanced, thereby reducing vulnerability to hunger in the face of hazards.
2. **Disaster Preparedness for Food Systems:** Integrating the Zero Hunger goal with hazard and DRR management emphasizes disaster preparedness in food systems. This involves developing early warning systems for impending hazards, establishing contingency plans for food supply and distribution, and strengthening local food storage and preservation capacity. By improving preparedness and response mechanisms, communities can better manage and ensure access to adequate nutritious food during and after disasters.
3. **Sustainable Agriculture and Resilient Livelihoods:** Integrating the Zero Hunger goal with hazard and DRR management entails promoting sustainable agriculture and resilient livelihoods. These include sustainable farming practices, land management, and water resource management to enhance agricultural productivity and reduce vulnerability to hazards. By strengthening local agricultural systems and promoting diversified livelihood opportunities,

communities can become more resilient to the impact of hazards, reduce food insecurity, and alleviate hunger.

4. **Nutrition in Emergency Response and Recovery:** Integrating the Zero Hunger goal with hazard and DRR management involves prioritizing nutrition in emergency response and recovery efforts. Disasters can disrupt access to nutritious food and exacerbate malnutrition and hunger. It is essential to ensure that emergency response includes provision of adequate and balanced nutrition, particularly for vulnerable groups such as children, pregnant women, and the elderly. Recovery efforts should focus on restoring and improving access to nutritious food and supporting community-level nutrition programs.
5. **Inclusive Approaches and Social Safety Nets:** Integrating the Zero Hunger goal with hazard and DRR management requires inclusive approaches and social safety nets. This involves addressing the underlying causes of hunger, such as poverty and inequality, while also considering the needs of vulnerable groups. Social safety nets, such as targeted cash transfer programs or school feeding initiatives, can help provide access to food and nutrition to those most affected by hazards and disasters.

Overall, integrating the goal of Zero Hunger (SDG 2) with the management of natural and technological hazards and disaster risk reduction aims to ensure food security, improve nutrition, and reduce hunger in vulnerable communities. By promoting sustainable agriculture, strengthening food systems, enhancing disaster preparedness, and incorporating nutritional considerations into emergency response and recovery, we can reduce vulnerability to hunger and achieve sustainable and resilient food systems in the face of hazards and disasters.

3.2.5 Goal 3: Good Health and Well-Being

The goal of "Good Health and Well-being" (SDG 3) in the context of integrating the Sustainable Development Goals (SDGs) with the management of natural and technological hazards and disaster risk reduction (DRR) focuses on promoting health, well-being, and resilient healthcare systems in the face of hazards and disasters. The goal of Good Health and Well-being is relevant to integrating SDGs with hazard and DRR management.

1. **Disaster Preparedness in Health Systems:** Integrating the Good Health and Well-being goal with hazard and DRR management involves strengthening disaster preparedness in healthcare systems. This includes developing emergency response plans, training healthcare professionals in disaster management, and ensuring the availability of essential medical equipment. By enhancing the preparedness of health systems, communities can respond better to health-related risks during and after disasters.
2. **Disease Prevention and Control:** Integrating the Good Health and Well-being goal with hazard and DRR management emphasizes disease prevention and control in disaster-affected areas. Disasters can increase the risk of disease outbreaks, injuries, and mental health problems. Efforts to integrate disease surveillance, immunization campaigns, and vector control measures into hazard and DRR management can help mitigate health risks and protect the affected populations.
3. **Accessible and Resilient Healthcare Services:** Integrating the Good Health and Well-being goal with hazard and DRR management involves ensuring accessible and resilient healthcare services. This includes promoting equitable access to healthcare, especially for vulnerable and marginalized groups. Strengthening the healthcare infrastructure, including hospitals, clinics, and health centers, to withstand hazards and disasters is crucial. In addition, integrating telemedicine and mobile health technologies can enhance healthcare delivery in disaster-affected areas.
4. **Mental Health and Psychosocial Support:** Integrating the Good Health and Well-being goal with hazard and DRR management recognizes the importance of mental health and psychosocial support in disaster-affected communities. Disasters can have profound psychological impacts, including increased stress, anxiety, and post-traumatic stress disorder (PTSD). Efforts to integrate mental health services, community support networks, and psychosocial interventions into hazard and DRR management can promote resilience and well-being in the affected populations.
5. **Health Education and Risk Communication:** Integrating the Good Health and Well-being goal with hazard and DRR management involves health education and risk

communication. This includes raising awareness about the health risks associated with hazards, promoting preventive measures, and providing accurate and timely information to affected communities. Effective risk communication can help empower individuals and communities to make informed decisions and take appropriate action to protect their health and well-being.

6. **Collaboration and Capacity Building:** Integrating Good Health and Well-being goal with hazard and DRR management requires collaboration and capacity building among various stakeholders. This includes partnerships between health authorities, emergency management agencies, community organizations, and other sectors. Capacity-building efforts focus on training healthcare professionals, community health workers, and volunteers in disaster response and recovery, as well as promoting cross-sectoral coordination to address the health needs of affected populations.

The integration of SDG 3 which is Good Health and Well-being with the management of natural and technological hazards and disaster risk reduction aims to promote resilient healthcare systems, disease prevention, mental health support, and equitable access to healthcare in disaster-affected areas. By strengthening disaster preparedness, disease control measures, healthcare infrastructure, and promoting health education and collaboration, we can enhance the health and well-being of communities in the face of hazards and disasters.

3.2.6 Goal 4: Quality Education

The goal of "Quality Education" (SDG 4) in the context of integrating the Sustainable Development Goals (SDGs) with the management of natural and technological hazards and disaster risk reduction (DRR) emphasizes the importance of inclusive and resilient education systems in disaster-prone areas. The goal of Quality Education is to integrate SDGs with hazard and DRR management.

1. **Continuity of Education during Disasters:** Integrating the Quality Education goal with hazard and DRR management involves ensuring continuity of education during and after disasters. Disasters can disrupt education systems, leading to school closures, student displacement, and loss of

educational infrastructure. By implementing strategies such as emergency education plans, alternative learning methods, and the use of technology, access to quality education can be maintained, even in the face of hazards.

2. **Inclusive Education for Vulnerable Groups:** Integrating the Quality Education goal with hazard and DRR management emphasizes inclusive education for vulnerable groups including children with disabilities, displaced populations, and those living in remote areas. Inclusive education ensures that all learners have equal opportunities to access quality education regardless of their background or circumstances. Efforts should be made to provide safe and supportive learning environments that accommodate the specific needs of vulnerable populations that are affected by hazards and disasters.
3. **Disaster Risk Reduction Education:** Integrating the Quality Education goal with hazard and DRR management involves incorporating disaster risk reduction education into the curriculum. This includes raising awareness of hazards, building resilience, and promoting preparedness among students, teachers, and the wider community. By integrating DRR education, learners can develop knowledge, skills, and attitudes that contribute to their own safety and resilience of their communities.
4. **Teacher Training and Capacity-Building:** Integrating the Quality Education goal with hazard and DRR management requires investing in teacher training and capacity-building. Teachers play a crucial role in ensuring quality education in disaster-prone areas. They need to be equipped with knowledge and skills related to disaster risk reduction, psychosocial support, and inclusive education practices. Training programs can enhance learner's capacity to address the specific challenges and needs of learners in disaster-affected contexts.
5. **Safe and Resilient Learning Environments:** Integrating the Quality Education goal with hazard and DRR management involves creating safe and resilient learning environments. This includes ensuring that the educational infrastructure, such as school buildings, are designed and constructed to withstand hazards. Additionally, safety protocols, evacuation plans, and early warning systems should be implemented to

protect students and staff during emergencies. The quality of education can be enhanced by promoting safe and resilient learning environments, and the well-being of learners can be safeguarded.

6. **Education as a Tool for Recovery:** Integrating the Quality Education goal with hazard and DRR management recognizes education as a tool for recovery and building back better. Education plays a crucial role in helping communities recover from disasters by providing a sense of normalcy, restoring social cohesion, and supporting opportunities for livelihood. By integrating post-disaster education strategies into recovery plans, communities can rebuild their education systems in a manner that is more resilient, inclusive, and responsive to future hazards.

In brief, integrating the goal of Quality Education (SDG 4) with the management of natural and technological hazards and disaster risk reduction involves ensuring the continuity of education, promoting inclusive and resilient learning environments, incorporating DRR education, investing in teacher training, and utilizing education as a tool for recovery. By prioritizing quality education in disaster-prone areas, we can empower learners, build community resilience, and contribute to sustainable development in the face of hazards and disasters.

3.2.7 Goal 5: Gender Equality

The goal of "Gender Equality" (SDG 5) in the context of integrating the Sustainable Development Goals (SDGs) with the management of natural and technological hazards and disaster risk reduction (DRR) focuses on promoting gender equality and empowering women and girls in disaster-prone areas. Here is how the goal of Gender Equality is relevant to integrating SDGs with hazard and DRR management.

1. **Gender-responsive Risk Assessment:** Integrating the Gender Equality goal with hazard and DRR management involves conducting gender-responsive risk assessments. This entails consideration of the specific vulnerabilities, needs, and capacities of women, girls, and other gender groups in disaster-prone areas. By recognizing and addressing gender disparities and power dynamics, risk assessments

can inform targeted interventions and ensure that risk-reduction measures are inclusive and effective.

2. **Women's Leadership and Participation:** Integrating the Gender Equality goal with hazard and DRR management emphasizes women's leadership and meaningful participation in decision-making processes. This includes women in the planning, implementation, and evaluation of risk-reduction strategies, policies, and programs. Women's active engagement ensures that their perspectives, experiences, and expertise are considered, leading to more comprehensive and gender-responsive disaster management approaches.
3. **Addressing Gender-based Violence:** Integrating the Gender Equality goal with hazard and DRR management involves addressing gender-based violence (GBV) in disaster contexts. Disasters can exacerbate the existing gender inequalities and increase the risk of GBV. Efforts should be focused on preventing and responding to GBV, providing safe spaces for survivors, and ensuring access to support services. By addressing GBV, an environment that promotes gender equality, safety, and well-being can be created.
4. **Access to Sexual and Reproductive Health Services:** Integrating the Gender Equality goal with hazard and DRR management includes ensuring access to sexual and reproductive health services in disaster-affected areas. Disasters can disrupt health systems and limit access to essential services, including reproductive healthcare. By providing comprehensive sexual and reproductive health services, including family planning, maternal healthcare, and access to contraceptives, we can empower women and girls, reduce maternal mortality rates, and promote their overall wellbeing.
5. **Gender-responsive Humanitarian Assistance:** Integrating the Gender Equality goal with hazard and DRR management involves delivering gender-responsive humanitarian assistance. This includes ensuring that relief and recovery efforts address the specific needs of women, girls, and other vulnerable groups. Access to clean water, sanitation facilities, hygiene products, and other essential supplies should be provided in a gender-sensitive manner, considering the unique needs and dignity of individuals.
6. **Data and Monitoring:** Integrating the Gender Equality goal with hazard and DRR management requires collecting and

analyzing sex-disaggregated data to monitor progress and inform decision-making. Gender-sensitive data allows for a better understanding of the differential impacts of hazards on women, men, and other gender groups. Monitoring progress on gender equality indicators helps to identify gaps and develop evidence-based interventions to promote gender equality in hazard and DRR management.

Being one of the most important sustainable development goal, SDG 5 reflects the status of gender equality and opportunity to everyone. The integration of SDG 5 with the management of natural and technological hazards and disaster risk reduction involves addressing gender disparities, promoting women's leadership and participation, addressing gender-based violence, ensuring access to sexual and reproductive health services, delivering gender-responsive humanitarian assistance, and collecting gender-sensitive data. By mainstreaming gender equality in hazard and DRR management, we can create more inclusive and resilient communities where women and girls are empowered, their rights are protected, and their contributions are valued.

3.2.8 Goal 6: Clean Water and Sanitation

The goal of "Clean Water and Sanitation" (SDG 6) in the context of integrating the Sustainable Development Goals (SDGs) with the management of natural and technological hazards and disaster risk reduction (DRR) focuses on ensuring access to clean water and sanitation facilities, particularly in disaster-prone areas. The goal of Clean Water and Sanitation is relevant to integrating SDGs with hazard and DRR management:

1. **Access to Safe Water during and after Disasters:** Integrating the Clean Water and Sanitation goal with hazard and DRR management involves ensuring access to safe water during and after disasters. Hazards, such as floods, storms, and earthquakes, can disrupt water supply systems and contaminate water sources, leading to waterborne diseases and health risks. Efforts should be focused on providing safe and reliable water sources, establishing emergency water storage and purification systems, and hygiene practices to prevent water-related diseases in disaster-affected areas.
2. **Sanitation Facilities and Hygiene Promotion:** Integrating the Clean Water and Sanitation goal with hazard and DRR

management includes promoting access to adequate sanitation facilities and promoting hygiene practices. Disasters can damage or destroy sanitation infrastructure, leading to poor hygiene and increased health risks. Efforts should focus on providing sanitation facilities such as toilets, handwashing stations, and waste management systems that are resilient to hazards. Hygiene promotion and behavioral change campaigns can further improve sanitation practices and reduce the risk of disease transmission.

3. **Water Resource Management and Climate Resilience:** Integrating the Clean Water and Sanitation goal with hazard and DRR management emphasizes water resource management and climate resilience. Climate change can affect water availability, quality, and distribution patterns, exacerbating water scarcity, and increasing the vulnerability of water resources to hazards. Integrating water resource management strategies that promote sustainable use, water conservation, and climate resilience can help mitigate the impacts of hazards and ensure the long-term availability of clean water in the long run.
4. **Inclusive and Equitable Access:** Integrating the Clean Water and Sanitation goal with hazard and DRR management involves ensuring inclusive and equitable access to clean water and sanitation services. Vulnerable groups such as women, children, the elderly, and people with disabilities are disproportionately affected by water and sanitation challenges during and after disasters. Efforts should be focused on addressing the specific needs and barriers faced by these groups, including accessibility, safety, and privacy considerations, to ensure that no one is left behind in accessing clean water and sanitation facilities.
5. **Water-Related Disaster Risk Reduction:** Integrating the Clean Water and Sanitation goal with hazard and DRR management includes incorporating water-related disaster risk reduction measures. This involves considering the impacts of hazards on water resources, infrastructure, and ecosystems and developing strategies to minimize risks and enhance resilience. Examples include watershed management, flood protection measures, early warning systems for water-related hazards, and ecosystem-based approaches to water-resource management.

6. **Data Monitoring and Governance:** Integrating the Clean Water and Sanitation goal with hazard and DRR management requires data monitoring and governance to track progress and inform decision-making. This includes collecting and analyzing data on water quality, access, and sanitation coverage to identify gaps and guide targeted interventions. Effective water governance, including stakeholder engagement, policy coordination, and institutional capacity building, are essential for sustainable water and sanitation management in disaster-prone areas.

Integrating the goal of Clean Water and Sanitation (SDG 6) with the management of natural and technological hazards and disaster risk reduction involves ensuring access to safe water and sanitation facilities, promoting hygiene practices, sustainably managing water resources, ensuring inclusive access, incorporating water-related disaster risk reduction measures, and establishing effective data monitoring and governance. By integrating clean water and sanitation considerations into hazard and DRR management, we can create more resilient and healthy communities, reduce health risks, and promote sustainable development in the face of hazards and disasters.

3.2.9 Goal 7: Affordable and Clean Energy

The goal of "Affordable and Clean Energy" (SDG 7) in the context of integrating the Sustainable Development Goals (SDGs) with the management of natural and technological hazards and disaster risk reduction (DRR) focuses on promoting access to reliable, sustainable, and affordable energy, particularly in disaster-prone areas. Here is how the goal of Affordable and Clean Energy is relevant to integrating SDGs with hazard and DRR management.

1. **Energy for Disaster Preparedness and Response:** Integrating the Affordable and Clean Energy goal with hazard and DRR management involves ensuring access to reliable energy sources for disaster preparedness and response efforts. Energy is essential for early warning systems, communication networks, emergency lighting, medical facilities, and other critical infrastructure during and after disasters. Promoting reliable and clean energy sources such

as renewable energy technologies can enhance the effectiveness of disaster preparedness and response measures.

2. **Energy Access in Disaster-Affected Areas:** Integrating the Affordable and Clean Energy goal with hazard and DRR management includes promoting energy access in disaster-affected areas. Disasters can damage energy infrastructure and disrupt access to electricity and other forms of energy, affecting the well-being and recovery of affected communities. Efforts should focus on restoring and strengthening energy systems, including decentralized and off-grid solutions, to ensure continuous access to affordable and clean energy in disaster-prone areas.
3. **Renewable Energy and Climate Resilience:** Integrating the Affordable and Clean Energy goal with hazard and DRR management emphasizes the promotion of renewable energy and climate resilience. Climate change can increase the frequency and intensity of hazards, thereby affecting the availability and reliability of traditional energy sources. By investing in renewable energy technologies, such as solar, wind, and hydro, communities can reduce their dependence on fossil fuels, enhance climate resilience, and mitigate the impact of disasters on energy infrastructure.
4. **Energy Efficiency and Disaster Mitigation:** Integrating the Affordable and Clean Energy goal with hazard and DRR management involves promoting energy-efficiency measures for disaster mitigation. Energy-efficient technologies and practices, such as energy-efficient buildings, lighting, and appliances, reduce energy consumption and greenhouse gas emissions, while enhancing resilience to hazards. By integrating energy-efficiency measures into hazard and DRR strategies, communities can lower their energy costs, reduce environmental impacts, and build a more sustainable and resilient energy infrastructure.
5. **Inclusive Energy Access and Gender Equality:** Integrating the Affordable and Clean Energy goal with hazard and DRR management requires promoting inclusive energy access and gender equality. Vulnerable groups, including women and marginalized communities, often face challenges in accessing affordable and clean energy services. Efforts should be focused on removing barriers, ensuring affordability, and providing targeted energy solutions to meet the specific

needs of these groups. Promoting gender equality in energy access involves recognizing and addressing gender disparities in decision-making, ownership, and benefits related to energy systems.

6. **Capacity Building and Partnerships:** Integrating the Affordable and Clean Energy goal with hazard and DRR management involves capacity building and fostering partnerships. Capacity-building efforts focus on enhancing the technical skills, knowledge, and institutional capacity to support the development and management of affordable and clean energy systems in disaster-prone areas. Partnerships among governments, private sector actors, civil society organizations, and international agencies are crucial for resource mobilization, technology transfer, and knowledge sharing to promote sustainable energy solutions.

In a nutshell integrating the goal of Affordable and Clean Energy (SDG 7) with the management of natural and technological hazards and disaster risk reduction involves ensuring access to reliable and sustainable energy for disaster preparedness and response, promoting energy access in disaster-affected areas, investing in renewable energy and energy efficiency, promoting inclusive energy access and gender equality, and building capacity and partnerships. By integrating affordable and clean energy considerations into hazard and DRR management, we can enhance resilience, reduce greenhouse gas emissions, and contribute to sustainable development in the face of hazards and disasters.

3.2.10 Goal 8: Decent Work and Economic Growth

The goal of "Decent Work and Economic Growth" (SDG 8) in the context of integrating Sustainable Development Goals (SDGs) with the management of natural and technological hazards and disaster risk reduction (DRR) focuses on promoting inclusive and sustainable economic growth, productive employment, and decent work opportunities in disaster-prone areas. Here is how the goal of Decent Work and Economic Growth is relevant to integrating SDGs with hazard and DRR management.

1. **Economic Resilience:** Integrating the Decent Work and Economic Growth goal with hazard and DRR management involves building economic resilience in disaster-prone areas. Hazards and disasters can disrupt economies, leading

to job losses, income reduction, and economic instability. Efforts should focus on promoting diversified economies, supporting small- and medium-sized enterprises (SMEs), and fostering entrepreneurship to enhance economic resilience and create decent work opportunities that can withstand the impacts of hazards.

2. **Job Creation and Livelihood Support:** Integrating the Decent Work and Economic Growth goal with hazard and DRR management includes promoting job creation and providing livelihood support in disaster-affected areas. Disasters can result in the loss of livelihoods and income sources. Efforts should be focused on creating employment opportunities, particularly in sectors such as renewable energy, sustainable agriculture, and infrastructure reconstruction. Providing support for micro-enterprises, skills development, and access to finance can help restore and enhance livelihoods in affected communities.
3. **Occupational Health and Safety:** Integrating the Decent Work and Economic Growth goal with hazard and DRR management emphasizes ensuring occupational health and safety in the workplace, including disaster-prone areas. Hazards and disasters can pose risks to worker health and safety, exacerbating vulnerabilities and inequalities. Efforts should focus on promoting safe working conditions, providing training on occupational hazards, and ensuring access to personal protective equipment. Protecting workers' rights and wellbeing contributes to decent work and sustainable economic growth.
4. **Social Protection and Social Dialogue:** Integrating the Decent Work and Economic Growth goal with hazard and DRR management involves strengthening social protection mechanisms and promoting social dialogue. Disasters can increase vulnerability, particularly for workers in informal sectors and marginalized groups. Efforts should focus on establishing social protection systems such as social insurance, safety nets, and social assistance programs to provide income support and access to essential services during and after disasters. Promoting social dialogue between employers, workers, and governments enhances cooperation and enables effective policy responses to support decent work and economic growth.

5. **Inclusive Employment and Gender Equality:** Integrating the Decent Work and Economic Growth goal with hazard and DRR management requires promoting inclusive employment opportunities and gender equality. Vulnerable groups, including women, youth, and persons with disabilities, often face barriers to accessing decent work and economic opportunities. Efforts should focus on eliminating discriminatory practices, promoting equal pay for equal work, and ensuring equal access to training and entrepreneurial support. Encouraging diverse and inclusive labor markets contributes to sustainable economic growth and resilience.
6. **Partnerships and Capacity Building:** Integrating the Decent Work and Economic Growth goal with hazard and DRR management involves fostering partnerships and capacity building. Collaborative efforts among governments, employers, trade unions, and civil society are essential for promoting decent work and sustainable economic growth. Capacity-building programs, technical assistance, and knowledge-sharing initiatives support the development of skills, entrepreneurship, and productive sectors in disaster-prone areas.

Integrating the goal of Decent Work and Economic Growth (SDG 8) with the management of natural and technological hazards and disaster risk reduction involves building economic resilience, creating decent work opportunities, ensuring occupational health and safety, promoting social protection, advancing gender equality, and fostering partnerships and capacity building. Integrating decent work and economic growth considerations into hazard and DRR management can enhance resilience, reduce inequalities, and contribute to sustainable development in the face of hazards and disasters.

3.2.11 Goal 9: Industry, Innovation, and Infrastructure

The goal of "Industry, Innovation, and Infrastructure" (SDG 9) in the context of integrating the Sustainable Development Goals (SDGs) with the management of natural and technological hazards and disaster risk reduction (DRR) focuses on promoting sustainable and resilient infrastructure development, fostering innovation, and enhancing industrialization in disaster-prone areas. Here is how the goal of Industry, Innovation, and Infrastructure is relevant to integrating SDGs with hazard and DRR management.

1. **Resilient Infrastructure:** Integrating the Industry, Innovation, and Infrastructure goal with hazard and DRR management involves developing resilient infrastructure systems. Hazards and disasters can damage critical infrastructure such as roads, bridges, energy networks, and communication systems, leading to disruptions in essential services and economic activities. Efforts should focus on incorporating disaster-resilient design principles, such as climate-proofing, early warning systems, and appropriate land-use planning, to ensure that infrastructure can withstand and recover from the impact of hazards.
2. **Innovation for Disaster Risk Reduction:** Integrating the Industry, Innovation, and Infrastructure goal with hazard and DRR management emphasizes fostering innovation for disaster risk reduction. Technological advancements and innovative solutions can enhance early warning systems, risk assessment tools, and response mechanisms. Encouraging research and development in areas such as remote sensing, data analytics, and artificial intelligence can support effective decision making and improve the overall resilience of infrastructure and communities.
3. **Access to Basic Services:** Integrating the Industry, Innovation, and Infrastructure goal with hazard and DRR management includes ensuring access to basic services in disaster-prone areas. Disasters can disrupt access to essential services including energy, water, sanitation, and healthcare facilities. Efforts should be focused on developing and improving infrastructure networks to provide reliable and affordable access to these services, particularly in vulnerable and remote areas. Sustainable and inclusive infrastructure development ensures that basic services are accessible to all, even during and after disasters.
4. **Sustainable Industrialization:** Integrating the Industry, Innovation, and Infrastructure goal with hazard and DRR management involves promoting sustainable industrialization practices. Industrial activities can contribute to environmental degradation and increase vulnerability to hazards. Efforts should focus on adopting cleaner and

more resource-efficient technologies, promoting circular economic principles, and integrating environmental considerations into industrial processes. Sustainable industrialization reduces the environmental impact of industries and enhances the resilience of economic sectors to hazards.

5. **Connectivity and Digital Infrastructure:** Integrating the Industry, Innovation, and Infrastructure goal with hazard and DRR management includes promoting connectivity and digital infrastructure in disaster-prone areas. Access to reliable and affordable internet connectivity supports information dissemination, early warning systems, and disaster response coordination. Efforts should focus on expanding broadband networks, enhancing digital literacy, and ensuring access to digital services for all, including marginalized communities. Digital infrastructure enables efficient communication and enhances the effectiveness of hazard and DRR management.
6. **Capacity Building and Partnerships:** Integrating the Industry, Innovation, and Infrastructure goal with hazard and DRR management requires capacity building and fostering partnerships. Strengthening the technical capacities of institutions, professionals, and communities in infrastructure development and risk management is crucial. Partnerships among governments, private sector entities, academic institutions, and civil society organizations can promote knowledge sharing, resource mobilization, and technology transfer for sustainable infrastructure development and hazard resilience.

The integration of the goal of Industry, Innovation, and Infrastructure (SDG 9) with the management of natural and technological hazards and disaster risk reduction involves developing resilient infrastructure, fostering innovation for risk reduction, ensuring access to basic services, promoting sustainable industrialization, enhancing connectivity and digital infrastructure, and building capacity through partnerships. Integrating sustainable and resilient infrastructure development practices into hazard and DRR management can enhance resilience, promote economic growth, and contribute to sustainable development in the face of hazards and disasters.

3.2.12 Goal 10: Reduced Inequality

Goal 10 of the United Nations' Sustainable Development Goals (SDGs) is "Reduced Inequality." This goal focuses on reducing inequalities within and among countries, including inequalities in income, opportunities, and access to resources. Integrating SDGs with the management of natural and technological hazards and disaster risk reduction is essential for effectively achieving Goal 10. On this section, we discuss how these areas intersect.

1. **Vulnerability and Inequality:** Natural and technological hazards disproportionately affect marginalized and vulnerable populations, exacerbating existing inequalities. Integrating SDGs with disaster risk reduction ensures that efforts are directed towards reducing vulnerabilities and addressing the specific needs of marginalized groups, such as women, children, persons with disabilities, and indigenous communities.
2. **Social Protection:** Disaster risk reduction can be aligned with SDG targets related to social protection, including poverty eradication, access to basic services, and social safety nets. By implementing inclusive and equitable social protection systems, communities can better cope with the impacts of hazards and reduce inequalities exacerbated by disasters.
3. **Access to Information and Participation:** Integrating SDGs with hazard and disaster management promotes the principles of transparency, accountability, and participation. Access to timely and accurate information, along with the meaningful participation of all stakeholders, including vulnerable groups, ensures that decision-making processes are inclusive and that the needs and concerns of marginalized communities are considered.
4. **Resilience and Capacity Building:** Disaster risk reduction efforts aligned with SDGs can focus on building resilience in vulnerable communities. This includes improving infrastructure, enhancing early warning systems, promoting sustainable livelihoods, and strengthening local capacities to respond to and recover from hazards. By addressing inequality in access to resources and opportunities, these efforts contribute to reducing vulnerability and enhancing resilience.

5. **International Cooperation:** Integrating SDGs with disaster risk reduction also emphasizes the importance of international cooperation and partnerships. This cooperation can facilitate the sharing of knowledge, expertise, and resources to support countries in implementing inclusive and equitable disaster risk reduction strategies, particularly for those most in need.

It is important to integrates SDGs with the management of natural and technological hazards and disaster risk reduction can help address the root causes of inequality and ensure that the most vulnerable populations are not left behind. By prioritizing inclusivity, social protection, resilience building, and international cooperation, inequalities can be reduced while effectively managing risks and building more sustainable and resilient societies.

3.2.13 Goal 11: Sustainable Cities and Communities

Goal 11 of the United Nations' Sustainable Development Goals (SDGs) is "Sustainable Cities and Communities." This goal aims to ensure that cities and human settlements are inclusive, safe, resilient, and sustainable. Integrating SDGs with the management of natural and technological hazards and disaster risk reduction is crucial for achieving Goal 11. Here, we discuss how these areas intersect.

Risk-Informed Urban Planning: Integrating SDGs with disaster risk reduction emphasizes the importance of risk-informed urban planning. This involves identifying and assessing hazards, vulnerabilities, and risks in urban areas, and incorporating this information into urban development plans. By considering disaster risks such as floods, earthquakes, or industrial accidents, cities can ensure that their infrastructure, housing, and services are designed to withstand and recover from such events, thereby creating safer and more resilient communities.

1. **Resilient Infrastructure:** SDG 11 promotes the development of sustainable and resilient infrastructure. Integrating disaster risk reduction strategies ensures that infrastructure projects consider potential hazards and incorporate measures to mitigate their impact. This includes the design of buildings, transportation systems, water and sanitation facilities, and energy networks to be more resilient and able to withstand and recover from natural and technological hazards.

2. **Inclusive Urban Development:** Integrating SDGs with disaster risk reduction emphasizes the inclusivity and needs of vulnerable groups in urban development. This involves ensuring that all communities, including low-income groups, women, children, persons with disabilities, and the elderly, have equal access to essential services such as housing, healthcare, education, and public spaces. Inclusive urban development reduces inequality and promotes social cohesion and resilience in cities.
3. **Climate Change Adaptation:** The integration of SDGs with disaster risk reduction recognizes the close relationship between climate change and natural hazards. Sustainable cities and communities need to address the challenges of climate change and adapt to its impacts. By incorporating climate change adaptation measures into urban planning, cities can reduce their vulnerability to extreme weather events, sea level rise, and other climate-related hazards.
4. **Community Engagement and Capacity Building:** Integrating SDGs with the management of hazards and disaster risk reduction involves actively engaging communities in decision-making processes. It encourages participatory approaches, community-based risk assessments, and empowerment of local actors. By involving communities, including marginalized groups, in planning and implementing risk-reduction strategies, cities can leverage local knowledge and build local capacities for disaster preparedness and response.
5. **Data and Information Management:** Effective management of hazards and disaster risk reduction requires reliable and up-to-date data and information. Integrating SDGs emphasizes the importance of data collection, analysis, and sharing to inform urban planning, risk assessments, early warning systems, and emergency response. By investing in data management systems, cities can improve their ability to understand and respond to hazards, thereby enhancing their resilience and reducing risks.

By integrating the SDGs with the management of natural and technological hazards and disaster risk reduction, cities can effectively address the challenges of creating sustainable and resilient communities. This approach ensures that urban development is inclusive, safe, and resilient while also considering the specific risks and vulnerabilities associated with natural and technological hazards.

3.2.14 Goal 12: Responsible Consumption and Production

Goal 12 of the United Nations' Sustainable Development Goals (SDGs) is "Responsible Consumption and Production." This goal aims to ensure sustainable consumption and production patterns by promoting resource efficiency, reducing waste generation, and minimizing the environmental impact of economic activities. Integrating SDGs with the management of natural and technological hazards and disaster risk reduction is crucial for effectively achieving Goal 12. Here is how these areas intersect:

1. **Waste Management and Hazard Reduction:** Integrating SDGs with disaster risk reduction emphasizes the importance of responsible waste management practices. Effective waste management reduces the environmental and health risks associated with hazardous materials such as chemicals and electronic waste, which can be a source of disaster risk. Implementing waste reduction, recycling, and proper disposal practices can help minimize the potential impacts of hazardous waste on communities and ecosystems.
2. **Sustainable Resource Management:** SDG 12 promotes sustainable resource management, including the efficient use of natural resources and reduction of resource consumption. Integrating this goal with disaster risk reduction encourages the adoption of sustainable production practices that minimize resource extraction, pollution, and the generation of hazardous materials. By reducing reliance on resource-intensive and environmentally harmful processes, businesses and industries can contribute to both responsible consumption and production and the reduction of disaster risks.
3. **Circular Economy Approach:** Integrating SDGs with the management of hazards and disaster risk reduction supports the transition to a circular economy. The aim of a circular economy is to minimize waste generation by promoting the reuse, recycling, and repurposing of materials. By adopting circular economy principles such as extended product lifecycles, sustainable design, and resource recovery, communities can reduce their vulnerability to natural and technological

hazards by minimizing the creation of waste and reducing the need for resource extraction.

4. **Climate Change Mitigation:** Responsible consumption and production are closely linked to efforts to mitigate climate change. Integrating SDGs with disaster risk reduction recognizes the interconnections between sustainable production practices, greenhouse gas emissions, and the potential for climate-related hazards. By adopting sustainable and low-carbon production methods, businesses and industries can contribute to climate-change mitigation, thereby reducing the frequency and intensity of climate-related disasters.
5. **Resilient Supply Chains:** Integrating SDGs with hazard management and disaster risk reduction emphasizes the importance of resilient supply chains. Sustainable and responsible consumption and production practices can enhance the resilience of supply chains by reducing dependence on vulnerable resources or locations and diversifying sourcing options. Resilient supply chains are better equipped to withstand and recover from disruptions caused by natural or technological hazards, thus ensuring the continuity of essential goods and services.
6. **Awareness and Education:** Integrating SDGs with disaster risk reduction recognizes the importance of raising awareness and promoting education on responsible consumption and production practices. By educating individuals, communities, and businesses about the environmental and social impacts of their consumption choices, integration efforts can foster behavior change towards more sustainable and responsible lifestyles. This awareness and education contributes to reducing the demand for environmentally harmful products and practices, thereby reducing disaster risks.

By integrating SDGs with the management of natural and technological hazards and disaster risk reduction, responsible consumption and production practices can be promoted effectively promoted. This integration ensures that economic activities are conducted sustainably with minimal impact on the environment and reduced exposure to disaster risks. It also encourages the adoption of circular economic principles, climate change mitigation measures, and the development of resilient supply chains, all of which contribute to creating a more sustainable and disaster-resilient society.

3.2.15 Goal 13: Climate Action

Goal 13 of the United Nations' Sustainable Development Goals (SDGs) is "Climate Action." This goal focuses on combating climate change and its impact through mitigation, adaptation, and resilience-building measures. Integrating SDGs with the management of natural and technological hazards and disaster risk reduction is critical for effectively achieving Goal 13. Here is how these areas intersect:

1. **Climate Change Mitigation and Hazard Reduction:** Integrating SDGs with disaster risk reduction recognizes the close link between climate change and the occurrence and intensity of natural and technological hazards. Climate change mitigation efforts, such as reducing greenhouse gas emissions and transitioning to renewable energy sources, have contributed to the reduction of disaster risks associated with climate-related events. By addressing the root causes of climate change, integration efforts can help reduce the frequency and severity of hazards, such as extreme weather events and sea-level rise.
2. **Adaptation and Resilience:** Integrating SDGs with the hazard management and disaster risk reduction emphasizes the importance of climate change adaptation and resilience-building measures. Climate change adaptation involves adjusting systems and practices to reduce vulnerability and increase resilience to climate change. By integrating adaptation measures into disaster risk reduction strategies, communities can better prepare for and respond to climate-related hazards and reduce associated risks and damages.
3. **Early Warning Systems:** Climate action and disaster risk reduction intersect in the development and strengthening of systems. Early warning systems for climate-related hazards, such as storms, floods, and droughts are crucial for effective disaster preparedness and response. Integrating SDGs with disaster risk reduction promotes the establishment of robust and inclusive early warning systems that provide timely and accurate information to vulnerable communities, enabling them to take appropriate actions and reduce the impacts of climate-related hazards.

4. **Sustainable Development Planning:** Integrating SDGs with climate action and disaster risk reduction promotes sustainable development planning at local, national, and global levels. This involves considering the impacts of climate change and disaster risks in development strategies, policies, and investment decisions. By incorporating climate resilience and disaster risk reduction into development planning, communities can build sustainable infrastructure, promote sustainable land-use practices, and ensure the long-term sustainability and resilience of their economies and ecosystems.
5. **Collaboration and Partnerships:** The integration of SDGs with climate action and disaster risk reduction emphasizes the importance of collaboration and partnerships at all levels. Addressing climate change and reducing disaster risks require collective action involving governments, businesses, civil society organizations, and local communities. Collaborative approaches foster knowledge sharing, capacity building, and resource mobilization, enabling effective climate action and resilience-building initiatives.
6. **Awareness and Education:** Integrating SDGs with the management of hazards and disaster risk reduction recognizes the role of awareness and education in climate action. Raising awareness of the causes and impacts of climate change, as well as promoting climate literacy, helps individuals and communities understand the importance of taking action. Education and awareness initiatives can empower people to make informed decisions, adopt sustainable practices, and support climate-change mitigation and adaptation efforts.

By integrating SDGs with the management of natural and technological hazards and disaster risk reduction, climate action can be pursued effectively. This integration ensures that efforts to combat climate change are aligned with disaster risk reduction strategies, promoting resilience, adaptation, and sustainable development. By addressing the impacts of climate change and reducing disaster risks, integration efforts can contribute to building a more sustainable, climate-resilient, and disaster-resilient future.

3.2.16 Goal 14: Life Below Water

Goal 14 of the United Nations' Sustainable Development Goals (SDGs) is "Life Below Water." This goal focuses on the conservation and sustainable use of oceans, seas, and marine resources for sustainable development. Integrating SDGs with the management of natural and technological hazards and disaster risk reduction is essential for effectively achieving Goal 14. Here is how these areas intersect:

1. **Marine Disaster Risk Reduction:** Integrating SDGs with disaster risk reduction recognizes the importance of addressing the risks and impacts of natural and technological hazards on marine ecosystems and coastal communities. This includes developing and implementing strategies to reduce the vulnerability of marine and coastal areas to hazards, such as storms, tsunamis, and oil spills. By incorporating disaster risk reduction measures into marine conservation and management efforts, integration helps to protect marine ecosystems and coastal communities from the impacts of disasters.
2. **Climate Change and Marine Ecosystems:** The integration of SDGs with hazard management and disaster risk reduction acknowledges the role of climate change in altering marine ecosystems. Climate change, including rising sea levels and ocean acidification, pose significant risks to marine life and coastal communities. By addressing climate change impacts, promoting climate resilience, and reducing greenhouse gas emissions, integration efforts can contribute to the preservation of marine ecosystems and the well-being of coastal communities.
3. **Sustainable Fisheries and Aquaculture:** Goal 14 emphasizes the sustainable use of marine resources. Integrating SDGs with hazard management and disaster risk reduction supports sustainable fisheries and aquaculture practices, reducing the vulnerability of these sectors to disasters. This includes implementing measures to prevent overfishing, promoting responsible fishing practices, protecting marine biodiversity, and ensuring long-term sustainability of fish stocks. By reducing the risks associated with unsustainable fishing practices, integration efforts can contribute to the conservation and sustainable use of marine resources.

4. **Pollution Prevention and Marine Environment Protection:** Integrating SDGs with disaster risk reduction emphasizes the importance of preventing and reducing marine pollution. This includes addressing pollution sources such as plastic waste, chemical contaminants, and oil spills, which can have severe impacts on marine ecosystems and coastal communities. By implementing pollution prevention measures, strengthening waste management systems, and promoting sustainable practices, integration efforts can help protect the marine environment and reduce the risks associated with pollution-related disasters.
5. **Ecosystem-Based Approaches:** The integration of SDGs with hazard management and disaster risk reduction promotes ecosystem-based approaches to marine conservation and management. Ecosystem-based approaches recognize the interconnectedness of marine ecosystems, coastal areas, and human well-being. By protecting and restoring coastal habitats, such as mangroves and coral reefs and implementing sustainable coastal zone management practices, integration efforts enhance the resilience of marine ecosystems and contribute to disaster risk reduction.
6. **Capacity Building and Collaboration:** Integrating SDGs with hazard management and disaster risk reduction emphasizes the importance of capacity building and collaboration among stakeholders. This includes promoting knowledge exchange, technology transfer, and collaboration among governments, scientific institutions, civil society, and local communities. By enhancing the capacity for marine conservation, disaster preparedness, and response, integration efforts contribute to the effective management of natural and technological hazards and achievement of Goal 14.

The integration of the SDGs with the management of natural and technological hazards, disaster risk reduction, and efforts to achieve Goal 14 are strengthened. This integration ensures that marine conservation and the sustainable use of marine resources are aligned with disaster risk reduction strategies, promoting the resilience of marine ecosystems and coastal communities. By addressing the impacts of hazards, pollution, and climate change, integration efforts can contribute to the preservation and sustainable management of life below water.

3.2.17 Goal 15: Life on Land

Goal 15 of the Sustainable Development Goals (SDGs) is "Life on Land." It focuses on protecting, restoring, and promoting the sustainable use of terrestrial ecosystems, sustainably managing forests, combating desertification, halting and reversing land degradation, and biodiversity loss.

Integrating SDG 15 with the management of natural and technological hazards and disaster risk reduction is crucial for ensuring the long-term sustainability and resilience of land ecosystems and communities that depend on them. The following are some key points to consider:

1. **Ecosystem Resilience:** Healthy and resilient terrestrial ecosystems can better withstand and recover from natural and technological hazards. By implementing effective disaster risk reduction strategies such as early warning systems, ecosystem-based approaches, and sustainable land management practices, we can enhance the resilience of ecosystems and minimize the negative impacts of disasters.
2. **Biodiversity Conservation:** Protecting and restoring biodiversity are essential for maintaining ecosystem balance and resilience. Biodiverse ecosystems are often more resistant to hazards and have a higher capacity for recovery. By conserving and restoring natural habitats, promoting sustainable land-use practices, and preventing biodiversity loss, we can reduce the vulnerability of ecosystems to disasters and support their recovery.
3. **Sustainable Land Management:** Unsustainable land use practices such as deforestation, overgrazing, and land degradation can increase the risk of natural hazards such as floods, landslides, and wildfires. Integrating SDG 15 with disaster risk reduction involves promoting sustainable land management approaches that prevent land degradation, restore degraded land, and enhance soil health. These practices can help reduce the likelihood and severity of hazards and improve the overall resilience of terrestrial ecosystems.
4. **Community Engagement and Capacity Building:** Successful integration of SDG 15 and disaster risk reduction requires active involvement and capacity building at the community level. Local communities should be engaged in decision-making processes, empowered with knowledge and skills, and supported in the implementation of

sustainable land management practices. This participatory approach ensures that local knowledge and needs are considered and enhances the effectiveness of hazard and risk reduction initiatives.

5. **Policy Integration:** Governments and policymakers play a crucial role in integrating SDG 15 and disaster risk reduction into national policies, plans, and strategies. Coherent and integrated policies can provide a framework for simultaneously addressing land degradation, biodiversity loss, and disaster risks. By aligning disaster risk reduction plans with sustainable land management strategies, governments can enhance the overall resilience of ecosystems and communities.

The integration and understanding of SDG 15 with the management of natural and technological hazards and disaster risk reduction requires a holistic and multi-sectoral approach. By recognizing the interlinkages between land ecosystems, biodiversity, and disaster risks, we can work towards sustainable development that ensures the long-term well-being of both people and the planet.

3.2.18 Goal 16: Peace and Justice Strong Institutions

Goal 16 of the Sustainable Development Goals (SDGs) is "Peace, Justice, and Strong Institutions." It aims to promote peaceful and inclusive societies, provide access to justice for all, and build effective, accountable, and inclusive institutions at all levels. Integrating SDG 16 with the management of natural and technological hazards and disaster risk reduction is essential for ensuring a comprehensive and inclusive approach to disaster risk management. The following conclusions can be drawn:

1. **Effective Governance:** Strong institutions and effective governance are crucial for disaster risk reduction. Integrating SDG 16 involves fostering transparent, accountable, and participatory governance structures that facilitate the coordination and implementation of disaster risk reduction efforts. This includes promoting the involvement of all stakeholders, including marginalized groups, in decision-making processes and ensuring access to justice and legal remedies related to disasters.

2. **Conflict Prevention and Peacebuilding:** Conflicts and violence can exacerbate the impact of natural and technological hazards, hindering effective disaster risk reduction. Integrating SDG 16 with disaster risk reduction requires efforts to prevent and resolve conflicts, build social cohesion, and promote peace. By addressing the root causes of conflict and promoting inclusive dialogue and cooperation, we can enhance the resilience of communities to hazards and reduce the risk of disasters.
3. **Social Inclusion and Equity:** Integrating SDG 16 involves ensuring that disaster risk reduction efforts are inclusive and leave no one behind. This requires addressing social inequalities, discrimination, and marginalization. Vulnerable and marginalized groups, including women, children, persons with disabilities, and indigenous peoples, should be actively engaged in decision-making processes, provided equal access to resources and information, and empowered to participate in disaster risk reduction activities.
4. **Data and Information Systems:** Effective disaster risk management relies on reliable and timely data and information systems. Integrating SDG 16 involves strengthening institutions' capacity to collect, analyze, and disseminate data related to hazards, vulnerabilities, and risks. This enables evidence-based decision making, supports early warning systems, and enhances preparedness and response efforts.
5. **Rule of Law and Legal Frameworks:** Integrating SDG 16 requires the development and implementation of robust legal frameworks and policies related to disaster risk reduction. These include laws and regulations that promote risk reduction, preparedness, response, and recovery. Strengthening the rule of law ensures accountability, protects the rights of the affected populations, and provides a framework for addressing the legal and governance aspects of disaster management.
6. **International Cooperation and Partnerships:** Achieving SDG 16 and integrating it with disaster risk reduction requires strong international cooperation and partnerships. This includes sharing knowledge, experiences, and best practices as well as providing technical and financial support to countries in need. Collaboration between governments, civil society organizations, the private sector, and international

organizations is essential for building strong institutions and promoting peace, justice, and effective global disaster risk reduction.

By integrating SDG 16 with the management of natural and technological hazards and disaster risk reduction, we can ensure that disaster risk reduction efforts are grounded in the principles of peace, justice, and inclusiveness. This approach helps create resilient societies that are better prepared to prevent, mitigate, and respond to disasters while upholding the rights and well-being of all individuals and communities.

3.2.19 Goal 17: Partnerships to Achieve the Goal

Goal 17 of the Sustainable Development Goals (SDGs) is "Partnerships for the Goals." It emphasizes the importance of global collaboration and partnerships in achieving the SDGs. Integrating SDG 17 with the management of natural and technological hazards and disaster risk reduction is crucial for enhancing the effectiveness and sustainability of disaster risk-management efforts. Here is how partnerships can contribute to this integration.

1. **Multi-Stakeholder Collaboration:** Integrating SDG 17 involves fostering collaboration and partnerships among governments, non-governmental organizations (NGOs), civil society, the private sector, academia, and other relevant stakeholders. By bringing together diverse perspectives, expertise, and resources, multi-stakeholder partnerships can promote innovative approaches to disaster risk reduction. Such collaborations can facilitate knowledge sharing, capacity building, and coordinated action, leading to more effective and comprehensive disaster-management strategies.
2. **Technology Transfer and Capacity Building:** Partnerships can facilitate the transfer of technology, knowledge, and expertise in disaster risk reduction. Developed countries, international organizations, and technology providers can support developing countries in building their capacity to monitor, assess, and respond to natural and technological hazards. Collaborative initiatives can help bridge this technological gap, enhance early warning systems, improve data analysis, and promote the use of innovative solutions for disaster risk reduction.

3. **Financial Partnerships:** Mobilizing sufficient financial resources is crucial for implementing effective disaster risk reduction measures. Integrating SDG 17 involves fostering partnerships to secure funding from various sources, including the public and private sectors, international financial institutions, and philanthropic organizations. Public–private partnerships, blended finance mechanisms, and innovative financing models can leverage resources to support risk-reduction investments, infrastructure development, and community resilience-building activities.
4. **Knowledge Sharing and Capacity Development:** Partnerships can facilitate the exchange of knowledge, best practices, and lessons learned in disaster risk reduction. By sharing experiences and expertise, stakeholders can learn from each other's successes and challenges, leading to improved practices and more efficient risk-reduction strategies. Capacity development programs, training initiatives, and knowledge-sharing platforms can strengthen the skills and competencies of the stakeholders involved in disaster risk management.
5. **Policy Coherence and Coordination:** Partnerships can contribute to policy coherence and coordination in disaster risk reduction. Governments can integrate disaster risk reduction objectives into their overall development agendas by aligning national policies, strategies, and frameworks with the SDGs. Partnerships can facilitate policy dialogue, coordination mechanisms, and joint planning processes to ensure coherence among the different sectors and stakeholders involved in disaster risk management.
6. **South-South Cooperation:** Integrating SDG 17 involves promoting South-South cooperation, where countries in the Global South share experiences, expertise, and resources to address common challenges in disaster risk reduction. This collaboration can enable knowledge exchange, technical assistance, and capacity-building support between countries that face similar hazards. South–South partnerships can foster mutual learning and enhance the resilience of participating countries to natural and technological hazards.

By fostering partnerships and collaborations, we can strengthen the integration of the SDGs with the management of natural and technological hazards and disaster risk reduction. These partnerships enable collective action, resource mobilization, and knowledge sharing, leading to more effective and sustainable approaches for reducing disaster risks and building resilient communities and ecosystems.

3.3 Conclusion

Integrating the Sustainable Development Goals (SDGs) with the management of natural and technological hazards and disaster risk reduction is vital for ensuring sustainable and resilient development. By incorporating the goals of the SDGs into disaster risk reduction efforts, we can address the underlying causes of vulnerabilities and build more effective and comprehensive strategies to reduce risks and enhance resilience.

Goal 15, "Life on Land," emphasizes the protection and sustainable use of terrestrial ecosystems, while Goal 16, "Peace and Justice Strong Institutions," focuses on promoting peaceful and inclusive societies and effective governance. These goals can be integrated by implementing sustainable land management practices, conserving biodiversity, involving communities in decision making, and addressing social inequalities and conflicts that contribute to vulnerability.

Goal 17, "Partnerships to achieve the Goal," underscores the importance of global collaboration and cooperation. By fostering partnerships among governments, civil society, the private sector, academia, and international organizations, we can leverage resources, share knowledge and best practices, and enhance coordination and policy coherence in disaster risk reduction.

Integrating SDGs with the management of natural and technological hazards and disaster risk reduction requires a holistic and multi-sectoral approach. This involves incorporating sustainable land management practices, promoting social inclusion and equity, strengthening institutions and governance structures, leveraging technology and financial resources, and fostering collaboration and partnerships at various levels.

By integrating the SDGs with disaster risk reduction, we can create a synergistic approach that addresses the interdependencies between environmental, social, economic, and governance factors. This integrated approach contributes to sustainable development, reduces the impact of hazards, protects ecosystems and biodiversity, promotes social justice, and builds resilient communities and institutions.

Through the integration of the SDGs and disaster risk reduction that we can effectively manage natural and technological hazards, reduce vulnerability, and create a more sustainable and resilient future for all.

References

Singh SK, Kanga S, Gulati B, Raiˇc M, Sajan B, Đurin B, Singh S (2023) Spatial and Temporal Analysis of Hydrological Modelling in the Beas Basin Using SWAT+ Model. *Water* 15, 3338. https://doi.org/10.3390/w1519333

Rani A, Gupta SK, Singh SK, Meraj G, Kumar P, Kanga S, Đurin B, Dogančić D (2023) Predicting Future Land Use Utilizing Economic and Land Surface Parameters with ANN and Markov Chain Models. *Earth* 4(3):728-751.

Ahmad T, Gupta SK, Singh SK, Meraj G, Kumar P, Kanga S (2023) Unveiling Nature's Resilience: Exploring Vegetation Dynamics during the COVID-19 Era in Jharkhand, India, with the Google Earth Engine. *Climate* 11(9):187.

Sharma M, Upadhyay RK, Tripathi G, Kishore N, Shakya A, Meraj G, Kanga S, Singh SK, Kumar P, Johnson BA, Thakur SN (2023) Assessing Landslide Susceptibility along India's National Highway 58: A Comprehensive Approach Integrating Remote Sensing, GIS, and Logistic Regression Analysis. *Conservation* 3(3):444-459.

Saqib N, Rai PK, Kanga S, Kumar D, Đurin B, Singh SK (2023) Assessment of Ground Water Quality of Lucknow City under GIS Framework Using Water Quality Index (WQI). *Water* 15(17):3048.

Choudhury U, Singh SK, Kumar A, Meraj G, Kumar P, Kanga S (2023) Assessing land use/land cover changes and urban heat island intensification: A case study of Kamrup Metropolitan District, Northeast India (2000–2032). *Earth* 4(3):503-521.

Rai NK, Mathur S, Singh SK, Tiwari M, Singh VK, Haque R, Tiwari S, Sharma LK (2023) Differential regulation of mitochondrial complex I and oxidative stress based on metastatic potential of colorectal cancer cells. *Oncol. Lett.* 26(1):1-1.

Debnath J, Sahariah D, Lahon D, Nath N, Chand K, Meraj G, Kumar P, Singh SK, Kanga S, Farooq M (2023) Assessing the impacts of current and future changes of the planforms of river Brahmaputra on its land use-land cover. *Geosci. Front.* 14(4):101557. Elsevier.

Kumar P, Singh SK, Gupta SK, Kanga S, Meraj G (2023) Uncovering the hydro-meteorological drivers responsible for forest fires utilizing geospatial techniques. *Theor. Appl. Climatol.* https://doi.org/10.1007/s00704-023-04497. Springer Nature.

Singh SK, Kotecha MJ, Kanga S, Pankhaniya SK, Agrawal S, Meraj G (2023) Urban Heat Island (UHI) Resilience Plan in Varying Climatic Conditions Using

Geospatial Approach: A Case Study of Rajkot City. In: *Climate Change, Agriculture and Society*, vol 1. pp 307-338.

Singh S, Singh SK, Prajapat DK, Pandey V, Kanga S, Kumar P, Meraj G (2023) Assessing the Impact of the 2004 Indian Ocean Tsunami on South Andaman's Coastal Shoreline: A Geospatial Analysis of Erosion and Accretion Patterns. *J Mar Sci Eng* 11(6):1134.

Rathore SS, Mathur SC, Mathur S, Sharma S, Singh SK (2023) Late Cretaceous–Early Paleocene Geodiversity of Fatehgarh Formation of Petroliferous Barmer Basin, Western Rajasthan, India: A Potential Geopark Site. *Geoheritage* 15:54.

Singh S, Meraj G, Kumar P, Singh SK, Kanga S, Johnson BA, Prajapat DK, Debnath J, Sahariah D (2023) Decoding Chambal River Shoreline Transformations: A Comprehensive Analysis Using Remote Sensing, GIS, and DSAS. *Water* 15(9):1793.

Tripathi G, Shakya A, Upadhyay RK, Singh SK, Kanga S, Pandey SK (2023) Landslide Susceptibility Mapping of Tehri Reservoir Region Using Geospatial Approach. In: *Climate Change Adaptation, Risk Management and Sustainable Practices in the Himalaya*, pp 135-156.

Sud A, Kanga R, Singh SK, Meraj G, Kanga S, Kumar P, Ramanathan AL, Bhardwaj V (2023) Simulating Groundwater Potential Zones in Mountainous Indian Himalayas—A Case Study of Himachal Pradesh. *Hydrology* 10(3):65.

Upadhyay RK, Tripathi G, Đurin B, Šamanović S, Cetl V, Kishore N, Sharma M, Singh SK, Kanga S, Wasim Md, Rai PK, Bhardwaj V (2023) Groundwater Potential Zone Mapping in the Ghaggar River Basin, North-West India, Using Integrated Remote Sensing and GIS Techniques. *Water* 15(5):961.

Lavane K, Kumar P, Meraj G, Han TG, Ngan LHB, Lien BTB, Ty TV, Thanh NT, Downes NK, Nam NGD, Minh HVT, Singh SK, Kanga S (2023) Assessing the effects of drought on rice yields in the Mekong Delta. *Climate* 11(1):13.

Nath N, Sahariah D, Meraj G, Debnath J, Kumar P, Lahon D, Chand K, Farooq M, Chandan P, Singh SK, Kanga S (2023) Land use and land cover change monitoring and prediction of a UNESCO world heritage site: Kaziranga eco-sensitive zone using cellular automata-Markov model. *Land* 12(1):151.

Kanga S, Upadhyay RK, Sharma M, Sharma S, Tripathi G, Singh SK (2023) Geotechnical Stability Assessment of Road-cut Slopes: A Case Study of Srinagar, Garhwal Himalaya, India. *Biological Forum – An International Journal* 15(5):1071-1080.

Gulati B, Sharma R, Kanga S, Singh SK, Sajan B, Meraj G, Kumar P, Ramanathan AL (2023) Unraveling the Relationship Between Stubble Burning and Air Quality Degradation in Punjab: A Temporal and Spatial Analysis (2019-2022). *J Clim Change* 9(2):43-53, IOS Press.

Tomar P, Singh SK, Kanga S, Pattanaik A (2023) Water Bodies mapping and monitoring using high-resolution satellite images. *Sustainability, Agri, Food and Environmental Research* 11.

Singh H, Meraj G, Singh S, Shrivastava V, Sharma V, Farooq M, Kanga S, Singh SK, Kumar P (2022) Status of Air Pollution during COVID-19-Induced Lockdown in Delhi, India. *Atmosphere* 13(12):2090.

Farooq M, Mushtaq F, Meraj G, Singh SK, Kanga S, Gupta A, Kumar P, Singh D, Avtar R (2022) Strategic Slum Upgrading and Redevelopment Action Plan for Jammu City. *Resources* 11(12):120.

4

Hydrological and Morphometric Study of the Girna River Basin, Maharashtra Using Remote Sensing and GIS Techniques

Pranaya Diwate[1]*, Ashish B Itolikar[2], Firoz Khan[3], Kunal Chinche[4] and Pranali Kathe[4]

[1]*Department of Geology, School of Basic and Applied Sciences, MGM University, Chhatrapati Sambhajinagar, Maharashtra, India*
[2]*School of Humanities and Engineering Sciences, MIT Academy of Engineering, Alandi Road, Pune, Maharashtra, India*
[3]*Department of Geology, HNB Garhwal Central University, Srinagar, Uttarakhand, India*
[4]*Centre for Climate Change and Water Research, Suresh Gyan Vihar University, Jaipur, Rajasthan, India*

Abstract

The present study aims to conduct a morphometric analysis of the Girna River Basin watershed, a tributary of the Tapi River in Maharashtra, India. A detailed morphometry analysis was performed using remote sensing and GIS using a Digital Elevation Model (Aster DEM) and SOI toposheets, and various drainage basin parameters such as Linear, Areal, and Relief aspects were also calculated. The Girna River Basin comprises 2,137 streams and has 6th order stream networks, in which 1,071, 526, 261, 125, 107, and 47 are the first-, second-, third-, fourth-, fifth-, and sixth-order streams, respectively. The Girna River Basin is 6th order stream with a circular shape, dendritic pattern, coarse drainage texture, and longer flow duration in lower areas. The bifurcation ratio lies between 1 and 2, indicating that the geological structure does not have a significant impact on the drainage patterns. The present study indicates various parameters of the Girna River Basin, which indicates that the basin has a high slope in the northwestern part, whereas most of the area has a low slope angle and a very low gradient ratio. The stream order helps understand the estimation of floods near the basin area because high

**Corresponding author*: pranayadiwate7@gmail.com

Shruti Kanga, Gowhar Meraj, Suraj Kumar Singh, Majid Farooq and M. S. Nathawat (eds.) Geospatial Technology for Natural Resource Management, (133–152) © 2024 Scrivener Publishing LLC

stream order rivers have a high chance of floods. Determining the water discharge flow speed and volume variables, such as Drainage Density and Stream Frequency, are very important parameters to determine in the river channels because higher values of both parameters cause a higher probability of flooding. These variables also cause a temporal variation in the speed at which water flows in the stream when the flood reaches its peak.

***Keywords*:** Hydrological parameters, morphometric analysis, Girna River Basin, remote sensing, GIS

4.1 Introduction

Morphometry is a quantitative and mathematical analysis of the size and shape of Earth resources such as lakes and rivers (Singh *et al.*, 2023; Rani *et al.*, 2023). The Morphometric study involves the drainage characteristics of the basin which consists of measuring the configuration of the channel network and areal and linear measurements (relief and gradient of the slope of the basin) (Ahmad *et al.*, 2023).

The main objective of morphometric study is to understand how the morphology of the channel network responds to various geological and geomorphological processes and the geohydrological behavior of drainage basins (Gulati *et al.*, 2023; Tomar *et al.*, 2023). This study also helps depict the prevailing climate, geomorphology, geology, and structural background of the catchment (Sharma *et al.*, 2023; Singh *et al.*, 2023). In addition, morphometric analysis plays a significant role in the understanding of watershed management, groundwater potential zones (Saqib *et al.*, 2023; Farooq *et al.*, 2023), soil physical properties, landform processes, basin management, and environmental assessment. A systematic approach for the analysis is necessary for the expansion of the drainage basin assessment of the area, e.g., the drainage line of the area describes the 3D geometry of the region with its evolutionary pattern. The geomorphological and geological history of the drainage basin, along with the slope angle and structural control, can be obtained from the study of the drainage systems. However, various drainage parameters need to be analyzed to evaluate the morphometric parameters, such as the stream order, basin area and perimeter, drainage channel length, bifurcation ratio (Rb), drainage density (Dd), relief ratio (Rh) and stream length ratio (RL). The morphometric parameters of the basin are significant elements in shaping the characteristics of watersheds or sub-watersheds; hence, the morphometric parameter is the foremost parameter to assess the development and formation of land surface processes (Choudhury *et al.*, 2023). Previous studies have used the conventional method for the analysis of drainage characteristics of several river basins and sub-basins. Similarly, Rai *et al.* (2023) used various geomorphic features related to morphometric parameters to evaluate the flow intensity and surface runoff of a drainage system. The stream order system eminently used for classifying stream networks is based on the number of upstream tributaries (Debnath *et al.*, 2023; Kumar *et al.*, 2023). A headwater stream is considered a 1st order stream and a segment downstream of the confluence of two 1st order streams is a 2nd order stream and so on. Remote sensing and GIS techniques are exact alternatives for performing quantitative and mathematical analyses of the size and shape of Earth resources, such as water and land, by providing a high-resolution aerial view of the study area.

The present study calculates morphometric parameters such as form factor, circularity ratio, elongation ratio, stream order, length, length ratio, bifurcation ratio, stream frequency drainage density, drainage texture, relief, relief ratio, slope, and gradient ratio. These parameters were estimated from the geospatial analysis of the DEM (ASTER) and SOI toposheets for the Girna River Basin. This study can be further useful for prioritizing watersheds for sustainable water resource planning for agricultural and drinking purposes.

4.2 Study Area

The Girna River is a tributary of the Tapi River and originates at the Kem Peak in the Western Ghats range of the Nashik District, with a latitude of 21° 07' 60"N and a longitude of 75° 19' 00"E. It flows through Maharashtra State in southern India. Girna originated from the name of Goddess Giraja (Parvati). It flows east across the Nashik and Jalgaon districts, flowing northward in the Jalgaon District to join the Tapi River. The Girna River Basin lies on the Deccan Plateau and its valley has fertile soils that are intensively farmed. The biggest dams on Girna are the Chankapur Dam, built by the British near Abhona in Kalwan tehsil, where the Sarpganga River joins the Girna River. Girna Dam was built in 1969. The total length of dam is 963.17 m, maximum height of dam is 54.56 m, and surface area is 60,040 km^2. The temperature in this area varied between 10 and 48 °C. The summer monsoon is the main source of precipitation in this region. The average annual rainfall is approximately 543 mm (Figure 4.1).

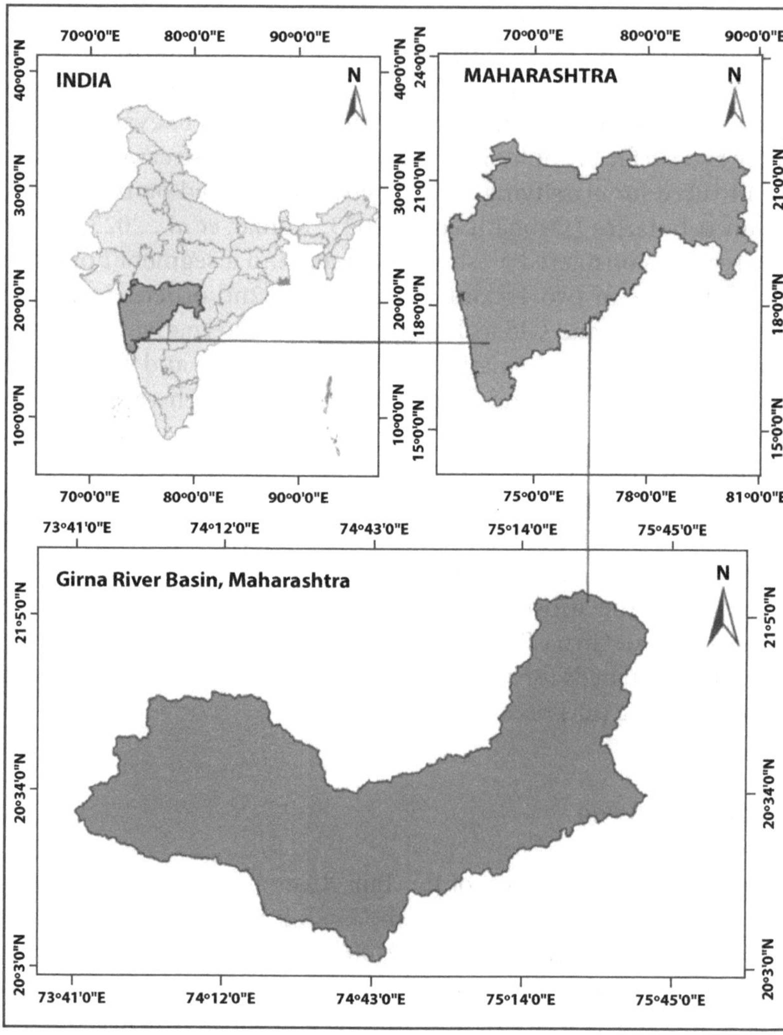

Figure 4.1 Location map of the Girna River Basin.

4.3 Database and Methodology

In this study, ASTER data were used to prepare a drainage map of the Girna River Basin. ASTER data of 30-meter resolution were used to determine the relief characteristics of the study area. The area and length of the tributaries in the basin were assessed using ArcGIS software. The stream order of the Girna River watershed was delineated by using the technique

Table 4.1 Formulae to calculate various parameters for the morphometric analysis.

Parameters	**Formulae**	**References**
Linear Aspects		
Stream Order (U)	The smallest permanent streams are called "first order". Two first order streams join to form a larger, second order stream; two second order streams join to form a third order, and so on. Smaller streams entering a higher-ordered stream do not change its order number.	Strahler (1964)
Stream Length (Lu)	The average length of streams of each of the different orders in a drainage basin tends closely to approximate a direct geometric ratio.	Horton (1945)
Stream Length Ratio (RL)	RL= Lu/(Lu-1)	
Bifurcation Ratio (Rb)	Rb = Nu/(Nu+1)	Horton (1932)
Areal Aspects		
Stream Frequency (Fs)	$Fs = \Sigma Nu/A$	Horton (1945)
Drainage Density (Dd)	Dd = Lu/A	Horton (1945)
Drainage Texture (T)	T = Dd x Fs	Smith (1950)
Elongation ratio (Re)	$Re = 1.128\surd A/L$	Schumm (1956)
Circularity ratio (Rc)	$Rc = 4\Pi A/P2$	Strahler 1964
Form factor (Ff)	Ff = A/L2	Horton (1945)
Relief Aspects		
Relief (R)	R = H – h	Hadley and Schumm (1961)
Relief Ratio (Rr)	Rr = R/L	Schumm (1963)
Slope (Sb)	Sb = (H-h)/L'	Mesa (2006)
Gradient Ratio (Gr)	Gr = (H-h)/L	

from Singh *et al.* (2023). Table 4.1 indicates the various standard formulae to explain the areal and linear aspects (Singh *et al.*, 2023; Rathore *et al.*, 2023). Relief aspects were evaluated using DEM from ASTER data with 30-meter resolution, and various formulae (Table 4.1) [14, 15].

4.4 Results and Discussion

The basin drainage network of the Girna River and its stream order are shown in Figure 4.2. The linear, relief, and areal parameters are the various morphometric factors classified as shown in Table 4.2. The study of morphometric influences in hydrology is very important to understand because it directly influences the extent of the peak and mean runoff, through which the maximum flood discharge per unit area is inversely associated with size (Singh *et al.*, 2023; Tripathi *et al.*, 2023).

4.4.1 Linear Aspects

The Girna River Basin drainage network was investigated to evaluate a variety of linear parameters such as stream order (Nu), stream length (Lu), mean stream length (Lu[1]), and bifurcation ratio (Rb), as shown in Table 4.2.

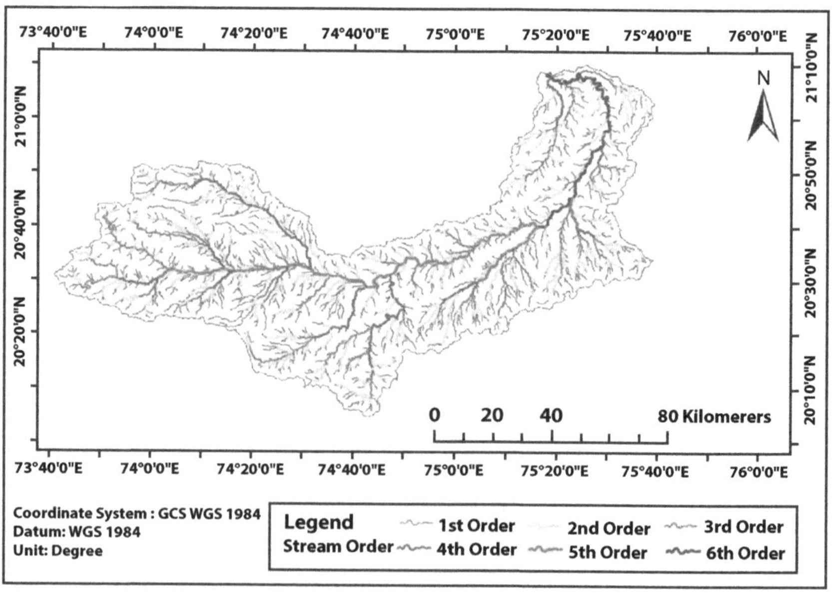

Figure 4.2 Stream order map of the Girna River Basin watershed.

Table 4.2 Linear parameter of the Girna River Basin.

Sr. no.	Parameter		Value
1	**Stream Order (Nu)**	1st order	1,071
		2nd order	526
		3rd order	261
		4th order	125
		5th order	107
		6th order	47
		Total	**2,137**
2	**Stream Length (Lu)**	1st Order	2,785,635.38
		2nd Order	1,390,133.27
		3rd Order	647,070.67
		4th Order	272,889.79
		5th Order	219,426.60
		6th Order	105,523.55
		Total	**5,420,679.284**
3	**Mean Stream Length (Lu^1)**	1st Order	2,600.96
		2nd Order	2,642.82
		3rd Order	2,479.19
		4th Order	2,183.11
		5th Order	2,050.71
		6th Order	2,245.18
4	**Mean Stream Length Ratio (RL)**		**2,366.99**
5	**Bifurcation Ratio (Rb)**	1st Order	2.03
		2nd Order	2.01
		3rd Order	2.08
		4th Order	1.16
		5th Order	2.27
		6th Order	____
6	**Mean Bifurcation Ratio (Rbm)**		**1.91**

4.4.1.1 *Stream Order (Nu)*

Lavane *et al.* (2023) proposed four distinct techniques for evaluating stream order from the basin. The Girna River Basin has been classified, which has been somewhat revised by Horton's system. According to Horton's system, the intersection of two first-order streams generates channel segments of second order, and then two second-order junctions join to develop a third order segment. A higher-order stream is formed when two streams of different orders join channels. Therefore, the Girna River watershed exhibited the sixth order as the highest in the basin (Figure 4.2).

4.4.1.2 *Stream Number (Nu)*

The total number of tributaries existing in every order is the stream number of any basin. It was observed that when an increase in the stream order decreased, the number of streams decreased in all sub-basins. The mountain regions have maximum first-order streams and a minimum in the lower part of the basin. The total number of tributaries of the streams in the Girna River Basin was 2,137, of which 1,071 are 1st order, 526 are 2nd order, 261 are 3rd order, 125 are 4th order, 107 are 5th order, and 47 are 6th order.

4.4.1.3 *Stream Length (Lu) and Mean/Average Stream Length (Lu^1)*

In the basin, the total length of the different stream sections of every single order is the stream length of that order, and the stream length in each order increases exponentially with increasing stream order. The total length of 1st order streams is 2,785.635 km, 2nd streams have a length of 1,390.133 km, 3rd streams have a length of 647.071 km, 4th streams have a length of approximately 272.889 km, 5th order streams have a length of 219.427 km, and 6th order streams have a length of approximately 105.524 km in the Girna River Basin. The average stream length is the ratio of the total length of a certain ordered stream to the total number of streams of similar order (Table 4.3).

4.4.1.4 *Stream Length Ratio (RL)*

The stream length ratio is the average length of the order 'u' segment to the mean length of order 'u − 1.' During this study, the stream length ratio of the entire Girna River Basin was 2,366.99 m (Table 4.4).

Table 4.3 Stream lengths of the Girna River Basin.

Stream order	Stream length (km)	Mean of all streams (km)
1st Order	2,785.635	903.447
2nd Order	1,390.133	
3rd Order	647.071	
4th Order	272.889	
5th Order	219.427	
6th Order	105.524	
Total = 5,420.679		

Table 4.4 Stream length ratio of the Girna River Basin.

Stream order	Mean stream length (M)Lu	Mean stream length ratio (RL)
1st Order	**2,600.96**	**2,366.99**
2nd Order	**2,642.82**	
3rd Order	**2,479.19**	
4th Order	**2,183.11**	
5th Order	**2,050.71**	
6th Order	**2,245.18**	

4.4.1.5 *Bifurcation Ratio (Rb)*

Strahler (1964) suggested that the bifurcation ratio is the ratio of the number of streams of order (u) to the number of streams of higher order (u + 1). Chow (1964) quantified that the bifurcation ratio values lie from 1 to 2, and the average bifurcation ratio (Rbm) of the Girna River watershed is 1.91 (Table 4.5) for watersheds where geological structures do not influence the drainage pattern (Sud *et al.*, 2023). The higher bifurcation ratio for the watershed is the outcome of the significant change in frequencies among successive orders and suggests a mature topography.

Table 4.5 Bifurcation ratio of the Girna River Basin.

Stream order	Number of streams	Bifurcation ratio (Rb)
1st order	1,071	2.03
2nd order	526	2.01
3rd order	261	2.08
4th order	125	1.16
5th order	107	2.27
6th order	47	___

4.4.2 Areal Aspects

The areal aspects are two-dimensional (2D) assets of a basin. It has the potential to demarcate a watershed area that provides water to each stream tributary. The total area of the Girna River watershed is 10,233.95 km^2. The aerial aspects of the drainage basin, such as stream frequency (Fs), drainage density (Dd), drainage texture (T), elongation ratio (Re), form factor ratio (Rf), and circularity ratio (Rc) were determined (Table 4.6).

4.4.2.1 Drainage Density (D_d)

The drainage density indicates the proximity of the spacing of the drainage networks. Therefore, the drainage density provides a quantitative amount of the mean length of the stream channel for the entire drainage basin. The lack of vegetation, impermeable subsurface materials, and mountainous relief contribute to the area's high drainage density. According to, high drainage density indicates a fine drainage texture, whereas low drainage density indicates a coarse drainage texture. The drainage density of the Girna River Basin was 0.52 km^{-1} signifying higher drainage densities (Table 4.6). This indicates that impermeable surface materials predominated in the watershed (Figure 4.3).

4.4.2.2 Stream Frequency (F_s)

Stream frequency is directly linked to lithological characteristics. The number of stream segments per unit area is called the stream frequency,

Table 4.6 Areal parameter of the Girna River Basin.

Sr. no.	Parameter	Value
1	Basin Area (sq.km)	10,233.94575
2	Basin Perimeter (km)	769.761
3	Form Factor (Ff)	3.48
4	Drainage density (Dd)	0.52
5	Drainage texture (Dt)	2.77
6	Texture Ratio (T)	1.39
7	Stream frequency (Fs) (no./sq.km)	0.2
8	Elongation ratio (Re)	4.86
9	Circulatory ratio (Rc)	0.21
10	Length of Overland Flow (Lg)	0.26
11	Infiltration Number (If)	0.104
12	Constant of Channel Maintenance (C)	1.92

drainage frequency, or channel frequency (Fs; Horton 1945). The total stream frequency of the Girna River watershed was 0.2 km^{-2} (Table 4.6) (Figure 4.4).

4.4.2.3 Drainage Texture (Dt)

The total number of stream segments of all orders in the watersheds per perimeter of the basin describes the drainage texture. Knowing the geomorphology or relative distances between the drainage lines is crucial. The infiltration capacity, underlying lithology, and relief aspect of the topography, as well as natural factors such as rainfall, climate, vegetation, type, stage, and relief of improvement of the rock and soil, affect drainage texture. According to Smith (1950), drainage texture can be divided into five distinct categories: very coarse has a value of less than 2, coarse has a value

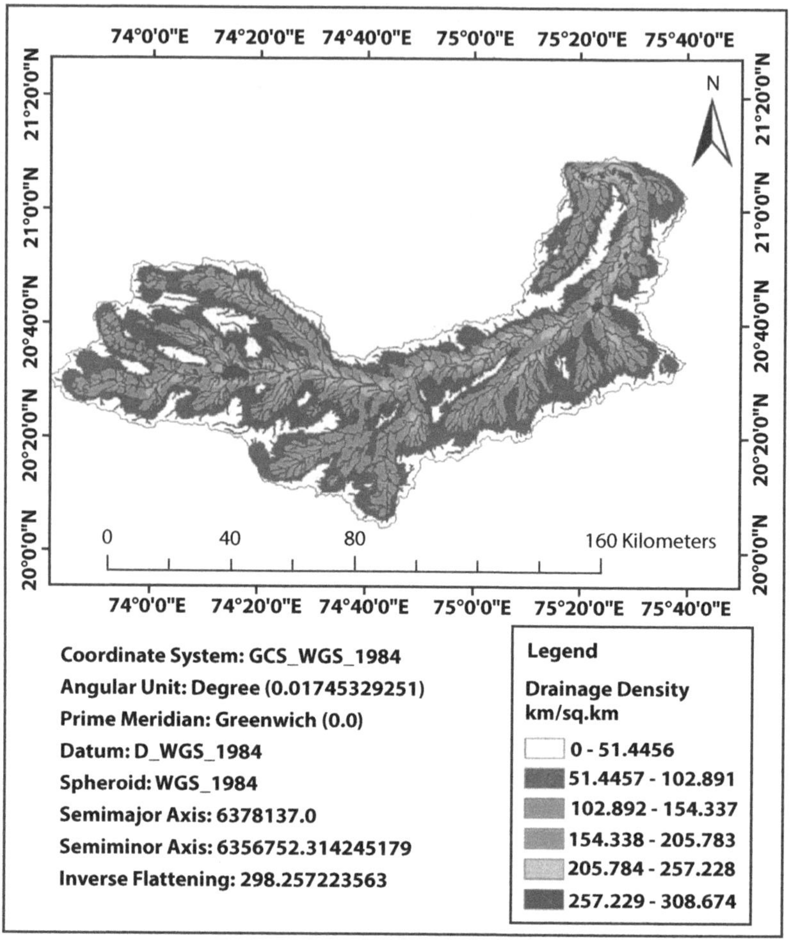

Figure 4.3 Drainage density map of the Girna River Basin watershed.

of 2 to 4, moderate has a value of 4 to 6, fine has a value of 6 to 8, and very fine has a value of less than 8. The Girna River Basin has a drainage texture of 2.77 (Table 4.6) (Upadhyay *et al.*, 2023).

4.4.2.4 *Elongation Ratio* (R_e)

The elongation ratio (Re) can be expressed as the ratio of the maximum basin length to the diameter of a circle with the same area as the basin. In a highly elongated shape, the value of Re (0.049) ranges from 0 to unity or 1.0, in a circular shape. The additional elongation values of 0.7, 0.7–0.8,

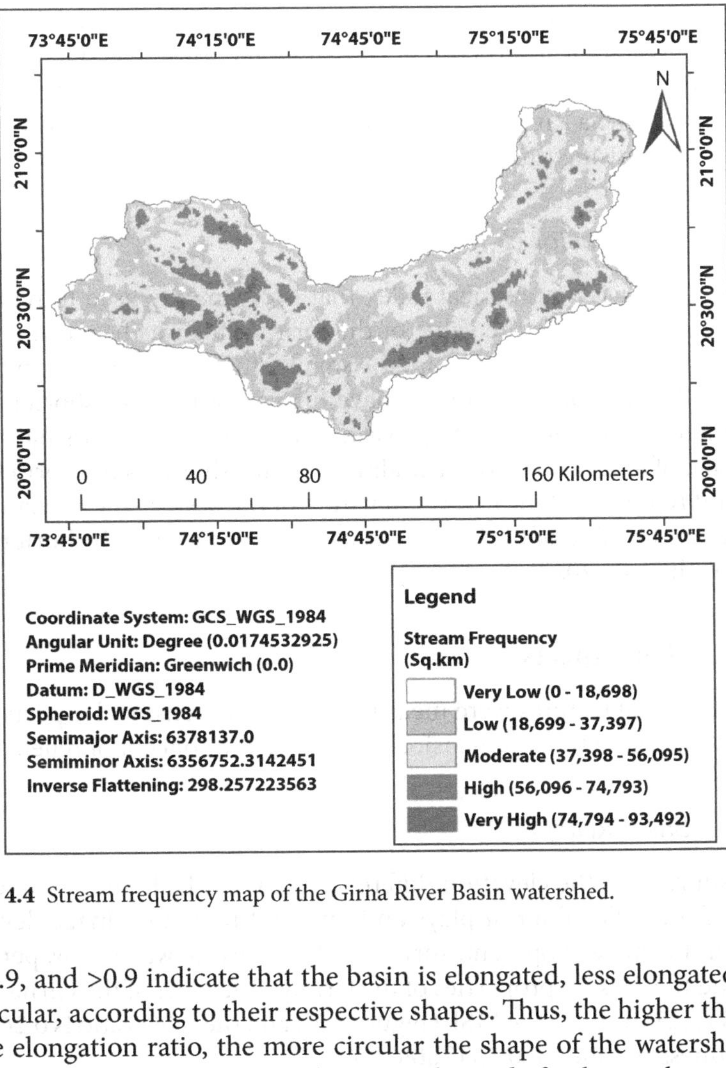

Figure 4.4 Stream frequency map of the Girna River Basin watershed.

0.8–0.9, and >0.9 indicate that the basin is elongated, less elongated, oval, or circular, according to their respective shapes. Thus, the higher the value of the elongation ratio, the more circular the shape of the watershed and reciprocal. Values close to 1.0, show very low relief, whereas between 0.6 and 0.8 are typically related to high relief and steep ground slope. The elongation ratio for the entire basin is 0.021. This shows that the Girna watershed is more circular in shape.

4.4.2.5 *Circularity Ratio (R_c)*

[19] defines the circularity ratio is defined as the ratio of the basin area to that of a circle with the same circumference as the perimeter of the basin. This ratio was comparable to the elongation ratio. The circularity ratio had

values ranging from 0 (in a line) to 1 (in a circle). The length and frequency of streams, geological structures, land use/land cover (LULC), climate, relief, and basin slope all affect the circulatory ratio. According to Table 4.6, Rc of the Girna River Basin was 0.21. The dendritic stage of a basin can be determined using this significant ratio.

4.4.2.6 *Form Factor (F_f)*

The form factor is a numerical index that is commonly used to represent different basin shapes. The value of the form factor was between 0.1 and 0.8. The more elongated the basin, the smaller the value of the form factor. Basins with high form factors 0.8 have high peak flows of shorter duration, whereas elongated drainage basins with low form factors had lower peak flows of longer duration. The alluvial basins showed a low form factor value and represented the elongated nature of the basins. The Girna watershed has an Ff value of 0.943, indicating a longer basin with shorter peak flows (Nath *et al.*, 2023).

4.4.3 Relief Aspects

Linear and areal features were measured as two-dimensional aspects lying on the plan. These three dimensions introduce the concept of relief.

4.4.3.1 *Basin Relief (R)*

The basin relief is the elevation difference between the highest and lowest points of the valley floor. It plays an important role in drainage development, landform development, surface and subsurface water flow, permeability, and the erosion properties of the terrain. The relief of the Girna River Basin is 1,428 m above the MSL (Figure 4.5), and the relief ratio is 0.26 with a ruggedness number of 0.73 (Table 4.7).

4.4.3.2 *Relief Ratio (Rr)*

The relief ratio is the ratio of basin relief to basin length. High values were found in mountainous areas, whereas low values were found in the plains and valleys. The height-to-length ratio, known as the relief ratio, can be used to compare the relative relief of any basin regardless of its scale or topography. According to, the relief ratio is the same as the right-angled

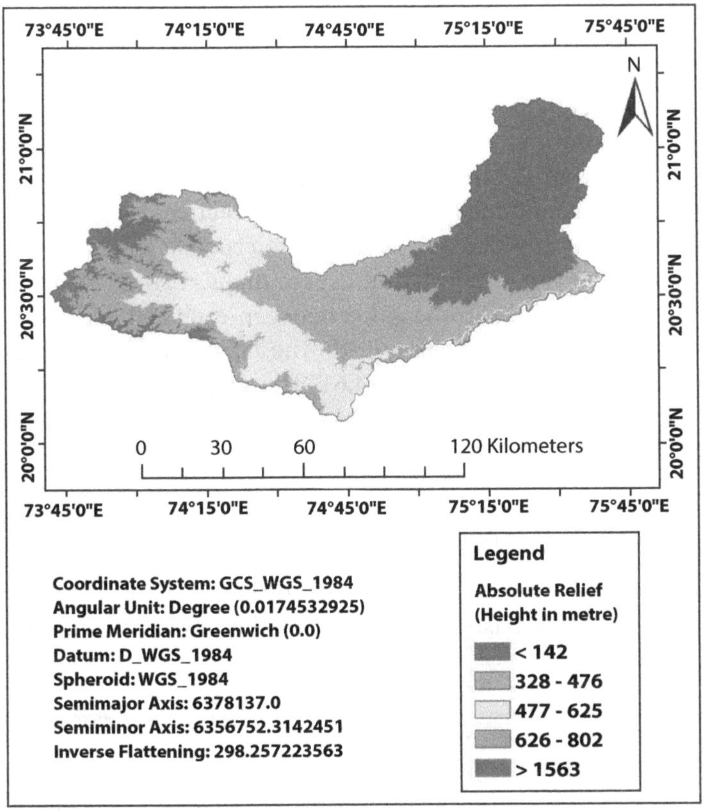

Figure 4.5 Absolute relief map of the Girna River Basin watershed.

Table 4.7 Relief parameter of the Girna River Basin.

Sr. no.	Parameters	Value
1	Basin Relief (R)	1,428
2	Relief Ratio (Rr)	0.26
3	Ruggedness Number (Rn)	0.73

triangle and tangent of the hypotenuse's angle of slope with respect to the horizontal. Thus, it measures the overall steepness of a drainage basin as an indicator of the intensity of the erosion processes operating on the slope of the basin. The relief ratio of the Girna watershed was 0.26 (Table 4.7).

4.4.3.3 *Slope*

A crucial parameter in geomorphic studies is the slope analysis. The climato-morphogenetic processes in the region with rocks of variable resistance control the slope elements. A slope map provides data for planning, settlement, agricultural mechanization, deforestation, engineering structure planning, and morph conservation practices (Kanga *et al.*, 2023). Understanding the distribution of slopes is crucial. The Aster DEM was used to create the slope, DEM, and aspect maps. Using this method, the maximum rate of change in the value between each cell and its neighbors was used to determine the slope grid. The Girna watershed has a slope ranging from 0° to 72.96° (Figure 4.6). The majority of the Girna River Watershed has a low slope, whereas the extremes to the north and south have high slopes.

4.4.3.4 *Gradient Ratio*

The total elevation drop from the source to the mouth of the trunk channels in each drainage basin was known as the gradient ratio. The Girna

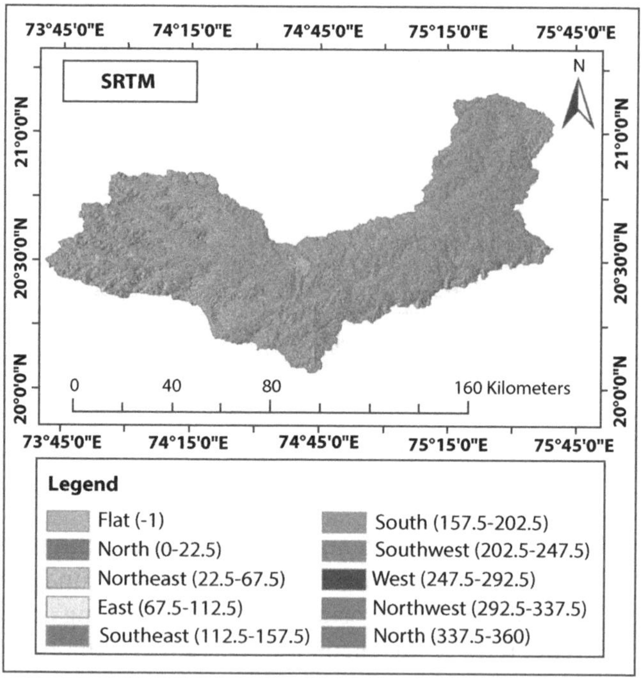

Figure 4.6 Gradient ratio map of Girna River Basin Watershed.

Table 4.8 Morphometric parameters of the Girna River Basin.

Sr. no.	Parameter value	Value
1	Basin Area (sq.km)	10,233.94575
2	Basin Perimeter (km)	769.761
3	Total length of streams (km)	5,420.67
4	Total number of streams (Nu)	2137
5	Mean Bifurcation ratio (Rbm)	1.91
6	Form Factor (Ff)	3.48
7	Drainage density (km/sq.km)	0.52
8	Drainage texture (no./km)	2.77
9	Texture Ratio (T)	1.39
10	Stream frequency (no./sq.km)	0.2
11	Elongation ratio (Re)	4.86
12	Circulatory ratio (Rc)	0.21
13	Length of Overland Flow (Lg)	0.26
14	Infiltration Number (If)	0.104
15	Constant of Channel Maintenance (C)	1.92
16	Maximum Height (H)	1,571
17	Minimum Height (h)	143
18	Basin Relief (R)	1,428
19	Relief Ratio (Rr)	0.26
20	Ruggedness Ratio (Rn)	0.73

River watershed has a gradient ratio of 0.053, which was very low in this study. This suggests that only a small amount of very fine sediments may be carried by slow-moving water and a higher-level stream bed (Table 4.8).

4.5 Conclusion

Dendritic patterns were revealed by morphometric analysis of the Girna River watershed drainage network, which also revealed a lack of structural control and a uniform texture. The Girna watershed has been classified as a six-order basin based on drainage orders. The study area has a drainage density (Dd) of 0.52 Km^{-1}, indicating that the surface materials are impermeable. The watershed drainage texture fell into the very coarse drainage texture category (>2). The basin's elongation and circularity ratios of 4.86 and 0.21, respectively indicate that the dendritic stage of the Girna Watershed is more circular. The slope of Girna was steep. Water network characteristics, such as the stream bifurcation ratio and discharge density, directly increase the valley's water discharge. Plans for watershed management and rainwater harvesting can benefit greatly from these studies. Because of their location on hilly terrain and the greater volume of water discharged through river tributaries in the basin, first- and second-order streams were ineffective for the construction of check dams in the study area. The Girna watershed also has a large number of lower-order streams. This study successfully demonstrated that remote sensing and geographic information systems (GIS) are extremely useful tools for basin morphometric analysis.

Acknowledgments

KC, PK, and PD thank the Director and Coordinator, C3WR, Suresh Gyan Vihar University, Jaipur, for providing the facilities necessary for this work.

References

Singh SK, Kanga S, Gulati B, Raiˇc M, Sajan B, Đurin B, Singh S (2023) Spatial and Temporal Analysis of Hydrological Modelling in the Beas Basin Using SWAT+ Model. *Water* 15, 3338. https://doi.org/10.3390/w1519333.

Rani A, Gupta SK, Singh SK, Meraj G, Kumar P, Kanga S, Đurin B, Dogančić D (2023) Predicting Future Land Use Utilizing Economic and Land Surface Parameters with ANN and Markov Chain Models. *Earth* 4(3):728-751.

Ahmad T, Gupta SK, Singh SK, Meraj G, Kumar P, Kanga S (2023) Unveiling Nature's Resilience: Exploring Vegetation Dynamics during the COVID-19 Era in Jharkhand, India, with the Google Earth Engine. *Climate* 11(9):187.

Sharma M, Upadhyay RK, Tripathi G, Kishore N, Shakya A, Meraj G, Kanga S, Singh SK, Kumar P, Johnson BA, Thakur SN (2023) Assessing Landslide Susceptibility along India's National Highway 58: A Comprehensive Approach Integrating Remote Sensing, GIS, and Logistic Regression Analysis. *Conservation* 3(3):444-459.

Saqib N, Rai PK, Kanga S, Kumar D, Đurin B, Singh SK (2023) Assessment of Ground Water Quality of Lucknow City under GIS Framework Using Water Quality Index (WQI). *Water* 15(17):3048.

Choudhury U, Singh SK, Kumar A, Meraj G, Kumar P, Kanga S (2023) Assessing land use/land cover changes and urban heat island intensification: A case study of Kamrup Metropolitan District, Northeast India (2000–2032). *Earth* 4(3):503-521.

Rai NK, Mathur S, Singh SK, Tiwari M, Singh VK, Haque R, Tiwari S, Sharma LK (2023) Differential regulation of mitochondrial complex I and oxidative stress based on metastatic potential of colorectal cancer cells. *Oncol. Let.* 26(1):1-1.

Debnath J, Sahariah D, Lahon D, Nath N, Chand K, Meraj G, Kumar P, Singh SK, Kanga S, Farooq M (2023) Assessing the impacts of current and future changes of the planforms of river Brahmaputra on its land use-land cover. *Geosci. Front.* 14(4):101557. Elsevier.

Kumar P, Singh SK, Gupta SK, Kanga S, Meraj G (2023) Uncovering the hydro-meteorological drivers responsible for forest fires utilizing geospatial techniques. *Theor. Appl. Climatol.* https://doi.org/10.1007/s00704-023-04497. Springer Nature.

Singh SK, Kotecha MJ, Kanga S, Pankhaniya SK, Agrawal S, Meraj G (2023) Urban Heat Island (UHI) Resilience Plan in Varying Climatic Conditions Using Geospatial Approach: A Case Study of Rajkot City. In: *Climate Change, Agriculture and Society*, vol 1. pp 307-338.

Singh S, Singh SK, Prajapat DK, Pandey V, Kanga S, Kumar P, Meraj G (2023) Assessing the Impact of the 2004 Indian Ocean Tsunami on South Andaman's Coastal Shoreline: A Geospatial Analysis of Erosion and Accretion Patterns. *J. Mar. Sci. Eng.* 11(6):1134.

Rathore SS, Mathur SC, Mathur S, Sharma S, Singh SK (2023) Late Cretaceous–Early Paleocene Geodiversity of Fatehgarh Formation of Petroliferous Barmer Basin, Western Rajasthan, India: A Potential Geopark Site. *Geoheritage* 15:54.

Singh S, Meraj G, Kumar P, Singh SK, Kanga S, Johnson BA, Prajapat DK, Debnath J, Sahariah D (2023) Decoding Chambal River Shoreline Transformations: A Comprehensive Analysis Using Remote Sensing, GIS, and DSAS. *Water* 15(9):1793.

Tripathi G, Shakya A, Upadhyay RK, Singh SK, Kanga S, Pandey SK (2023) Landslide Susceptibility Mapping of Tehri Reservoir Region Using Geospatial Approach. In: *Climate Change Adaptation, Risk Management and Sustainable Practices in the Himalaya*, pp 135-156.

Sud A, Kanga R, Singh SK, Meraj G, Kanga S, Kumar P, Ramanathan AL, Bhardwaj V (2023) Simulating Groundwater Potential Zones in Mountainous Indian Himalayas—A Case Study of Himachal Pradesh. *Hydrology* 10(3):65.

Upadhyay RK, Tripathi G, Đurin B, Šamanović S, Cetl V, Kishore N, Sharma M, Singh SK, Kanga S, Wasim Md, Rai PK, Bhardwaj V (2023) Groundwater

Potential Zone Mapping in the Ghaggar River Basin, North-West India, Using Integrated Remote Sensing and GIS Techniques. *Water* 15(5):961.

Lavane K, Kumar P, Meraj G, Han TG, Ngan LHB, Lien BTB, Ty TV, Thanh NT, Downes NK, Nam NGD, Minh HVT, Singh SK, Kanga S (2023) Assessing the effects of drought on rice yields in the Mekong Delta. *Climate* 11(1):13.

Nath N, Sahariah D, Meraj G, Debnath J, Kumar P, Lahon D, Chand K, Farooq M, Chandan P, Singh SK, Kanga S (2023) Land use and land cover change monitoring and prediction of a UNESCO world heritage site: Kaziranga eco-sensitive zone using cellular automata-Markov model. *Land* 12(1):151.

Kanga S, Upadhyay RK, Sharma M, Sharma S, Tripathi G, Singh SK (2023) Geotechnical Stability Assessment of Road-cut Slopes: A Case Study of Srinagar, Garhwal Himalaya, India. *Biol. Forum.* 15(5):1071-1080.

Gulati B, Sharma R, Kanga S, Singh SK, Sajan B, Meraj G, Kumar P, Ramanathan AL (2023) Unraveling the Relationship Between Stubble Burning and Air Quality Degradation in Punjab: A Temporal and Spatial Analysis (2019-2022). *J. Clim. Change* 9(2):43-53, IOS Press.

Tomar P, Singh SK, Kanga S, Pattanaik A (2023) Water Bodies mapping and monitoring using high-resolution satellite images. *Sustain. Agri. Food Environ. Res.* 11.

Singh H, Meraj G, Singh S, Shrivastava V, Sharma V, Farooq M, Kanga S, Singh SK, Kumar P (2022) Status of Air Pollution during COVID-19-Induced Lockdown in Delhi, India. *Atmosphere* 13(12):2090.

Farooq M, Mushtaq F, Meraj G, Singh SK, Kanga S, Gupta A, Kumar P, Singh D, Avtar R (2022) Strategic Slum Upgrading and Redevelopment Action Plan for Jammu City. *Resources* 11(12):120.

5

A Geospatial Analysis of the Effect of Waste Disposal on Groundwater Quality in Ife North Local Government Area, Osun State, Nigeria

J. O. Nwaezeigwe* **and T. A. Babatunde**

Department of Geography, Obafemi Awolowo University, Ile-Ife, Nigeria

Abstract

Indiscriminate waste disposal is a major environmental challenge in developing countries, especially because waste often ends downstream. This study investigated the vulnerability of selected groundwater sources in semi-rural communities in the Ife North Local Government Area of Osun State, Nigeria, to contamination by adjacent dumpsites. The specific objective was to assess the quality of the groundwater in the area. The data used were the results of a laboratory investigation of selected water samples from sampled groundwater sources and information about the use of groundwater by the residents. The results showed that the groundwater sources close to the dumpsites were more contaminated than the sources farther from the dumpsites, and that many (>70%) residents complained of water-related illnesses, although they rarely report such to the hospitals because of the probable distrust of the healthcare system. The study concluded that groundwater sources around the dumpsites were contaminated and unfit for potable use in the area.

Keywords: Groundwater quality, solid waste disposal, physico-chemical properties, heavy metals, GIS

**Corresponding author*: revauyeh@oauife.edu.ng

Shruti Kanga, Gowhar Meraj, Suraj Kumar Singh, Majid Farooq and M. S. Nathawat (eds.) Geospatial Technology for Natural Resource Management, (153–176) © 2024 Scrivener Publishing LLC

5.1 Introduction

Water plays a vital role in environmental factors in all forms of life and in the socioeconomic development of the human population. Approximately 70% of the Earth is filled with water, making water a fundamental asset for the world economy, as well as a precondition for humans, creatures, and vegetation (Singh *et al.*, 2023; Rani *et al.*, 2023). Groundwater is used domestically, agriculturally and industrially. It is distinguished from other water bodies for two main reasons: it has slow movement within rocks, which makes it highly susceptible to pollution. In addition, the chemical properties of groundwater are dependent on the physiochemical composition of the rock containing the groundwater (Rani *et al.*, 2023). Wastes are materials that are not considered useful and are generated at different levels during the extraction of raw materials, processing of raw materials into intermediate or final products, consumption of final products, and other human activities. When waste is not well disposed, it has an adverse effect on the soil of the area where it is dumped and on the surrounding underground water of the area. Waste contaminates water to varying degrees, and some wastes are more hazardous than others.

There are different ways of taking care of the waste generated by humans, which can be broadly classified into traditional and emerging methods (Ahmad *et al.*, 2023; Gulati *et al.*, 2023). Traditional methods include open burning, composting, landfill, and incineration. Open burning, which is the burning of waste generated in open spaces, has been adopted in different countries over time. This method is practiced because it is the cheapest and fastest method for waste care. This method is still mostly practiced by low-level and developing countries (Saqib *et al.*, 2023; Tomar *et al.*, 2023). The emerging methods that have taken over previous ways of managing waste include waste reduction, waste recycling, and waste treatment. These types of waste management believe that all wastes are not useless and can still be reused for another function if they are processed again (Sharma *et al.*, 2023; Singh *et al.*, 2023).

When waste is not well managed and disposed of properly, it affects the health status of the residents of the area. Poor waste disposal has been reported to affect the environment and groundwater, causing different health hazards; for example, at the Gazipur municipal dumping site, Delhi was found to be prone to groundwater contamination through leaching action (Choudhury *et al.*, 2023; Farooq *et al.*, 2023). Waste management is a problem in Nigeria. It manifests in the form of piles of disposed heaps of uncovered waste and illegal dumpsites on major roads and street corners.

Open dumps are a major cause of environmental degradation and public health concerns in many developing countries, including Nigeria. These waste dumps may contain a mixture of general waste and toxic, infectious, or radioactive wastes, and are susceptible to burning and exposure to scavengers. This has been a major problem in Nigeria, with the Osun state inclusive. Solid waste, if not taken into account appropriately, constitutes wellbeing perils, causes hostile scent, and contaminates underground water sources when rain falls (Rai *et al.*, 2023; Debnath *et al.*, 2023). This study examined solid waste dumped around well points in the study area using geospatial techniques and laboratory analysis. This is because the waste management system in the study area has not been sufficiently effective and there have been growing environmental problems due to the increasing population, the increasing number of small-scale industries in the area, and some other factors that cause the waste disposed in the study area to increase; wastes are dumped along the roads and in the bushes, as well as on rivers and drainage areas. The indiscriminate dumping of refuses along roads and drainages is likely to have an effect on the water, which in turn affects the health of people living in the study area. In this regard, this study evaluated the impacts of waste disposal on groundwater quality to ascertain its fitness for household use in the Ife North Local Government Area of Osun State. While, the specific objective was to examine the quality of groundwater in the area. The main null hypothesis is that dumpsites near groundwater sources do not exhibit significant contamination effects on the surrounding groundwater quality in the study area.

5.2 Study Area

Ife North is a Local Government Area (LGA) in Osun State, in the south western part of Nigeria. Its headquarter is in Ipetumodu, a town to the north of the area. The LGA is on the outskirts of the town of Ile-Ife, and the major language spoken is Yoruba. Ife North lies between latitudes 7°28′N and 7°31′N and longitudes 4°27′E and 4°28′E (Figure 5.1). The local government has a total land area of 889 km^2. It lies at the intersection of the roads from Ibadan to Ife and from Ife Central to Ede. It comprises Ipetumodu (the local government headquarter), Edunabon, Moro, Yakoyo, Asipa, Akinlalu, and Isope (which fused with Ipetumodu). According to the 2006 National Census, Ife North Local Government Area had a total population of 153,274.

Ife North Local Government is the home of the Centre for Distance Learning of the Obafemi Awolowo University. It also includes two tertiary

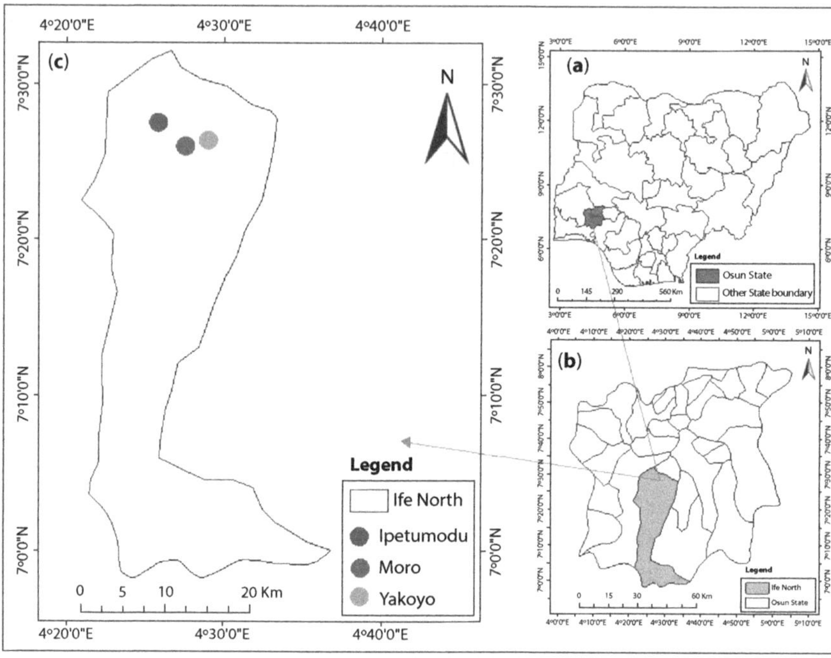

Figure 5.1 Map of the study area.

institutions and a secondary school. The major economic activities in the area are farming and the public sector, which are government-owned secondary schools (Jeje, 2014). Ife North has an undulating topography which is characterized by ridges, regolith covered hills and a few rock outcrops. It is also underlain by a basement complex of Nigerian rocks, composed mostly of granitic and metamorphic rocks (Kumar *et al.*, 2023; Singh *et al.*, 2023).

5.3 Materials and Methods

5.3.1 Data

The target population was residents of the local government who generated waste in the area. The primary and secondary data were used in this study. Three towns (Ipetumodu, Moro, and Yakoyo) were sampled as the populous communities in the local government area. Information about the type of waste, the ways in which waste is disposed, and waste collection points were collected through participant observation and interviews. Information was also gathered from the groundwater points closest to the waste collection points through field observations. Water samples were collected in sterilized 250 ml bottles from the groundwater points close to the waste dumpsites, numbered from 1 to 9 against their locations and taken to the Centre for Research and Development Laboratory, Obafemi Awolowo University Ile-Ife to assess the chemical quality of the groundwater. The samples were spread across each community with the samples taken far from each other (Figure 5.2) and at three locations: between 15 and 45 m, between 50 and 100 m, and above 100 m because the distance to the dumpsite and concentration of heavy metals were been examined. Coordinates (x, y) of the locations from which water samples were obtained were taken with a global positioning system (GPS) handheld device at ±3 m accuracy and in the UTM zone 31N, the zone of the study area. The secondary data were a base map of the area from ArcGIS 10 online, where the roads were digitized to obtain the road network of the area.

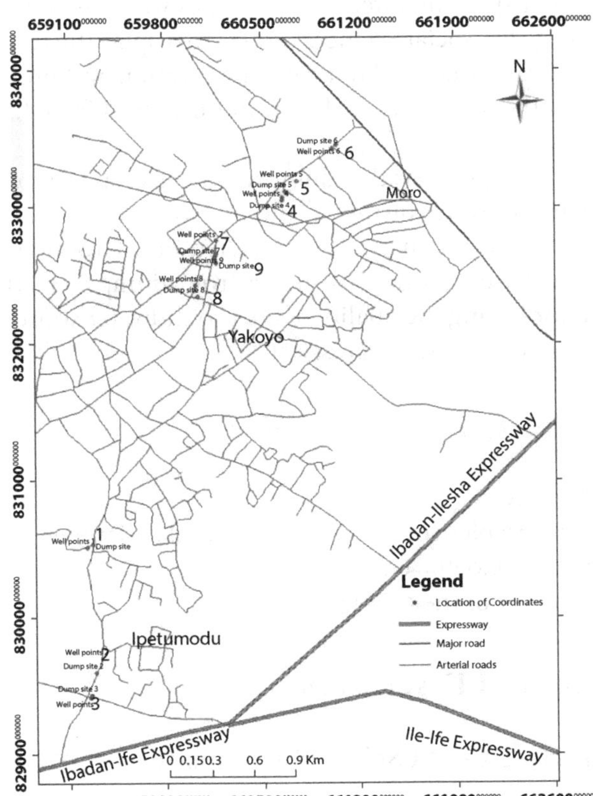

Figure 5.2 Map showing the location of the sampled points (1–9).

5.3.2 Data Analysis

Water samples were taken to the laboratory to determine the concentrations of selected heavy metals (lead (Pb) copper (Cu), iron (Fe)) and anions (sulfate (SO_4^{2-})) that are associated with leachates from dumpsites and are known to cause severe health implication when consumed in contaminated water (Singh *et al.*, 2023; Rathore *et al.*, 2023) they are inorganic compounds that can pollute groundwater. The concentrations of Pb, Cu, and Fe were determined using an Atomic Absorption Spectrophotometer at their respective wavelengths, whereas sulfate was determined using spectrophotometric methods, as described in APHA (Singh *et al.*, 2023). Prior to laboratory analysis, the pH, conductivity, and total dissolved solids (TDS) of the water samples were obtained *in situ* using pen-type pH/conductivity and TDS meters. To ensure quality control, the water samples were analyzed within 7 h after they were taken into the laboratory, and the instruments were properly calibrated following standard laboratory protocols.

Descriptive statistics and inferential statistics were carried out using the Statistical Package for Social Sciences (SPSS) to display the results. ArcGIS, a geospatial software, was used to assess the location and proximity of the sample points to the dumpsites to determine any effect from the dumpsites at a 25 m radius.

To examine the hypothesis that dumpsites near groundwater sources do not affect the groundwater quality of the Ife North local government area of Osun state, regression analysis was carried out. Groundwater sources were considered based on their distance from dumpsites. The chemical variables used for testing the quality of the groundwater sources were lead, iron, copper, and sulfate. The regression is:

$$Y = a + bx + c \tag{5.1}$$

Where Y = Distance,

x = chemical variables (iron, lead, copper, and sulfate)

b = Standardized Coefficients (Beta)

c = standard error of the estimate.

5.4 Results and Discussion

5.4.1 Characteristics of Selected Dumpsites

In Moro, the wastes generated were mainly household wastes consisting of food remnants, paper waste, and plastic bottles, which are mainly caused by

Figure 5.3 A typical vunerable groundwater source in the study area.

the presence of students from the Centre for Distance Learning and in other areas. The major result of the waste generated here is air pollution, which causes a foul odor in the environment. In Ipetumodu, household waste is generated from residential areas; small-scale industrial establishments such as filling stations, block-making industries, and retail outlets (e.g. cement mini-depots) generate waste such as plastic kegs, leakage from fuel tanks, and cement dust due to damaged goods. Most of the waste generated in this study is toxic. For example, leakage from fuel tanks can seep into water bodies and fuel tanks in the community. The type of waste generated in Akinlalu and Edunabon is mainly household waste and agricultural waste because the communities are residential towns, and the majority of the residents are farmers. In Yakoyo, Isope, and Asipa, the waste generated was mainly household waste.

Most households in Moro, Yakoyo, Isope, Akinlalu, Edunabon, Asipa, and some parts of Ipetumodu dispose of the waste generated at the back of the house collectively, with the waste burned at the point of collection. The residue was left at this site to decay. Most of the waste that is burned is collected and burned near groundwater sources (Figure 5.3). At some locations in Ipetumodu, waste is collected in drums by waste collectors and disposed of in a refuse dump. Households that use waste collectors pay a certain amount to waste collecting agencies to collect waste on their behalf.

5.4.2 Uses of Groundwater

At the study area, it was observed that the main source of water are the groundwater sources, therefore, the groundwater sources are used for domestic activities like cooking, bathing, washing, etc. Also, due to the presence of some small scale industries in the area, water is also used for some other processes such as by block making industries.

5.4.3 Quality of Selected Groundwater

To check the quality of the water, water samples were taken from groundwater points in three communities with three different divisions from the groundwater points and the dumpsites: between 15 and 45 m, between 50 and 100 m, and above 100 m.

The measurement of pH is related to the acidity or alkalinity of water. Acidic water can corrode the metal pipes. Meanwhile, alkaline water is disinfected. Figure 5.4 shows the pH readings at different sample points. The readings are 6.6 at sample point 1, 6.6 at sample point 2, 6.7 at sample point 3, 6.65 at sample point 4, 6.5 at sample point 5, 6.8 at sample point 6, 6.5 at sample point 7, 6.8 at sample point 8, and 6.9 at sample point 9. The WHO standard for suitable drinking water quality is between the values of 6.5 and 8.5. Therefore, all sample points fall within the WHO standard for the pH reading of potable water.

Figure 5.5 shows the conductivity readings at different sample points. The conductivities measured at different points were 550 μS/cm at sample point 1, 500 μS/cm at sample point 2, 250 μS/cm at sample point 3, 400 μS/cm at sample point 4, 445 μS/cm at sample point 5, 390 μS/cm at sample point 6, 322 μS/cm at sample point 7, 442 μS/cm at sample point 8, and 450 μS/cm at sample point 9. The WHO standard for suitable drinking water quality is ≤1,000 μS/cm. However, these results indicate that all sample points fall within the WHO standard for conductivity reading of potable water.

Total Dissolved Solids (TDSs) are inorganic matter and small amounts of organic matter present as solutions in water. Figure 5.6 shows the TDS readings at different sample points. The readings are 385 mg/L at sample point 1, 350 mg/L at sample point 2, 175 mg/L at sample point 3, 280 mg/L at sample point 4, 311.5 mg/L at sample point 5, 273 mg/L at sample point 6, 225.4 mg/L at sample point 7, 309.4 mg/L at sample point 8, and 315 mg/L at sample point 9. The WHO standard for potable water TDS is ≤300 ≥600. This indicates that all sample points fall within the WHO standard for TDS reading of potable water.

Lead (Pb)

The lead readings at different sample points and distances were 0.012 ppm at sample point 1, 0.009 ppm at sample point 2, 0.015 ppm at sample point 3, 0.018 ppm at sample point 4, 0.011 ppm at sample point 5, 0.007 ppm at

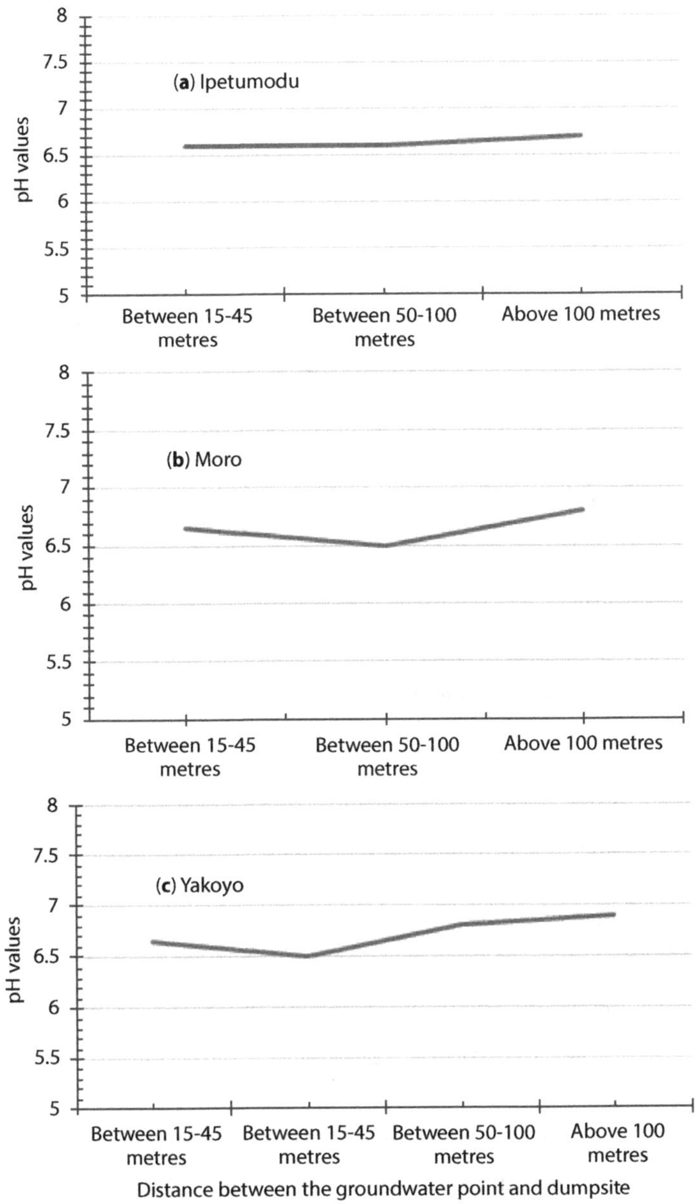

Figure 5.4 pH at different distances in (a) Ipetumodu, (b) Moro, and (c) Yakoyo.

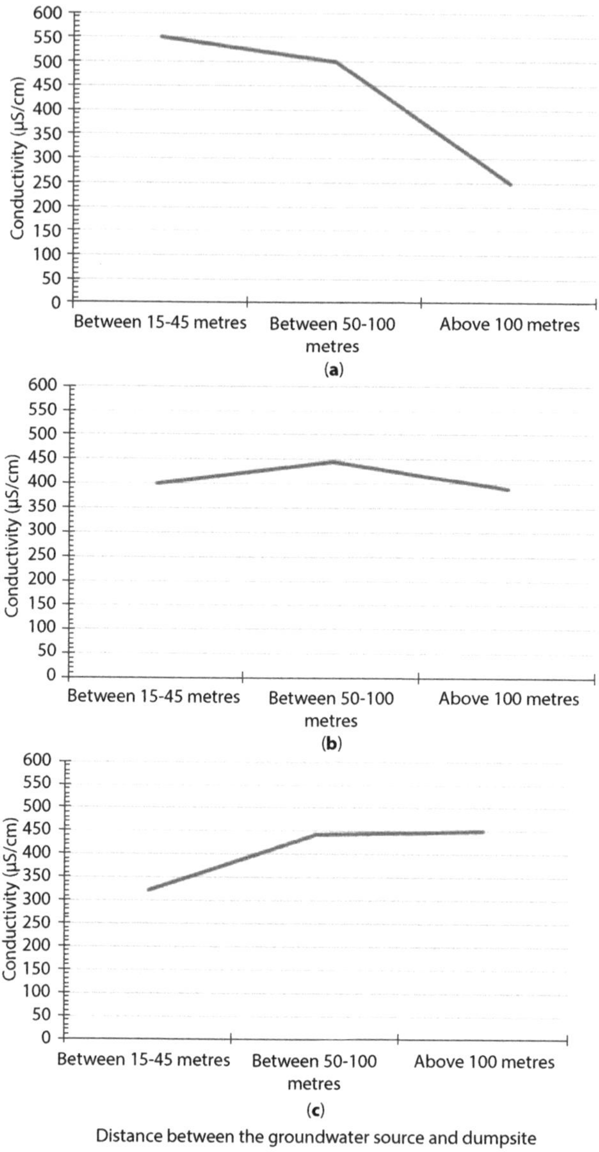

Figure 5.5 Conductivity at different distances in (a) Ipetumodu, (b) Moro, and (c) Yakoyo.

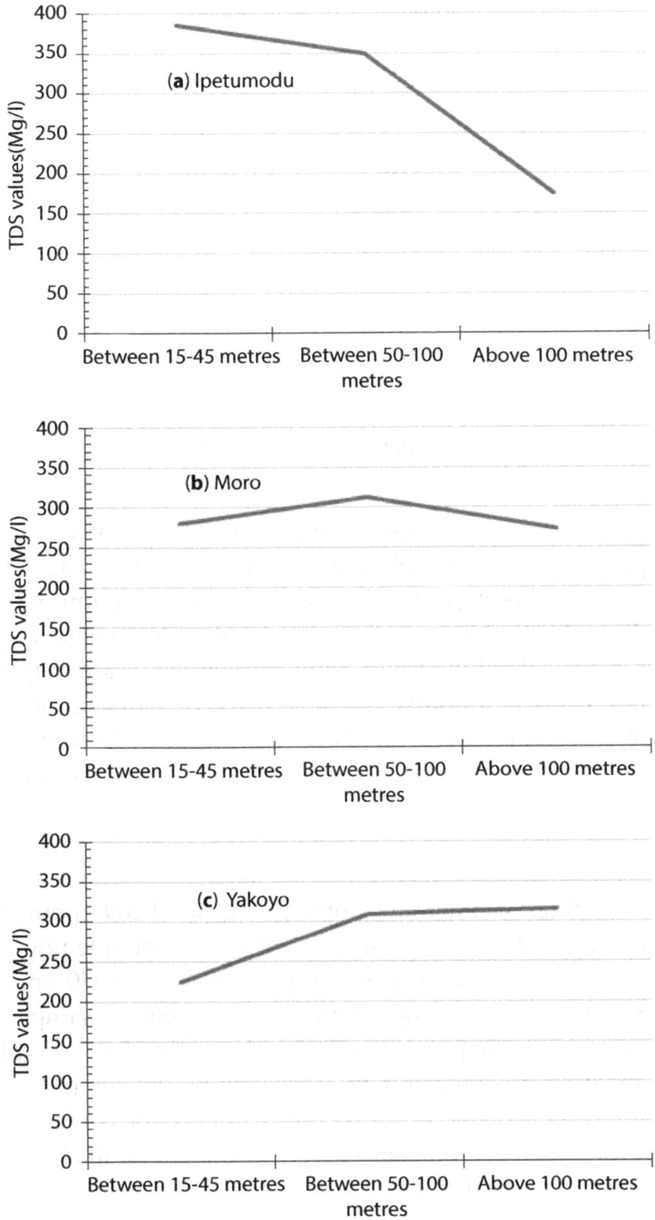

Figure 5.6 TDS at different distances in (a) Ipetumodu, (b) Moro, and (c) Yakoyo.

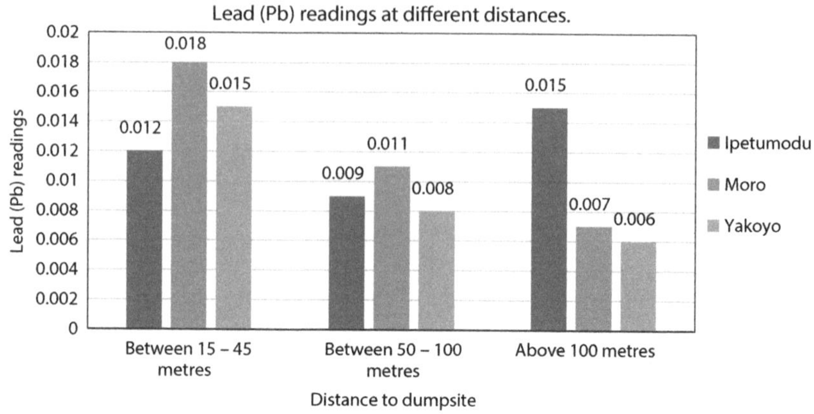

Figure 5.7 Concentration of lead at different distances in the different towns.

sample point 6, 0.015 ppm at sample point 7, 0.008 ppm at sample point 8, and 0.006 ppm at sample point 9, respectively (Figure 5.7). The WHO standard for suitable drinking water quality for lead is a maximum 0.01 ppm. These results indicate that not all sampled points had suitable lead concentrations in their drinking water. Sampled points 1–3 are in Ipetumodu, points 4–6 in Moro, and points 7–9 in Yakoyo. The concentration of lead decreased with distance from the dumpsite, except in Ipetumodu, which connotes that there are other factors contributing to lead concentration in the area.

Copper

The copper readings at different sample points are 0.0094 ppm at sample point 1, 0.072 ppm at sample point 2, 0.066 ppm at sample point 3, 0.082 ppm at sample point 4, 0.070 ppm at sample point 5, 0.070 ppm at sample point 6, 0.085 ppm at sample point 7, 0.055 ppm at sample point 8, and 0.080 ppm at sample point 9, respectively (Figure 5.8). Sample points 1–3 are in Ipetumodu, points 4–6 in Moro, and points 7–9 in Yakoyo. The WHO standard for suitable drinking water quality for copper is a maximum 2.0 ppm. This result indicates that the sampled points had a suitable copper concentration in drinking water.

Iron

The iron readings at the different sample points were 0.108 ppm at sample point 1, 0.119 ppm at sample point 2, 0.087 ppm at sample point 3, 0.088 ppm at sample point 4, 0.093 ppm at sample point 5, 0.095 ppm at sample point 6, 0.110 ppm at sample point 7, 0.101 ppm at sample point 8, and

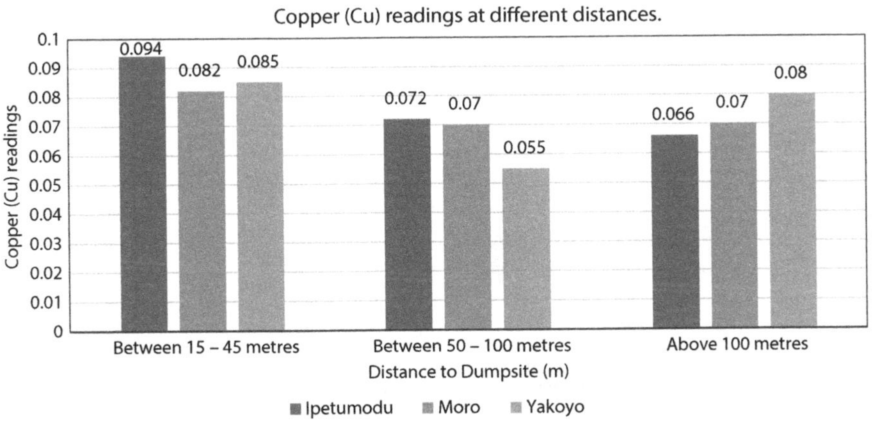

Figure 5.8 Concentration of copper in groundwater at different distances from the dumpsites.

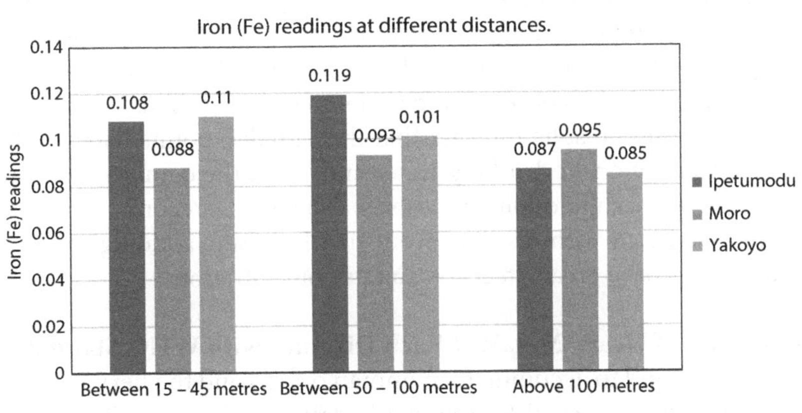

Figure 5.9 Amount of iron concentration at different distances from the dumpsite.

0.085 ppm at sample point 9, respectively (Figure 5.9). The WHO standard for suitable drinking water quality for iron concentration is a maximum of 0.300 ppm. Based on these results, the sample points were suitable for the iron concentration in drinking water.

Sulfate

The sulfate readings at different sample points are 331.71 ppm at sample point 1, 321.34 ppm at sample point 2, 272.25 ppm at sample point 3, 396.41 ppm at sample point 4, 311.55 ppm at sample point 5, 274.45 ppm at sample

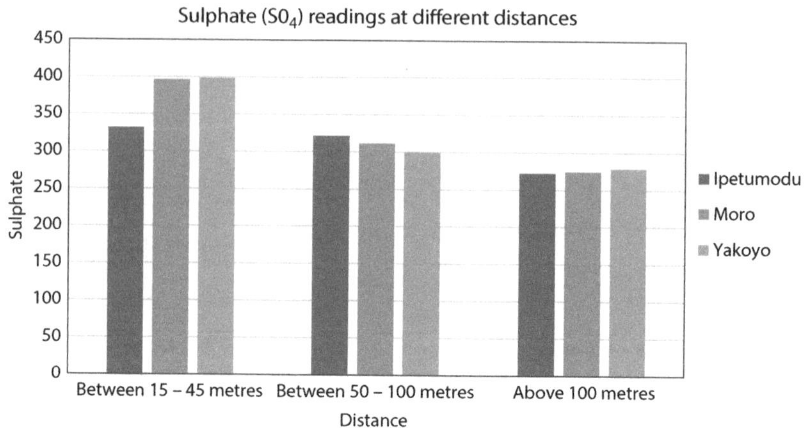

Figure 5.10 Amount of sulfate concentration in the groundwater at different distances.

point 6, 398.75 ppm at sample point 7, 299.67 ppm at sample point 8, and 278.74 ppm at sample point 9, respectively (Figure 5.10). Sample points 1–3 are in Ipetumodu, points 4–6 are in Moro, and points 7–9 are in Yakoyo. The WHO standard for suitable drinking water quality for sulfate is a maximum of 400 ppm. These results indicate that the sampled points had a suitable sulfate concentration in drinking water and, thus, the community in general. Although the sulfate concentration was below the recommended maximum, it decreased with distance from the dumpsite, alluding to the fact that dumpsites have great effects on sulfate concentration.

Comparison of Heavy Metals at Each Distance with WHO Standard

Heavy metals tend to accumulate in human organs and the nervous system and interfere with normal functions (Tripathi *et al.*, 2023). The concentration of heavy metals for each water sample parameter compared with the WHO standard for water quality is shown in Table 5.1. For lead, the maximum concentration that potable water should have is 0.01 ppm. At locations between 15 and 45 m, at sample points 1, 4, and 7. These concentrations are higher than the standard and did not meet the WHO water quality standard for lead. At distances between 50 and 100 m, at sample points 2, 5, and 8. At this distance, it was only at sample point 2 (0.011 ppm) that the concentration did not meet the WHO standard for lead concentration. At distances greater than 100 m, the concentrations at sample points 3, 6, and 9. At these distances, it was only at sample point 3 (0.015 ppm) that the concentration was greater than the set standard for lead concentration in water by the WHO. This means that water is not fit for

Table 5.1 The comparison between WHO (WHO, 2009) standard and the concentration of the metals at different locations.

	WHO standard (ppm)	Locations	Between 15 and 45 m (ppm)	Between 50 and 100 m (ppm)	Above 100 m (ppm)
Lead (Pb)	0.0100	Ipetumodu	0.012	0.009	0.015
		Moro	0.018	0.011	0.007
		Yakoyo	0.015	0.008	0.006
Copper (Cu)	2.000	Ipetumodu	0.094	0.072	0.066
		Moro	0.082	0.070	0.070
		Yakoyo	0.085	0.055	0.080
Iron (Fe)	0.300	Ipetumodu	0.108	0.119	0.087
		Moro	0.088	0.093	0.095
		Yakoyo	0.110	0.101	0.085
Sulphate (SO_4)	400.00	Ipetumodu	331.71	321.34	272.25
		Moro	396.41	311.55	274.45
		Yakoyo	398.75	299.67	278.74

use. Young children, infants, and fetuses are particularly vulnerable to lead because the physical and behavioral effects of lead occur at lower exposure levels (Sud *et al.*, 2023).

For copper, the WHO standard for the maximum concentration of potable water should be 2.0 ppm. At locations between 15 and 45 m, the concentrations are 0.094 ppm at sample point 1, 0.082 ppm at sample point 4, and 0.085 ppm at sample point 7. These concentrations met the set standards for copper water quality. At distances between 50 and 100 m, the concentrations are 0.072 ppm at sample point 2, 0.070 ppm at sample point 5, and 0.055 ppm at sample point 8. These values meet up with the standard set for copper water quality by the WHO. At locations with a distance greater than 100 m, the concentrations were 0.066 ppm at sample point 3, 0.070 ppm at sample point 6, and 0.080 ppm at sample point 9. These values meet the standards set for copper water quality by WHO. When copper exceeds the normal standard, its consumption can cause nausea, vomiting, diarrhea, gastric (stomach) complaints, and headaches. Long-term exposure for several months and years can cause liver damage and death (Upadhyay *et al.*, 2023).

For iron, the maximum concentration of water should be 0.300 ppm; at distances between 15 and 45 m, the concentrations were 0.108 ppm at sample point 1, 0.088 ppm at sample point 4, and 0.110 ppm at sample point 7. These concentrations are not more than the set standard for water quality, which indicates that the water meets the standard for the concentration of iron in water. At distances between 50 and 100 m, the concentrations were 0.119 ppm at sample point 2, 0.093 ppm at sample point 5, and 0.101 ppm at sample point 8; the concentrations of iron in the water were not more than the standard set for maximum water quality, indicating that the water meets the standards for iron water quality by the WHO. For distances greater than 100 m, the concentrations were 0.087 ppm at sample point 3, 0.095 ppm at sample point 6, and 0.085 ppm at sample point 9; the concentrations of iron in the water were not more than the standard set for maximum water quality, which indicates that the water meets the standard for iron concentration in water quality as defined by the WHO. Water with a high concentration of iron has a metallic taste, which makes it unpleasant to drink. When used for cooking, it darkens what it is used for and can cause vomiting (Lavane *et al.*, 2023).

For sulfate, the maximum concentration a water should have is 400 ppm; the sulfate concentrations for distances between 15 and 45 m are 331.71 ppm at sample point 1, 396.41 ppm at sample point 4, and 398.75 ppm at sample point 7, these parameters meet up with the standard for sulfate water quality, for distances between 50 and 100 m, 321.34 ppm at sample point 2, 311.55 ppm at sample point 5, and 299.67 ppm at sample point 8, these concentrations meet up with the standard for water quality for sulfate. This result is similar to observations of (Nath *et al.*, 2023) For distances that are above 100 m, the concentrations are 272.25 ppm at sample point 3, 274.45 ppm at sample point 6, and 278.74 ppm at sample point 9, these concentrations meet up with the standard for water quality for sulfate. When sulfate is high in water, it may have a laxative effect that can lead to dehydration, which is of special concern for infants (Kanga *et al.*, 2023).

Relationship between Location of Groundwater Sources and the Dumpsites

Collected samples were representative of each community. The samples were spread across each community, with samples taken far from each other (Figure 5.2). Samples were collected at different distances to determine the effects of waste at different distances. ArcGIS 10 was used to digitize the road map of the area from the base map of the area.

The buffering of the groundwater points in Moro (Figure 5.11) at a 25 m area of influence shows that at well point 4, there is a dumpsite that falls within 25 m from the well point; therefore, there will be an influence of the dumpsite on the well point. At well point 5, there is no dumpsite that falls within 25 m from the well point; therefore, there is not much influence on the well point. At well point 6, there is no dumpsite that falls within 50 m from the well point; therefore, the dumpsite will not have much influence on the well point.

The buffering of the groundwater points in Ipetumodu (Figure 5.12) in a 25 m area of influence shows that at well point 1, there is a dumpsite that

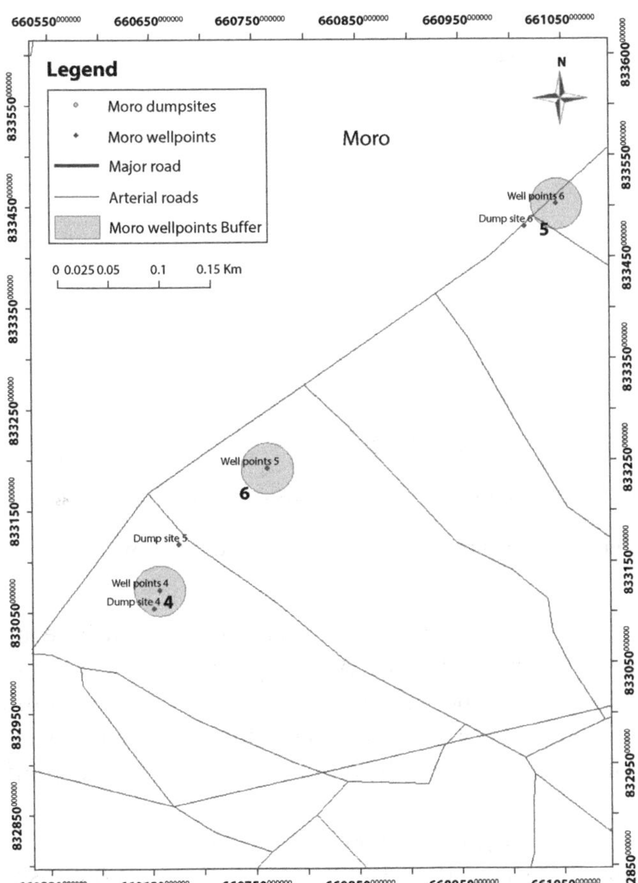

Figure 5.11 The buffering of the wellpoints in Moro at 25 m distance.

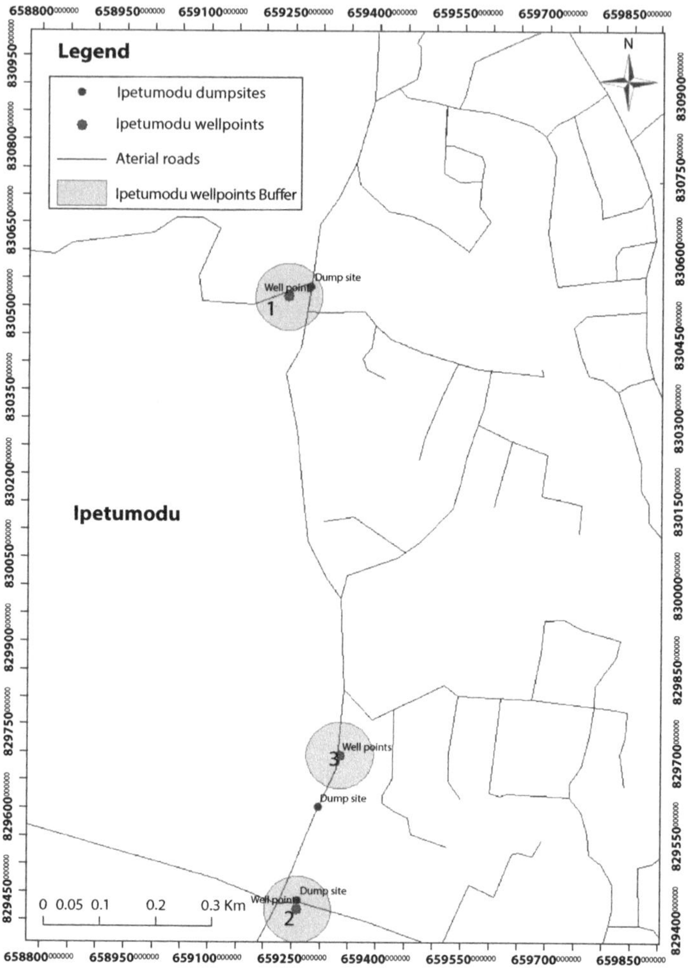

Figure 5.12 The buffering of the wellpoints in Ipetumodu at 25 m distance.

falls within 25 m from the well point; therefore, there will be an influence of the dumpsite on the well point. At well point 2, there is a dumpsite that falls within 25 m from the well point; therefore, there is an influence on the well point. At well point 3, there is no dumpsite that falls within 25 m from the well point; therefore, the dumpsite will not be much influence on the well point.

The buffering of the groundwater points in Yakoyo (Figure 5.13) in a 25 m area of influence shows that at well point 7, there is a dumpsite that falls within 25 m from the well point; therefore, there will be an influence of the

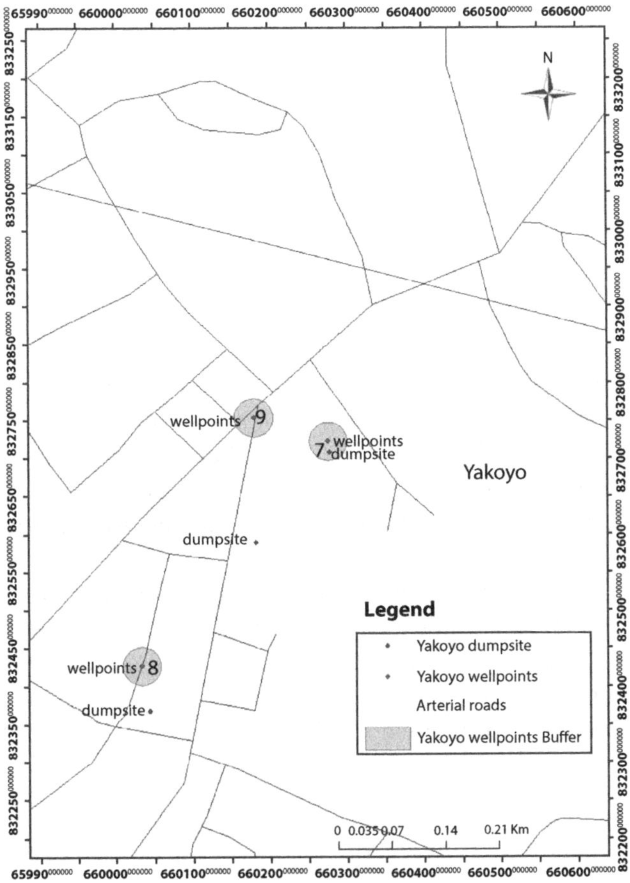

Figure 5.13 The buffering of the wellpoints in Yakoyo at 25 m distance.

dumpsite on the well point; at well point 8, there is no dumpsite that falls within 25 m from the well point; therefore, there will be less influence on the well point. At well point 9, the dumpsite is not close to the well point 25 m from the well point; therefore, the dumpsite will not be much influence on the well point.

Changes in Concentrations of Selected Variables Away From Dumpsites

The results of the regression analyses as stated in Equation 5.1 is shown in Table 5.2.

The results of the analysis of variations/changes in the concentrations of selected variables as one moves away from the targeted source of pollutants (dumpsites) indicate that the ions (Fe, Cu, Pb, and SO_4^{2-}) were

Table 5.2 Change in concentration of selected chemical variables away from dumpsite.

Variable	Trend (a + bx + c)	R^2
Iron	298.81 – 0.54x +46.48	0.29
Copper	191.741 – 0.375x + 51.47	0.14
Lead	$160.151 - 0.664x + 41.28$	0.44
Sulfate	$342.097 - 0.805x + 32.76$	0.65

reduced at different rates. The coefficient of determination (R^2) suggests that the rate of decrease in SO_4^{2-} and Pb was much better explained by distance ($R^2 = 0.65$ and 0.44, respectively) than Fe or Cu. The results suggest possible diffuse sources of Fe, Cu, and Pb in the area, implying that the heavy metals in the groundwater need further investigation to identify other sources.

In general, this study rejects the main hypothesis that dumpsites near groundwater sources do not have an effect on the groundwater quality of the Ife north local government area of Osun State and accepts the alternative hypothesis that dumpsites near groundwater sources affect the groundwater quality of the Ife north local government area of Osun State.

Conclusion

Waste is harmful to human health. This study examined the wastes generated close to groundwater sources and their effects on the quality of water from groundwater sources. Although the proximity of the dumpsite to groundwater sources has an adverse influence on them, waste generated close to the dumpsite is not the only variable that can affect the quality of the groundwater. There are other variables that could contaminate and reduce the quality of the water, including the geology of the rock that the water passes through and the weather conditions that can affect the level of the water.

This study also examined the types of waste generated in the Ife north local government area of Osun state and discovered that the wastes were mainly household wastes and wastes from farmlands. Waste disposal methods in different places affect the quality of the surrounding water. If the waste generated is not properly treated, it will cause a nuisance to the dwellers of the area. This research recommends that the dumpsite close to the wells should be moved away from the wells to improve the quality of water and for better health status. In addition, the residents of the area should be sensitized about the benefits of using a collective waste dumping system, where all the wastes are treated in a place far away from the groundwater source for drinking.

References

Singh SK, Kanga S, Gulati B, Raiˇc M, Sajan B, Đurin B, Singh S (2023) Spatial and Temporal Analysis of Hydrological Modelling in the Beas Basin Using SWAT+ Model. *Water* 15, 3338. https://doi.org/10.3390/w1519333

Rani A, Gupta SK, Singh SK, Meraj G, Kumar P, Kanga S, Đurin B, Dogančić D (2023) Predicting Future Land Use Utilizing Economic and Land Surface Parameters with ANN and Markov Chain Models. *Earth* 4(3):728-751.

Ahmad T, Gupta SK, Singh SK, Meraj G, Kumar P, Kanga S (2023) Unveiling Nature's Resilience: Exploring Vegetation Dynamics during the COVID-19 Era in Jharkhand, India, with the Google Earth Engine. *Climate* 11(9):187.

Sharma M, Upadhyay RK, Tripathi G, Kishore N, Shakya A, Meraj G, Kanga S, Singh SK, Kumar P, Johnson BA, Thakur SN (2023) Assessing Landslide Susceptibility along India's National Highway 58: A Comprehensive Approach Integrating Remote Sensing, GIS, and Logistic Regression Analysis. *Conservation* 3(3):444-459.

Saqib N, Rai PK, Kanga S, Kumar D, Đurin B, Singh SK (2023) Assessment of Ground Water Quality of Lucknow City under GIS Framework Using Water Quality Index (WQI). *Water* 15(17):3048.

Choudhury U, Singh SK, Kumar A, Meraj G, Kumar P, Kanga S (2023) Assessing land use/land cover changes and urban heat island intensification: A case study of Kamrup Metropolitan District, Northeast India (2000–2032). *Earth* 4(3):503-521.

Rai NK, Mathur S, Singh SK, Tiwari M, Singh VK, Haque R, Tiwari S, Sharma LK (2023) Differential regulation of mitochondrial complex I and oxidative stress based on metastatic potential of colorectal cancer cells. *Oncology Letters* 26(1):1-1.

Debnath J, Sahariah D, Lahon D, Nath N, Chand K, Meraj G, Kumar P, Singh SK, Kanga S, Farooq M (2023) Assessing the impacts of current and future changes of the planforms of river Brahmaputra on its land use-land cover. *Geosci. Front.* 14(4):101557. Elsevier.

Kumar P, Singh SK, Gupta SK, Kanga S, Meraj G (2023) Uncovering the hydrometeorological drivers responsible for forest fires utilizing geospatial techniques. *Theor. Appl. Climatol.* https://doi.org/10.1007/s00704-023-04497. Springer Nature.

Singh SK, Kotecha MJ, Kanga S, Pankhaniya SK, Agrawal S, Meraj G (2023) Urban Heat Island (UHI) Resilience Plan in Varying Climatic Conditions Using Geospatial Approach: A Case Study of Rajkot City. In: *Climate Change, Agriculture and Society*, vol 1. pp 307-338.

Singh S, Singh SK, Prajapat DK, Pandey V, Kanga S, Kumar P, Meraj G (2023) Assessing the Impact of the 2004 Indian Ocean Tsunami on South Andaman's Coastal Shoreline: A Geospatial Analysis of Erosion and Accretion Patterns. *J. Mar. Sci. Eng.* 11(6):1134.

Rathore SS, Mathur SC, Mathur S, Sharma S, Singh SK (2023) Late Cretaceous–Early Paleocene Geodiversity of Fatehgarh Formation of Petroliferous Barmer Basin, Western Rajasthan, India: A Potential Geopark Site. *Geoheritage* 15:54.

Singh S, Meraj G, Kumar P, Singh SK, Kanga S, Johnson BA, Prajapat DK, Debnath J, Sahariah D (2023) Decoding Chambal River Shoreline Transformations: A Comprehensive Analysis Using Remote Sensing, GIS, and DSAS. *Water* 15(9):1793.

Tripathi G, Shakya A, Upadhyay RK, Singh SK, Kanga S, Pandey SK (2023) Landslide Susceptibility Mapping of Tehri Reservoir Region Using Geospatial Approach. In: *Climate Change Adaptation, Risk Management and Sustainable Practices in the Himalaya*, pp 135-156.

Sud A, Kanga R, Singh SK, Meraj G, Kanga S, Kumar P, Ramanathan AL, Bhardwaj V (2023) Simulating Groundwater Potential Zones in Mountainous Indian Himalayas—A Case Study of Himachal Pradesh. *Hydrology* 10(3):65.

Upadhyay RK, Tripathi G, Đurin B, Šamanović S, Cetl V, Kishore N, Sharma M, Singh SK, Kanga S, Wasim Md, Rai PK, Bhardwaj V (2023) Groundwater Potential Zone Mapping in the Ghaggar River Basin, North-West India, Using Integrated Remote Sensing and GIS Techniques. *Water* 15(5):961.

Lavane K, Kumar P, Meraj G, Han TG, Ngan LHB, Lien BTB, Ty TV, Thanh NT, Downes NK, Nam NGD, Minh HVT, Singh SK, Kanga S (2023) Assessing the effects of drought on rice yields in the Mekong Delta. *Climate* 11(1):13.

Nath N, Sahariah D, Meraj G, Debnath J, Kumar P, Lahon D, Chand K, Farooq M, Chandan P, Singh SK, Kanga S (2023) Land use and land cover change monitoring and prediction of a UNESCO world heritage site: Kaziranga eco-sensitive zone using cellular automata-Markov model. *Land* 12(1):151.

Kanga S, Upadhyay RK, Sharma M, Sharma S, Tripathi G, Singh SK (2023) Geotechnical Stability Assessment of Road-cut Slopes: A Case Study of Srinagar, Garhwal Himalaya, India. *Biol. Forum* 15(5):1071-1080.

Gulati B, Sharma R, Kanga S, Singh SK, Sajan B, Meraj G, Kumar P, Ramanathan AL (2023) Unraveling the Relationship Between Stubble Burning and Air Quality Degradation in Punjab: A Temporal and Spatial Analysis (2019-2022). *J. Clim. Change* 9(2):43-53, IOS Press.

Tomar P, Singh SK, Kanga S, Pattanaik A (2023) Water Bodies mapping and monitoring using high-resolution satellite images. *Sustain. Agri Food Environ. Res.* 11.

Singh H, Meraj G, Singh S, Shrivastava V, Sharma V, Farooq M, Kanga S, Singh SK, Kumar P (2022) Status of Air Pollution during COVID-19-Induced Lockdown in Delhi, India. *Atmosphere* 13(12):2090.

Farooq M, Mushtaq F, Meraj G, Singh SK, Kanga S, Gupta A, Kumar P, Singh D, Avtar R (2022) Strategic Slum Upgrading and Redevelopment Action Plan for Jammu City. *Resources* 11(12):120.

World Health Organisation (WHO) (2009) Guidelines for drinking water quality, 3rd ed., Geneva, p. 540.

Jeje JA (2014) Assessment of Heavy Metals of Boreholes and Hand Dug Wells in Ife North. *Int. J. Sci. Technol.*

6

Enhancing Sustainable Natural Resource Management Through Innovative Use of Waste Materials in Concrete Production

Saurabh Singh[1], Suraj Kumar Singh[2*], Sujeet Kumar[3], Bhartendu Sajan[2] and Gowhar Meraj[4]

[1]*Department of Civil Engineering, Suresh Gyan Vihar University, Jaipur, India*
[2]*Centre for Climate Change and Water Research, Suresh Gyan Vihar University, Jaipur, India*
[3]*Department of Mechanical Engineering, Vivekananda Global University, Jaipur, India*
[4]*Department of Ecosystem Studies, Graduate School of Agricultural and Life Sciences, The University of Tokyo, Tokyo, Japan*

Abstract

The global demand for construction materials significantly impacts natural resource depletion, with sand extraction notably affecting river ecosystems and biodiversity. Innovations in construction materials focusing on sustainability are crucial to mitigate these impacts. This chapter explores the use of various waste materials as partial substitutes for natural sand in concrete production, aiming to enhance resource management and environmental sustainability. This study aims to evaluate the feasibility of using multiple waste materials, including marble powder, glass fines, and fly ash, as alternative aggregates in concrete. The primary goal is to assess their potential in reducing the reliance on natural sand, thus promoting sustainable natural resource management. An experimental setup was conducted where concrete samples incorporated different percentages (10%, 20%, 30%) of waste materials as partial substitutes for natural sand. Mechanical properties such as compressive strength, tensile strength, and durability tests, including water absorption and chloride ion permeability, were systematically evaluated. Results indicated that concrete mixes containing up to 20% marble powder, 15% glass fines, and 25% fly ash achieved enhanced mechanical and durability properties

**Corresponding author*: suraj.kumar@mygyanvihar.com

Shruti Kanga, Gowhar Meraj, Suraj Kumar Singh, Majid Farooq and M. S. Nathawat (eds.) Geospatial Technology for Natural Resource Management, (177–188) © 2024 Scrivener Publishing LLC

compared to conventional concrete. For instance, concrete with 20% marble powder showed a 5% increase in compressive strength after 28 days, signifying improved material efficiency and reduced environmental impact. The study confirms that integrating various waste materials in concrete production can significantly contribute to natural resource conservation and waste reduction. Optimal replacement levels were identified, balancing material properties with sustainability benefits. These findings encourage further exploration into innovative mix designs and underscore the necessity of adopting sustainable practices in the construction industry to manage natural resources effectively. Future research should focus on the long-term performance and economic analysis of these alternative materials to integrate them fully into construction standards.

***Keywords*:** Marble powder waste, sustainable construction, concrete production, eco-friendly materials, natural resource management, water contamination, environmental impact

6.1 Introduction

The incessant global demand for construction materials profoundly affects the depletion of natural resources, notably impacting river ecosystems and biodiversity due to extensive sand extraction [1, 2]. Such practices necessitate urgent innovations in construction materials to enhance sustainability and conserve natural resources. This chapter delves into the utilization of various waste materials, including marble powder, glass fines, and fly ash, as alternative aggregates in concrete. The focus is on minimizing reliance on natural sand, thereby promoting more sustainable natural resource management and reducing the environmental footprint of the construction industry. The exploration of these waste materials as partial substitutes aims to address the dual challenges of environmental degradation and resource scarcity. By integrating recycled waste into concrete production, this approach seeks to not only mitigate the adverse impacts on ecosystems and biodiversity but also enhance the overall sustainability of construction materials [3]. This initiative aligns with global sustainability goals and reflects an increasing commitment to responsible resource management within the construction sector. The integration of such practices represents a crucial step towards the development of more resilient and sustainable construction methodologies, emphasizing the efficient use of available resources and the reduction of environmental impacts. Singh *et al.* (2022) address the severe ecological consequences of excessive river sand mining—a key component traditionally used in concrete—which leads to river bank erosion and habitat destruction [4]. By experimenting with copper slag, quarry dust, foundry sand, and sawdust as substitutes for river sand, the study emphasizes reducing dependency on river sand, thus promoting more sustainable resource management practices [5]. Firoozi *et al.* (2023) extend this approach by incorporating waste materials, such as crushed glass and plastic fibers, into concrete. This not only utilizes waste that would otherwise contribute to landfill but also reduces the demand for virgin raw materials. The successful integration of these recycled materials highlights a dual benefit: enhancing concrete's structural properties and supporting circular economic practices within the construction industry [6]. Kumar *et al.* (2020) focus on the integration of glass fibers into concrete to improve its mechanical properties and durability. The use of glass fiber reinforced concrete (GFRC) demonstrates how alternative materials can provide enhanced functionality while reducing the need for traditional, often more resource-intensive materials such as steel for reinforcement. This approach not only conserves natural resources but also offers long-term sustainability benefits through the extended lifespan and durability of constructed structures [7]. Jagarapu (2016) explores the integration of glass fibers into concrete, which apart from enhancing the structural properties, also reduces potential damage during incidents like fires. The incorporation of glass fibers, which are inherently strong and fire-resistant, highlights an innovative use of materials that contribute to safer, more durable building practices, while also promoting the reuse of industrial by-products [8]. İskender and Karasu (2018) trace the evolution of GFRC from the 1940s, highlighting significant advancements such as the inclusion of zirconium dioxide in the 1960s to enhance its durability. Their study discusses the versatility and potential of GFRC in modern construction, emphasizing its suitability for creating lightweight, strong, and durable structures, thus promoting resource efficiency and reducing the environmental footprint associated with traditional construction materials [9]. Sharma and Mehta (2023) investigate the potential of incorporating glass fibers in concrete mixes that use recycled coarse aggregates. Their findings suggest that using up to 70% recycled aggregates can effectively replace traditional materials without compromising structural integrity, particularly when enhanced with glass fibers. This study underscores the importance of reusing materials as a way to manage natural resources wisely and reduce construction waste [10]. Devi *et al.* (2022) provide a comprehensive review of the use of glass fiber in concrete, emphasizing its impact on improving concrete's mechanical properties and its role in the production of new structural elements. By facilitating the use of lighter, yet strong materials, GFRC can significantly reduce the quantity of raw materials needed, thereby conserving natural resources and promoting sustainability in construction practices [11]. Annamaneni and Pedarla (2023) focus on the specific advantages of using short, alkali-resistant glass fibers in concrete to enhance its mechanical and fracture properties. Their study highlights the material's potential to reduce concrete's inherent fragility and improve its durability, which directly contributes to longer-lasting structures and reduces the need for frequent replacements or repairs, thereby optimizing resource use [12]. Jedulius (2018) explores the mechanical enhancements brought by integrating glass fibers into concrete, specifically targeting the material's tensile strength, ductility, and crack resistance. This approach not only extends the life of concrete structures but also demonstrates the potential of glass fibers as a sustainable alternative to steel, avoiding issues like corrosion and thus extending the material's service life and sustainability [13]. Anas *et al.* (2022) review the overall benefits and challenges of integrating fibers into concrete. They discuss how fiber-reinforced concrete (FRC) can overcome the limitations of traditional concrete by enhancing tensile strength and durability while also managing shrinkage cracks more effectively [14].

Together, these studies paint a comprehensive picture of how modern concrete technology can align with natural resource management goals. They illustrate a shift towards more sustainable construction practices that prioritize the efficient use of resources, leverage recycled materials, and innovate within the industry to reduce environmental footprints. This cohesive approach not only addresses the immediate needs of the construction industry but also contributes to broader environmental sustainability objectives by managing natural resources more effectively and responsibly.

6.2 Data Collection and Methodology

The global construction industry faces significant challenges due to the depletion of natural resources, particularly river sand, which is widely used in concrete production. The extraction of river sand has detrimental effects on river ecosystems and biodiversity. This study seeks to address these concerns by exploring the potential of marble powder, glass fines, and fly ash as partial substitutes for natural sand in concrete, aiming to improve sustainability and resource management in construction practices.

The primary objective of this research is to assess the feasibility of using marble powder, glass fines, and fly ash as alternative aggregates in concrete mixes. The study aims to determine the optimal substitution rates that maintain or enhance the mechanical properties and durability of concrete while reducing the reliance on natural sand.

The materials selected for this study include:

Natural sand: standard grade sand used as the control variable in concrete mixes.
Marble powder: a byproduct of marble processing considered waste material.
Glass fines: recycled glass ground into fine particles.
Fly Ash: a byproduct from coal combustion in power plants.
Cement: ordinary portland cement (OPC) conforming to IS 12269.
Aggregates: coarse aggregates conforming to IS 383.
Water: potable water meeting the standards set by IS 456.

Each material was chosen based on its potential environmental benefits and its physical properties similar to those of natural sand.

6.2.1 Experimental Design

The experimental design involves creating concrete mixes with varying percentages of marble powder, glass fines, and fly ash replacing natural sand. These mixes are tested for their mechanical properties and durability to establish a comprehensive understanding of the performance of these alternative materials in concrete applications.

6.2.2 Mix Proportions

The concrete mixes are prepared according to IS 10262 recommendations. The substitution of natural sand with marble powder, glass fines, and fly ash is set at different percentages to explore a range of potential outcomes. The specific proportions are detailed in Table 6.1.

The experimental setup and procedures include the preparation, curing, and testing of concrete specimens to evaluate their mechanical properties and durability.

6.2.3 Preparation of Specimens

Concrete mixes are prepared by accurately weighing the specified amounts of cement, aggregates, water, and substitute materials. These components are thoroughly mixed using a mechanical mixer to achieve a homogeneous mixture. The mixing process ensures uniform distribution of the substitute

Table 6.1 Composition of concrete mixes.

Mix ID	Cement (kg)	Marble powder (kg)	Glass fines (kg)	Fly ash (kg)	Natural sand (kg)	Aggregates (kg)	Water (L)
Base	50	0	0	0	50	75	25
MP10	50	5	0	0	45	75	25
MP20	50	10	0	0	40	75	25
GF10	50	0	5	0	45	75	25
GF15	50	0	7.5	0	42.5	75	25
FA10	50	0	0	5	45	75	25
FA20	50	0	0	10	40	75	25

materials within the concrete matrix. After mixing, the concrete is poured into standard molds for specimen preparation. The molds are then vibrated to eliminate air pockets and ensure compaction.

6.2.4 Curing Process

Once the concrete has set sufficiently (typically after 24 hours), the specimens are demolded carefully to avoid any damage. The demolded specimens are transferred to a curing tank where they are fully submerged in water maintained at room temperature. This curing process is essential for the hydration of cement and development of strength. The curing periods are set at 7, 14, and 28 days, allowing for comprehensive strength development analysis.

6.2.5 Testing of Specimens

The compressive strength of the concrete specimens is measured at 7, 14, and 28 days using a hydraulic press, following the procedures outlined in IS 516. This test determines the ability of the concrete to withstand axial loads. The tensile strength is determined at the same intervals using a tensile testing machine, as per IS 5816. This test evaluates the concrete's resistance to tension and its overall flexibility. This test measures the amount of water absorbed by the concrete specimens over a specified period, providing an indication of the concrete's porosity and potential durability.

Conducted to assess the concrete's resistance to chloride ion penetration, which is critical for structures exposed to marine environments or deicing salts. The test follows standard procedures to evaluate the permeability of concrete to chloride ions, which can cause corrosion of embedded steel.

6.3 Results and Analysis

6.3.1 Compressive Strength and Tensile Strength at 28 Days

Figure 6.1 shows that the concrete mixes with marble powder (MP10 and MP20) and glass fines (GF15) exhibit higher compressive strengths compared to the base mix. Specifically, MP20 demonstrates the highest compressive strength at 45 MPa, followed by GF15 and MP10. This indicates that the addition of 20% marble powder and 15% glass fines significantly enhances the compressive strength of the concrete, likely due to their material properties that contribute positively to the concrete matrix.

The Figure 6.2 highlights the tensile strength results. Similar to the compressive strength, mixes with marble powder (MP10 and MP20) and glass fines (GF15) show improved tensile strength over the base mix. The highest tensile strength is observed in MP20 at 5.4 MPa, which suggests that marble powder not only enhances compressive strength but also contributes to

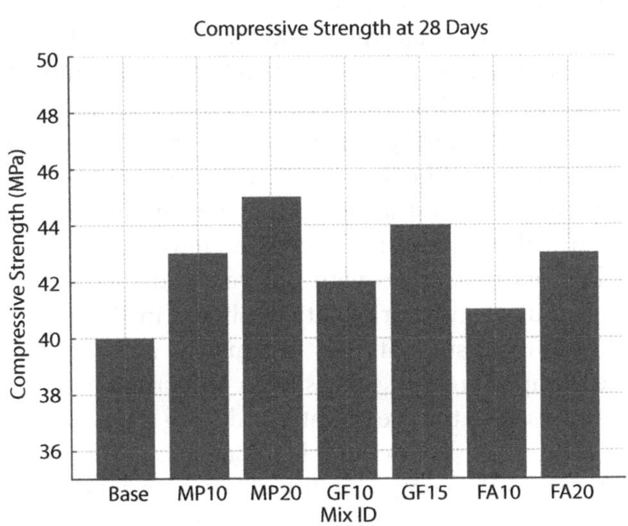

Figure 6.1 Compressive strengths at 28 days.

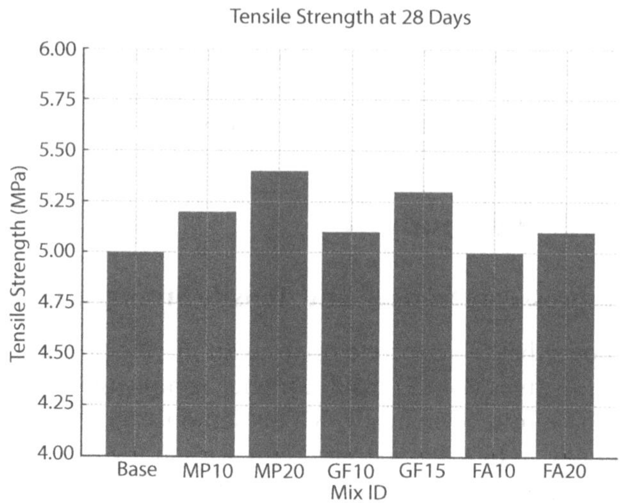

Figure 6.2 Tensile strengths at 28 days.

better tensile properties. This enhancement in tensile strength is crucial for applications where concrete might experience tensile forces.

6.3.2 Workability Results from Slump and Compaction Factor Tests

The workability of concrete is a crucial factor in construction, influencing both the ease of concrete placement and its final strength and durability. Two common tests used to measure workability are the Slump Test and the Compaction Factor Test. The results from these tests provide insights into the concrete's behavior during mixing, transportation, and placing. The slump test results indicate the fluidity and consistency of the concrete mix. Higher slump values suggest a more fluid mix that can be easily placed and compacted without requiring excessive effort or vibration. From the results, it's observed that mixes with natural sand (Base) have the highest slump, indicating good workability as shown in Figure 6.3. However, as the percentage of waste materials like marble powder and glass fines increases, the slump decreases, suggesting a reduction in workability. This decrease may necessitate the use of water-reducing admixtures to maintain adequate workability for practical applications.

This test measures the compactibility of concrete, which is particularly important for achieving dense and durable concrete. The compaction factor results range from 0.92 to 0.95, with higher values indicating better compactibility. The results show in Figure 6.4 that the base mix and those with a lower percentage of waste materials exhibit slightly higher compaction factors, reflecting better potential for achieving well-compacted concrete structures. Mixes with higher waste material content, although slightly

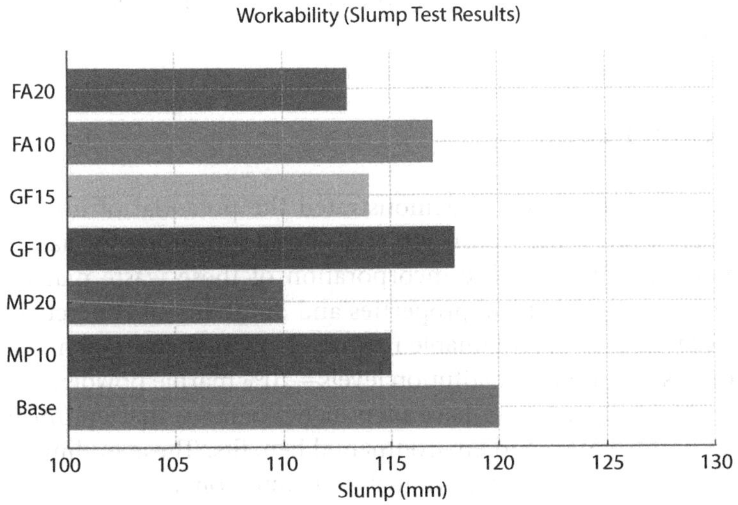

Figure 6.3 Workability (Slump Test results).

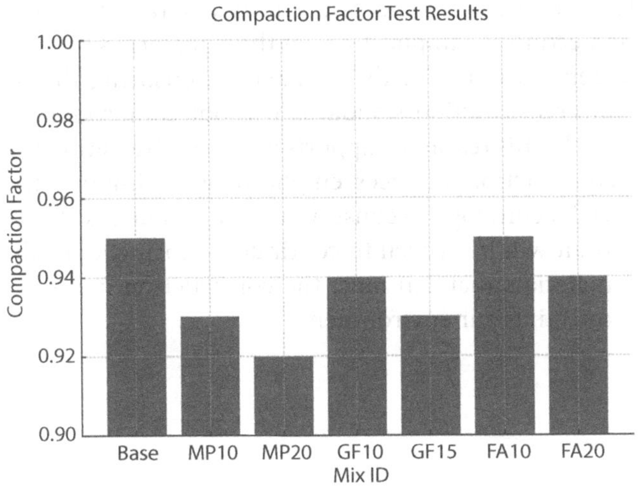

Figure 6.4 Compaction Factor Test results.

lower in compaction factor, still demonstrate adequate compactibility, which can be optimized with adjustments in mix design.

Both tests together reveal that while the addition of waste materials enhances certain mechanical properties of the concrete, they also affect its fresh properties, such as reducing workability and slightly altering compatibility. To ensure that concrete with high percentages of waste materials remains workable and compactible, careful consideration must be given to mix design, possibly including the adjustment of water content or the addition of specific admixtures.

6.4 Conclusion

This study has conclusively demonstrated the potential of using marble powder, glass fines, and fly ash as viable partial substitutes for natural sand in concrete production. The incorporation of these waste materials not only boosts the mechanical properties and durability of concrete but also plays a critical role in sustainable natural resource management. Through the careful selection of substitution levels—20% marble powder, 15% glass fines, and 25% fly ash—we have identified a balance that optimizes both material performance and environmental benefits. These findings suggest that integrating such alternative materials into concrete can significantly contribute to the construction industry's sustainability goals by reducing dependence on natural sand, which is both environmentally detrimental and unsustainable in the long term. However, while the short-term benefits are clear, further research is necessary to understand the long-term performance and economic implications of these materials fully. Such studies will help to refine the use of these materials, ensuring that they can be integrated into construction standards without compromising quality or safety. Ultimately, this research supports a broader transition towards more sustainable construction practices, emphasizing the importance of innovation in material technology to conserve natural resources effectively. As we move forward, it will be crucial to continue exploring and validating new sustainable materials that can meet the construction industry's demands while also protecting our environment.

References

1. P. Pitchaiah, 'Impacts of Sand Mining on Environment–A Review', *Int. J. Geoinf. Geol. Sci.*, vol. 4, pp. 1–6, Feb. 2017, doi: 10.14445/23939206/IJGGS-V4I1P101.
2. M. Bendixen *et al.*, 'Sand, gravel, and UN Sustainable Development Goals: Conflicts, synergies, and pathways forward', *One Earth*, vol. 4, pp. 1095–1111, Aug. 2021, doi: 10.1016/j.oneear.2021.07.008.
3. S. Singh, S. K. Singh, S. Kanga, P. K. Shrivastava, P. Kumar and G. Meraj, 'Hydrological dynamics of the Kalisindh and Parbati Rivers: An integrated analysis in the context of the Eastern Rajasthan Canal Project (ERCP)', *Results Eng.*, p. 101960, 2024.
4. 'Water | Free Full-Text | Decoding Chambal River Shoreline Transformations: A Comprehensive Analysis Using Remote Sensing, GIS, and DSAS'. Accessed: Apr. 02, 2024. [Online]. Available: https://www.mdpi.com/2073-4441/15/9/1793
5. S. Singh, S. K. Singh, R. Kumar, A. Shrama and S. Kanga, 'Finding Alternative to River Sand in Building Construction', 2022, Accessed: May 24, 2024. [Online]. Available: https://catalog.lib.kyushu-u.ac.jp/ja/recordID/6625713/?repository=yes
6. A. A. Firoozi, A. A. Firoozi and D. Oyejobi, 'ENHANCING CONCRETE PERFORMANCE BY UTILIZING CRUSHED GLASS AND WASTE BOTTLE PLASTIC FIBERS FOR IMPROVED STRENGTH AND FLEXURAL PROPERTIES', *J. Teknol.*, vol. 85, no. 6, Art. no. 6, Sep. 2023, doi: 10.11113/jurnalteknologi.v85.20328.
7. D. Kumar, L. K. Rex, V. S. Sethuraman, V. Gokulnath and B. Saravanan, 'High performance glass fiber reinforced concrete', *Mater. Today: Proc.*, vol. 33, pp. 784–788, Jan. 2020, doi: 10.1016/j.matpr.2020.06.174.
8. D. C. K. Jagarapu, 'Experimental Studies on Glass Fiber Concrete', *Am. J. Eng. Res.*, vol. 5, pp. 100–104, Jan. 2016.
9. M. İskender and B. Karasu, 'Glass Fibre Reinforced Concrete (GFRC)', *El-Cezeri J. Sci. Eng. (EJCSE)*, vol. 5, pp. 136–162, Feb. 2018, doi: 10.31202/ecjse.371950.
10. S. Kumar Sharma and S. Mehta, 'Experimental study on effect of glass fibre on concrete using recycled aggregate', *Mater. Today: Proc.*, Mar. 2023, doi: 10.1016/j.matpr.2023.03.241.
11. Ch. Devi, D. S. Vijayan, R. Nagalingam and S. Arvindan, 'A review of the implementations of glass fiber in concrete technology', *Mater. Today: Proc.*, vol. 62, pp. 2010–2015, Jan. 2022, doi: 10.1016/j.matpr.2022.02.293.

12. K. K. Annamaneni and K. Pedarla, 'Compressive and flexural behavior of glass fiber-reinforced concrete', *J. Phys.: Conf. Ser.*, vol. 2423, no. 1, p. 012025, Jan. 2023, doi: 10.1088/1742-6596/2423/1/012025.
13. J. G, 'Experimental Study on Glass Fibre Concrete', *Int. J. Eng. Res. Technol.*, vol. 4, no. 8, Jul. 2018, doi: 10.17577/IJERTCONV4IS08017.
14. M. Anas, M. Khan, H. Bilal, S. Jadoon and M. N. Khan, 'Fiber Reinforced Concrete: A Review', *Eng. Proc.*, vol. 22, no. 1, Art. no. 1, 2022, doi: 10.3390/engproc2022022003.

7

Dynamics of Land Use/Land Cover of Watershed Changes in Kolhapur District Maharashtra

K.S. Kumbhar[1], Rishikesh Golekar[2] and Pranaya Diwate[3*]

[1]*Department of Geology, Nowrosjee Wadia College, Pune, Maharashtra, India*
[2]*Khare Dhere Bhosale College, Guhagar, Ratnagiri, Maharashtra, India*
[3]*Department of Geology, School of Basic and Applied Sciences, MGM University, Chhatrapati Sambhajinagar, Maharashtra, India*

Abstract

Planning and observing of land use land cover (LULC) dynamics is imperative for economical turn of events, arranging and the executives. This study attempts to track the LULC patterns of the watershed in Kolhapur District, Maharashtra. It is based on remote sensing (RS) and geographic information systems (GIS). Using a Maximum Likelihood Classifier (MLC), supervised classification was used to create LULC maps of the watershed. This study uses multispectral satellite data to demonstrate the dynamics of Land Use land cover (LULC) in the Watershed of Kolhapur District, Maharashtra. There are six major LULC classes: Settlement, Agriculture, Fallow land, Water body, Open land, and Forest. The identification of accounts for most watershed land use. In this study, agriculture was the major land-use pattern of the district. Agricultural activities were observed to be higher in the eastern part of the study area i.e., in watersheds KR 55, KR 63, and KR 64 than in watersheds KR 66 and KR 71. Deciduous forests also occur in patches in these watersheds, with degraded forests found intermittently in smaller patches. Land with and without scrubs was observed in large patches throughout the study area. Small patches of barren rocky waste land were observed in watersheds KR 66, KR 71, and KR 77.

**Corresponding author*: pranayadiwate7@gmail.com

Shruti Kanga, Gowhar Meraj, Suraj Kumar Singh, Majid Farooq and M. S. Nathawat (eds.) Geospatial Technology for Natural Resource Management, (189–206) © 2024 Scrivener Publishing LLC

Keywords: Watershed, Kolhapur, agriculture, land use, land cover

7.1 Introduction

It is now acknowledged that satellite-based data can serve as a foundation for the generation of useful information for LULC, even though efforts to incorporate various interpretation methods into LULC mapping began in the middle of the 1970s. Globally, a variety of LULC mapping and change detection methods have been developed and implemented over the past few decades. Globally, various land-use categories are rapidly replacing land cover (Singh *et al.*, 2022).

In the existing situation of development, natural resources are being exploited indiscriminately, leading to the degradation of the environmental ecosystem, and causing global warming, floods, and droughts worldwide (Rani *et al.*, 2023). This situation calls for the implementation of a channel to restore and conserve the existing ecosystems. The ecosystem plays a vital role for present and future human generations; therefore, this is a cause of concern for land managers, environmental scientists, and decision-makers (Singh *et al.*, 2023). Sustainable development is associated with achieving desired growth for economic or social reasons, along with caring for the environment and maintaining a good quality of life. In the process of development, geo-environmental components may degrade and therefore need to be studied with sufficient awareness and proper replacement measures (Ahmad *et al.*, 2023). In this approach, many researchers, globally, have concentrated their research on the development and evolution of appropriate methodologies for geo-environmental assessment studies (Sharma *et al.*, 2023).

Land use land cover studies have become increasingly important as the nation plans to overcome the problems of uncontrolled development and deteriorating environmental quality. Land use data are needed to analyze environmental processes and problems that must be understood if living conditions and standards are to be improved or maintained at current levels (Saqib *et al.*, 2023). There is no ideal classification for land use and land cover, and it is unlikely that one could ever be developed (Choudhury *et al.*, 2023). There are different perspectives on the classification process, and the process itself tends to be subjective. There is no logical reason to expect that a detailed inventory should be adequate for more than a short time, since land use and land cover patterns change according to demands for natural resources (Rai *et al.*, 2023; Kumar *et al.*, 2023). Each classification is made to suit the needs of the user, and a few users will be satisfied with an inventory that does not meet most of their needs (Debnath *et al.*, 2023; Singh *et al.*, 2023; Rathore *et al.*, 2023).

Land use refers to the use of the landscape for different purposes, such as development, conservation, or mixing. Land cover is natural, planted vegetation, or manmade. Agriculture is the main source of occupation for most of the population in the study area (Singh *et al.*, 2023). The proportion of cultivated area varied from one watershed to another. Cropping patterns are an important aspect of studies, particularly relevant to agriculture (Tripathi *et al.*, 2023; Upadhyay *et al.*, 2023). Population growth and changes in socio-economic conditions lead to changes in cropping patterns. Urbanization and industrialization are also important contributory factors responsible for changing land use (Sud *et al.*, 2023). The main objective of this study to assess irrigation development in the watershed of Kolhapur District through land use/land cover analysis.

7.2 Study Area

The Krishna and Panchganga Rivers are the two major rivers dissecting the southern part of the Deccan Volcanic Province (DVP) in the Kolhapur district. Several tributaries join these rivers and form drainage networks. Some of these are contaminated by rapid industrial and urban development along the banks. The rivers Dudhaganga, Vedganga, and Panchganga have always attracted the attention of geoscientists and environmentalists for a long time. The area lies in the region of the seismic zone adjacent to the Koyana Valley. The seismic activity in the recent past has indicated that this area is seismically active and unstable, as previously thought. The indirect signatures of these activities conflicted in the form of different geological, geomorphological, structural, environmental, and other related features, along with subsurface abnormalities.

A detailed scientific approach to this area provides useful information about these activities in the geological past in different forms and signatures.

The study area includes different watershed areas from Shirol, Gadhinglaj, Bhudargad, and Radhanagari Talukas located in the Kolhapur district, Maharashtra (Figure 7.1, Table 7.1). The study area is bounded by latitude 16° 04’33” to 16°45’04” N and 73°50’10” to 74°41’52” E longitude. The study area was included in Toposheet Nos. 47 H/15, 16, 47 L/3, 4, 47 L/6, 8, 9, and 10 of the Survey of India, and covers an area of approximately 830 km^2 (Figure 7.1).

Physiographically, the area is an undulating terrain, a typical trappean country with hills in western parts at Bhudargad and Radhanagari Tahsils, moderately undulating topography, and a flat area in Gadhinglaj Tahsil,

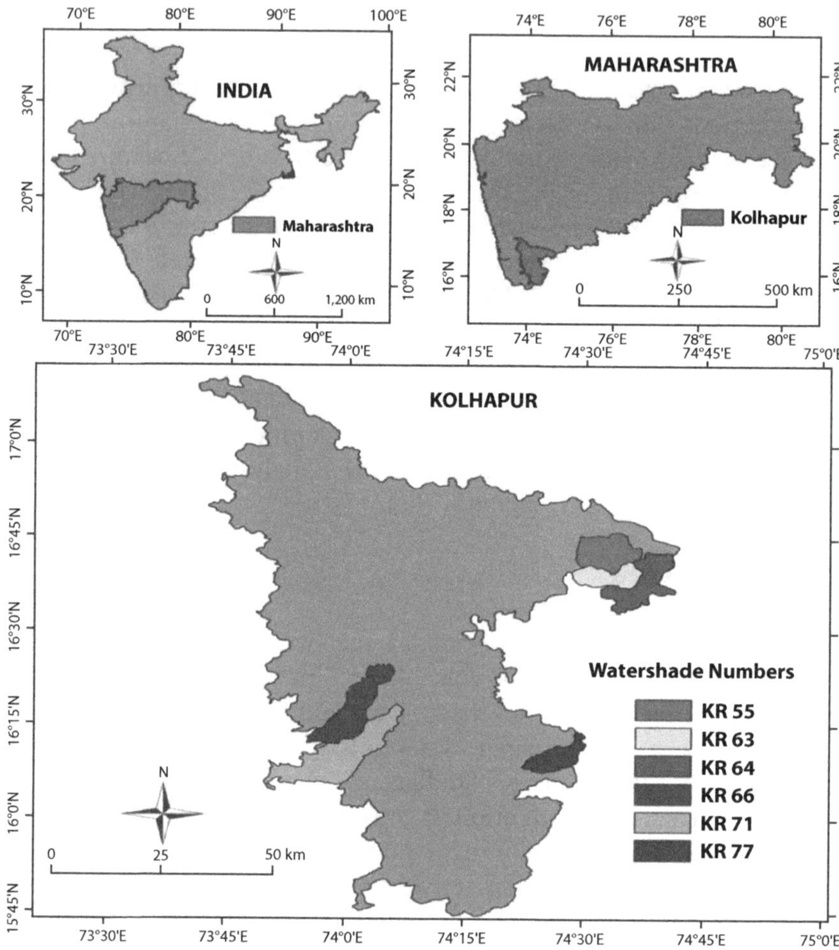

Figure 7.1 Location map of the study area.

which is an almost flat flood plain of the Krishna River in the Shirol Tahsil, which is the eastern part of the study area.

7.3 Methodology

Landsat imagery (March 2012) was used to study the land use and land cover patterns in the study area. The land use land cover map of the study area was prepared using IRS-P6–LISS-III satellite image data by executing the maximum likelihood classification algorithm technique over the

Table 7.1 Talukawise distribution of watersheds in study area.

Sr. no.	Taluka	WS no	Mini WS no	Area (Ha)	Villages
1	Sirol	KR 55	3/4, 5/5	84.75	9
2	Sirol	KR 63	1/2, 3/3	64.27	8
3	Sirol	KR 64A	1/2, 2/2	53.97	8
4	Sirol	KR 64B	1/2, 2/2	53.26	6
5	Sirol	KR 64C	1/2, 2/2	39.88	4
6	Radhanagari	KR 66	1 to 8/13	194.44	22
7	Bhudargad	KR 71	1 to 12/12	257.22	29
8	Gadhinglaj	KR 77	1 to 4/4	82.3	19
				830.09	**105**

false color composite (FCC). These FCC were prepared from geometrically corrected IRS-P6–LISS-III images, as shown in Figures 7.1–7.9. The entire classification process of the satellite image was performed using ArcGIS 9.0 software (Lavane *et al.*, 2023).

Digital image classification helped identify, delineate, and map the land use/land cover into several classes. Image classification was performed using the maximum likelihood classifier decision rule in the supervised classification method. The overall accuracy is 86%. Land use/land cover changes were determined using the post-classification change detection method, and land use/land cover statistics derived from the datasets were computed and compared for quantification of change.

7.4 Results and Discussion

As a result of anthropogenic activity on the planet and changes in the landscape, the natural environment has undergone significant changes. These anthropogenic influences on shifting patterns of land use are a primary component of many current environmental concerns, as land use and land cover changes are becoming recognized as key drivers of environmental change. The classified images were categorized into different LULC patterns (Kanga *et al.*, 2023; Nath *et al.*, 2023).

Despite their similarities, land use and land cover have distinct meanings. Land use reflects human decisions about how land will be used and is an aspect of human activity. The vegetative characteristics or man-made structures on the surface of the land are referred to as the land cover. For instance, land cover may have changed following a timber harvest, but the area's use will not change if seedlings are planted or natural regeneration takes place, and it will continue to be used to produce timber. Estimates of land use and land cover are frequently produced using a variety of approaches. Surveys based on field observations or enumeration typically determine land use, whereas remote sensing techniques or aerial photography interpretations typically determine land cover. Land cover is the observed biophysical cover on the Earth's surface, whereas land use is characterized by the arrangements, activities, and inputs that people perform in a particular type of land cover to produce, change, or maintain it (Gulati *et al.*, 2023; Farooq *et al.*, 2022).

The surface of the Earth has been altered or transformed by humans since agriculture was developed approximately 10,000 years ago. The biosphere has been altered more by agricultural production than by any other land use type (Tomar *et al.*, 2023; Singh *et al.*, 2022). Only 5% of the world's land, including parks and preserves, is left unmanaged, while it is estimated that 50% of the land is used for agriculture and animal production. Agriculture has been the primary means by which humans have changed their lands. Globally, 20% of forests and woodlands, 1% of grasslands and pastures (although most grasslands have been converted to pastures), and 466% of croplands have been lost over the past 300 years. Even though land-use practices vary widely worldwide, the result is typically the same: the purchase of natural resources to meet immediate human requirements, frequently at the expense of deteriorating environmental conditions.

Changes in land cover also affect regional climates through changes in surface energy and water balance. In addition, humans have altered the hydrological cycle to supply irrigation, industry, and domestic consumption with fresh water. In addition, fertilizer and atmospheric pollution-derived anthropogenic nutrient inputs to the biosphere now exceed natural sources and have a significant impact on water quality, as well as coastal and freshwater ecosystems. Through the destruction, alteration, and fragmentation of habitats, land use has also contributed to a decline in biodiversity; water and soil degradation; and excessive use of native species. Despite its success in increasing food production, modern agriculture has caused significant harm to the environment. In many places, for instance, the use of fertilizers has deteriorated the quality of water. Additionally, some irrigated lands have experienced severe salinization, resulting in an annual loss of approximately 1.5 million hectares of arable land worldwide, and an estimated $11 billion in production. Soil erosion decreases fertility, and overgrazing may affect up to 40% of croplands worldwide. By diminishing the services provided by pollinators, particularly bees, the destruction of native habitats also affects agricultural production. Essentially, modern agricultural land use practices may sacrifice ecosystem services, many of which are crucial to agriculture, in exchange for short-term increases in food production.

The main land use and land cover categories found in the study area are Settlement/Built upland, Agricultural land, Fallow land, Water bodies, Open land, and Forest.

7.4.1 Watershed KR 55

The land use patterns of watersheds KR 55, KR 63, and KR 64, which are in Shirol Tahsil, were studied. The land use pattern of watershed KR 55 is shown in Table 7.2 and Figures 7.2, 7.3. It is observed that out of 120.97 km^2 Area 11.45 km^2 is used for settlement which comprises to 9.47%. Agriculture is 50.78 km^2. Percentage-wise, this category uses more land area, which is 41.99%. Followed by fallow land having 40.95 km^2. The open land in the study area is 16.47 km^2, which is 13.62%. The watershed water body occupies only 1.27 km^2, which is 1.05%. There was no forest cover in the watershed.

Table 7.2 Land use statistics of watershed KR 55.

Sr. no.	Land use category	Area in km^2	Percentage (%)
1	Settlement	11.45	9.47
2	Agriculture	50.78	41.99
3	Fallow land	40.95	33.87
4	Water body	1.27	1.05
5	Open land	16.47	13.62
6	Forest	0.00	0.00
	Total	120.92	100.00

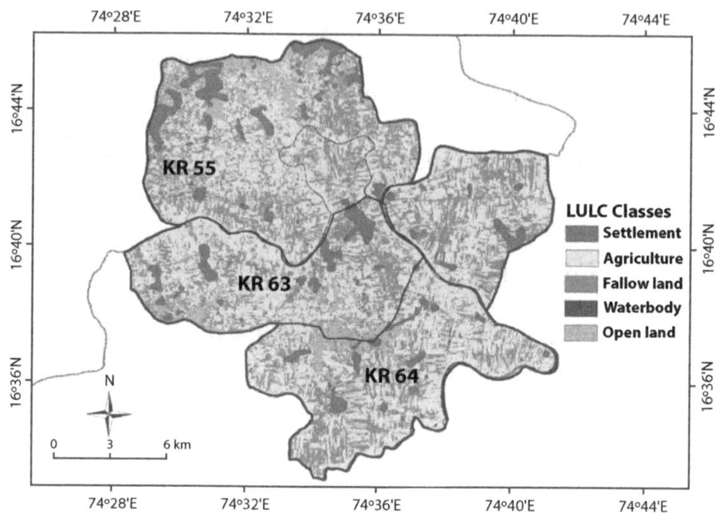

Figure 7.2 Land use land cover pattern in watersheds KR 55, KR 63, and KR 64.

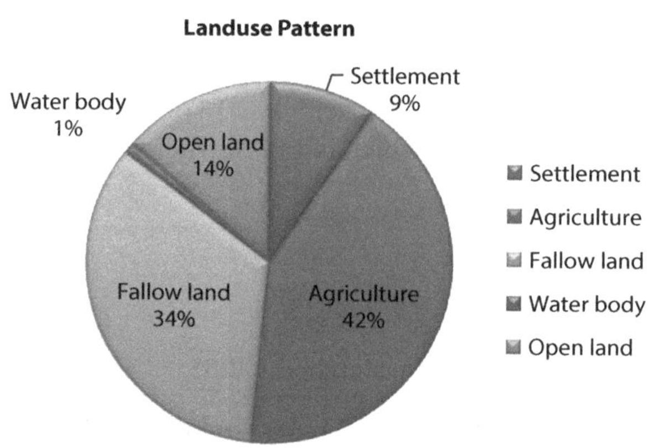

Figure 7.3 Land use pattern in watershed KR 55.

7.4.2 Watershed KR 63

The land use pattern of watershed KR 63 is shown in Table 7.3 and Figures 7.2, 7.4. It was observed that out of 78.67 km^2 area agriculture is spread over 37.12 km^2 comprising 47.18% of the watershed area. Fallow land covers 24.87 km^2 constituting 31.61%. This is followed by open land which covers 9.16 km^2 constitutes 11.64%. Settlement is spread over 6.50 km^2 area and the water body cover only 1.02 km^2. There was no forest cover in the watershed.

Table 7.3 Land use statistics of watershed KR 63.

Sr. no.	Land use category	Area in km²	Percentage (%)
1	Settlement	6.50	8.26
2	Agriculture	37.12	47.18
3	Fallow land	24.87	31.61
4	Water body	1.02	1.30
5	Open land	9.16	11.64
6	Forest	0.00	0.00
	Total	**78.67**	**100.00**

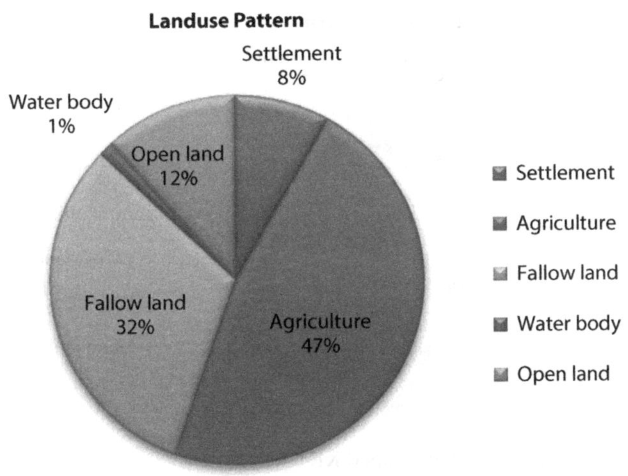

Figure 7.4 Land use pattern in watershed KR 63.

7.4.3 Watershed KR 64

The land use pattern in watershed KR 64 is shown in Table 7.4 and Figures 7.2, 7.5. The total area of the watershed is 134.69 km². Agriculture is spread over 66.02 km² area constituting 49% of the total area followed by fallow land over 48.68 km², which comes to 36%. Open land in the watershed is 11.23 km² and settlements are spread over 6.40 km² constituting about 5%. Open land covers over 11.23 km² constituting 8% of the watershed area. There was no forest cover in the watershed.

Table 7.4 Land use statistics of watershed KR 64.

Sr. no.	Land use category	Area in km^2	Percentage (%)
1	Settlement	6.40	4.75
2	Agriculture	66.02	49.02
3	Fallow land	48.68	36.14
4	Water body	2.36	1.75
5	Open land	11.23	8.34
6	Forest	0.00	0.00
	Total	**134.69**	**100.00**

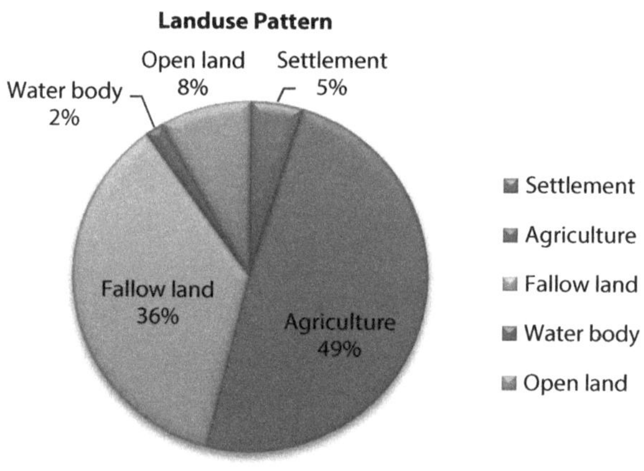

Figure 7.5 Land use pattern in watershed KR 64.

7.4.4 Watershed KR 66

The land use pattern of watershed KR 66, which is in Bhudargad Tahsil, was studied. The land use pattern of watershed KR 66 is shown in Table 7.5 and Figures 7.6, 7.7. It is observed that out of 165.13 km^2 area 81.59 km^2 is open land which constitutes 49.41% this means almost half the area of the watershed is unused. This is followed by forest, which covers an area of 56.7 km^2 this constitutes 34.35%. Agriculture is 14.43 km^2. Percentage-wise, this is 8.74%. Fallow land having 5.69 km^2, i.e., 3.45%. Water body occupies 3.89 km^2, area i.e., 2.36%, and settlement was only 2.8 km^2 area which is 1.70% of the watershed area.

Table 7.5 Land use statistics of watershed KR 66.

Sr. no.	Land use category	Area in km^2	Percentage (%)
1	Settlement	2.81	1.70
2	Agriculture	14.43	8.74
3	Fallow land	5.69	3.45
4	Water body	3.89	2.36
5	Open land	81.59	49.41
6	Forest	56.72	34.35
	Total	**165.13**	**100.00**

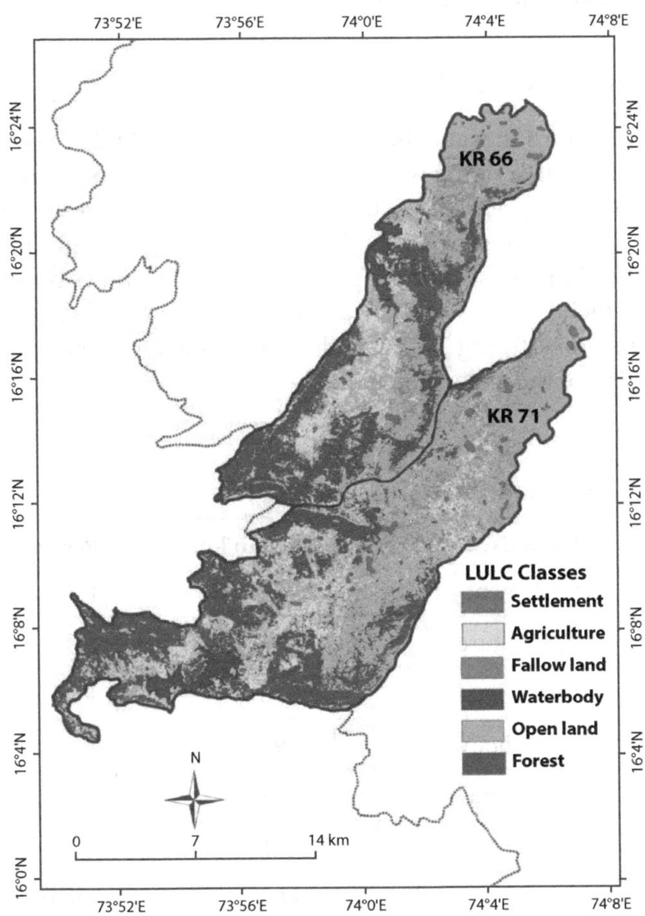

Figure 7.6 Land use land cover pattern in watersheds KR 66 and 71.

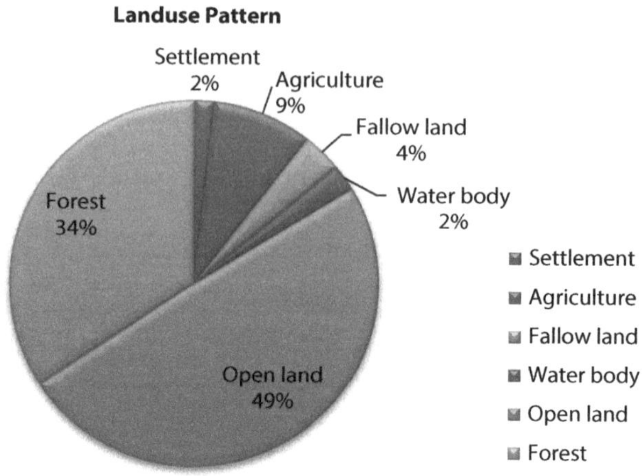

Figure 7.7 Land use pattern in watershed KR 66.

7.4.5 Watershed KR 71

The land use pattern in watershed KR 71 is shown in Table 7.6 and Figures 7.8, 7.9. The total area of the watershed is 249.57 km^2. Open land is 125.56 km^2 area constituting 50% of the total area, followed by forest cover of over 77.41 km^2, which comes to 31%. This is followed by an agricultural area covering 25.15 km^2 constituting 10.08%, and fallow land covers an area of 11.18 km^2, which is 4.48% of the watershed area. The waterbody is spread over 5.79 km^2 constituting 2.32%, and the settlement covers 4.48 km^2 constituting 1.80% of the watershed area.

Table 7.6 Land use statistics of watershed KR 71.

Sr. no.	Land use category	Area in km^2	Percentage (%)
1	Settlement	4.48	1.80
2	Agriculture	25.15	10.08
3	Fallow land	11.18	4.48
4	Water body	5.79	2.32
5	Open land	125.56	50.31
6	Forest	77.41	31.02
	Total	**249.57**	**100.00**

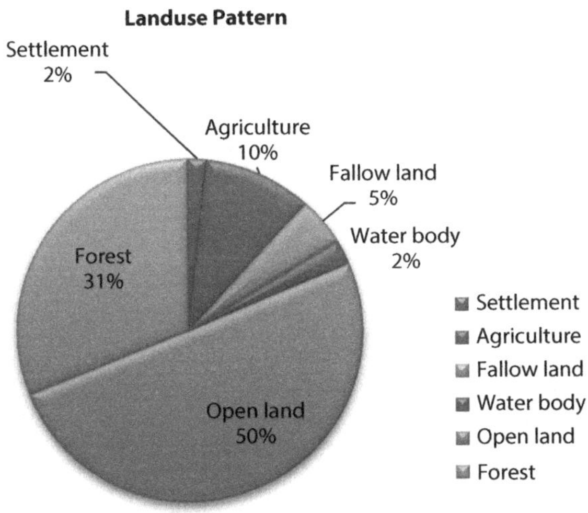

Figure 7.8 Land use pattern in watershed KR 71.

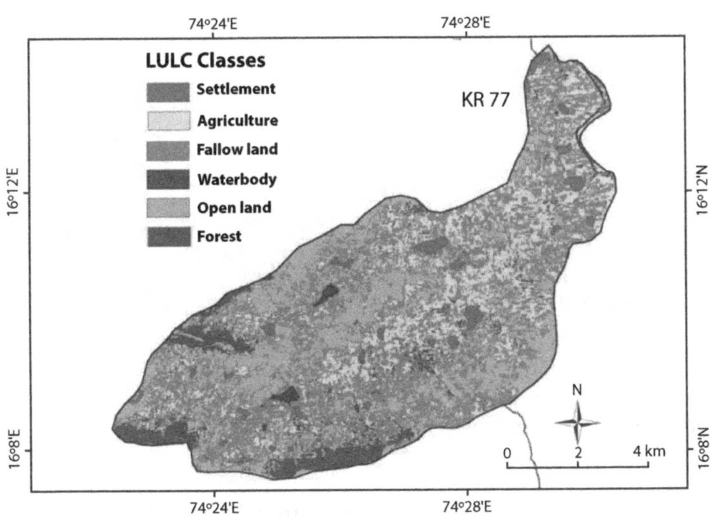

Figure 7.9 Land use land cover pattern in watershed KR 77.

7.4.6 Watershed KR 77

The land use pattern of watershed KR 77 is in the Gadhinglaj Tahsil. The land use pattern of watershed KR 77 is shown in Table 7.7 and Figure 7.10. It is observed that out of 78.97 km^2 Area 2.21 km^2 is used for settlement,

Table 7.7 Land use statistics of watershed KR 77.

Sr. no.	Land use category	Area in km^2	Percentage (%)
1	Settlement	2.21	2.80
2	Agriculture	13.79	17.46
3	Fallow land	37.53	47.52
4	Water body	0.78	0.99
5	Open land	18.74	23.73
6	Forest	5.92	7.50
7	**Total**	**78.97**	**100.00**

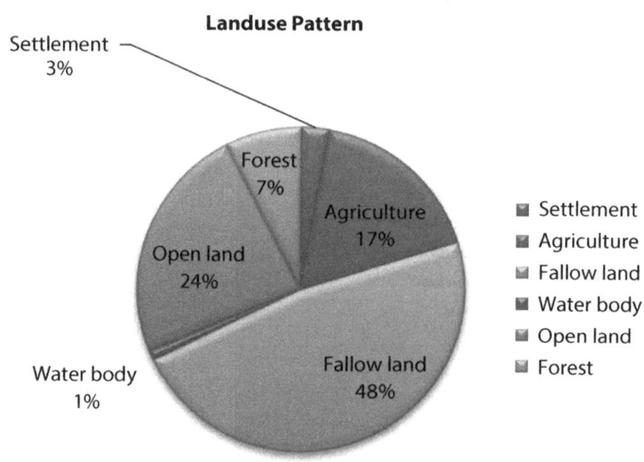

Figure 7.10 Land use pattern in watershed KR 77.

which comprises to 2.80% of the total area. Agriculture is 13.79 km^2. Percentage-wise. this category used 17.46% of the total area. Fallow land is having an area of 37.53 km^2, i.e., 47.52% of the watershed area, followed by Open land spread over an area 18.74 km^2, this is 23.73% of the watershed area. Forest cover is spread over 5.92 km^2 area constituting 7.50% of the watershed area. In the watershed water body occupies only 0.78 km^2 area which is 0.99% of the watershed area.

7.5 Conclusion

Agriculture is the major land use pattern in the district. Agricultural activities are observed to be more in the eastern part of the study area i.e., in watersheds KR 55, KR 63, and KR 64 as compared to the watersheds KR 66 and KR 71. This may be due to hilly areas with forest cover. The eastern part of the KR 77 watershed is cultivated more than the western hilly region. Sugarcane is the main cash crop in all the watersheds. Crops are cultivated in both seasons along the rivers. Thick forest cover, categorized as evergreen forest, is present adjoining the western boundary of watersheds KR 66 and KR 71. Deciduous forest also occurs in patches in these watersheds with degraded forest found intermittently in smaller patches. Land with or without scrub is observed in large patches all over the study area. Small patches of barren rocky waste land are observed to be present in watersheds KR 66, KR 71, and KR 77.

References

Singh, S., Singh, S. K., Kumar, R., Shrama, A., & Kanga, S. (2022). *Finding Alternative to River Sand in Building Construction.*

Singh, S., Meraj, G., Kumar, P., Singh, S. K., Kanga, S., Johnson, (2023). Decoding Chambal River Shoreline Transformations: A Comprehensive Analysis Using Remote Sensing, GIS, and DSAS. *Water*, 15(9), 1793.

Singh SK, Kanga S, Gulati B, Raiˇc M, Sajan B, Đurin B, Singh S (2023) Spatial and Temporal Analysis of Hydrological Modelling in the Beas Basin Using SWAT+ Model. *Water* 15, 3338. https://doi.org/10.3390/w1519333.

Rani A, Gupta SK, Singh SK, Meraj G, Kumar P, Kanga S, Đurin B, Dogančić D (2023) Predicting Future Land Use Utilizing Economic and Land Surface Parameters with ANN and Markov Chain Models. *Earth* 4(3):728-751.

Ahmad T, Gupta SK, Singh SK, Meraj G, Kumar P, Kanga S (2023) Unveiling Nature's Resilience: Exploring Vegetation Dynamics during the COVID-19 Era in Jharkhand, India, with the Google Earth Engine. *Climate* 11(9):187.

Sharma M, Upadhyay RK, Tripathi G, Kishore N, Shakya A, Meraj G, Kanga S, Singh SK, Kumar P, Johnson BA, Thakur SN (2023) Assessing Landslide Susceptibility along India's National Highway 58: A Comprehensive Approach Integrating Remote Sensing, GIS, and Logistic Regression Analysis. *Conservation* 3(3):444-459.

Saqib N, Rai PK, Kanga S, Kumar D, Đurin B, Singh SK (2023) Assessment of Ground Water Quality of Lucknow City under GIS Framework Using Water Quality Index (WQI). *Water* 15(17):3048.

Choudhury U, Singh SK, Kumar A, Meraj G, Kumar P, Kanga S (2023) Assessing land use/land cover changes and urban heat island intensification: A case study of Kamrup Metropolitan District, Northeast India (2000–2032). *Earth* 4(3):503-521.

Rai NK, Mathur S, Singh SK, Tiwari M, Singh VK, Haque R, Tiwari S, Sharma LK (2023) Differential regulation of mitochondrial complex I and oxidative stress based on metastatic potential of colorectal cancer cells. *Oncol. Lett.* 26(1):1-1.

Debnath J, Sahariah D, Lahon D, Nath N, Chand K, Meraj G, Kumar P, Singh SK, Kanga S, Farooq M (2023) Assessing the impacts of current and future changes of the planforms of river Brahmaputra on its land use-land cover. *Geosci. Front.* 14(4):101557. Elsevier.

Kumar P, Singh SK, Gupta SK, Kanga S, Meraj G (2023) Uncovering the hydro-meteorological drivers responsible for forest fires utilizing geospatial techniques. *Theor. Appl. Climatol.* https://doi.org/10.1007/s00704-023-04497. Springer Nature.

Singh SK, Kotecha MJ, Kanga S, Pankhaniya SK, Agrawal S, Meraj G (2023) Urban Heat Island (UHI) Resilience Plan in Varying Climatic Conditions Using Geospatial Approach: A Case Study of Rajkot City. In: *Climate Change, Agriculture and Society*, vol 1. pp 307-338.

Singh S, Singh SK, Prajapat DK, Pandey V, Kanga S, Kumar P, Meraj G (2023) Assessing the Impact of the 2004 Indian Ocean Tsunami on South Andaman's Coastal Shoreline: A Geospatial Analysis of Erosion and Accretion Patterns. *J. Mar. Sci. Eng.* 11(6):1134.

Rathore SS, Mathur SC, Mathur S, Sharma S, Singh SK (2023) Late Cretaceous–Early Paleocene Geodiversity of Fatehgarh Formation of Petroliferous Barmer Basin, Western Rajasthan, India: A Potential Geopark Site. *Geoheritage* 15:54.

Singh S, Meraj G, Kumar P, Singh SK, Kanga S, Johnson BA, Prajapat DK, Debnath J, Sahariah D (2023) Decoding Chambal River Shoreline Transformations: A Comprehensive Analysis Using Remote Sensing, GIS, and DSAS. *Water* 15(9):1793.

Tripathi G, Shakya A, Upadhyay RK, Singh SK, Kanga S, Pandey SK (2023) Landslide Susceptibility Mapping of Tehri Reservoir Region Using Geospatial Approach. In: *Climate Change Adaptation, Risk Management and Sustainable Practices in the Himalaya*, pp 135-156.

Sud A, Kanga R, Singh SK, Meraj G, Kanga S, Kumar P, Ramanathan AL, Bhardwaj V (2023) Simulating Groundwater Potential Zones in Mountainous Indian Himalayas—A Case Study of Himachal Pradesh. *Hydrology* 10(3):65.

Upadhyay RK, Tripathi G, Đurin B, Šamanović S, Cetl V, Kishore N, Sharma M, Singh SK, Kanga S, Wasim Md, Rai PK, Bhardwaj V (2023) Groundwater Potential Zone Mapping in the Ghaggar River Basin, North-West India, Using Integrated Remote Sensing and GIS Techniques. *Water* 15(5):961.

Lavane K, Kumar P, Meraj G, Han TG, Ngan LHB, Lien BTB, Ty TV, Thanh NT, Downes NK, Nam NGD, Minh HVT, Singh SK, Kanga S (2023) Assessing the effects of drought on rice yields in the Mekong Delta. *Climate* 11(1):13.

Nath N, Sahariah D, Meraj G, Debnath J, Kumar P, Lahon D, Chand K, Farooq M, Chandan P, Singh SK, Kanga S (2023) Land use and land cover change monitoring and prediction of a UNESCO world heritage site: Kaziranga eco-sensitive zone using cellular automata-Markov model. *Land* 12(1):151.

Kanga S, Upadhyay RK, Sharma M, Sharma S, Tripathi G, Singh SK (2023) Geotechnical Stability Assessment of Road-cut Slopes: A Case Study of Srinagar, Garhwal Himalaya, India. *Biol. Forum* 15(5):1071-1080.

Gulati B, Sharma R, Kanga S, Singh SK, Sajan B, Meraj G, Kumar P, Ramanathan AL (2023) Unraveling the Relationship Between Stubble Burning and Air Quality Degradation in Punjab: A Temporal and Spatial Analysis (2019-2022). *J. Clim. Change* 9(2):43-53, IOS Press.

Tomar P, Singh SK, Kanga S, Pattanaik A (2023) Water Bodies mapping and monitoring using high-resolution satellite images. *Sustain. Agri, Food Environ. Res.* 11.

Singh H, Meraj G, Singh S, Shrivastava V, Sharma V, Farooq M, Kanga S, Singh SK, Kumar P (2022) Status of Air Pollution during COVID-19-Induced Lockdown in Delhi, India. *Atmosphere* 13(12):2090.

Farooq M, Mushtaq F, Meraj G, Singh SK, Kanga S, Gupta A, Kumar P, Singh D, Avtar R (2022) Strategic Slum Upgrading and Redevelopment Action Plan for Jammu City. Resources 11(12):120.[11] Singh, S., Singh, S. K., Prajapat, D. K., Pandey, V., Kanga, S., Kumar, P., & Meraj, G. (2023). Assessing the Impact of the 2004 Indian Ocean Tsunami on South Andaman's Coastal Shoreline: A Geospatial Analysis of Erosion and Accretion Patterns. *J. Mar. Sci. Eng.*, 11(6), 1134.

8

Formulation and Mapping of GIS-Based Smart Village Plan Using Drone Imagery

Suraj Kumar Singh[1], Shruti Kanga[2]*, Sayeed Afridi[1], Bhartendu Sajan[2], Saurabh Singh[3] and Gowhar Meraj[4]

[1]*Centre for Climate Change and Water Research, Suresh Gyan Vihar University, Jaipur, India*
[2]*Department of Geography, School of Environment and Earth Sciences, Central University, Punjab, India*
[3]*Department of Civil Engineering, Suresh Gyan Vihar University, Jaipur, India*
[4]*Department of Ecosystem Studies, Graduate School of Agricultural and Life Sciences, The University of Tokyo, Tokyo, Japan*

Abstract
Village maps can give geospatial data and information to become an essential tool in regional development planning. The is a study aimed at producing geospatial data and information on the resources of the villages, based on the reference directives of energy, knowledge, agriculture, healthcare, technology to achieve the desired goal related to the development of the village. The study was carried out at Dungarpur Reelka village, India, as it lacks education, health, employment, water supply, sewage, gas, agriculture, waste collection, transportation, environment and social, etc., at the village level. Raster data consisting of images collected from unmanned vehicles with high resolution were analyzed, and various thematic layers like land use/landcover, digital elevation model, slope, aspect, base map were generated. Spatial analysis was conducted using the licensed ArcGIS 10.5 software with limited field checks and community interviews. The space information in the village resource and border map can be viewed as a breakthrough step and might be used as a pilot by other communities.

Keywords: Rural development, smart village, drone image, GIS, facilities

**Corresponding author*: shruti.kanga@cup.edu.in

Shruti Kanga, Gowhar Meraj, Suraj Kumar Singh, Majid Farooq and M. S. Nathawat (eds.) Geospatial Technology for Natural Resource Management, (207–236) © 2024 Scrivener Publishing LLC

8.1 Introduction

The population growth of the urban population has been relatively quick, and the percentage of the Indian population living in urban areas has more than doubled from 14% at the time of independence to 31.8% in 2011. (Census of India). This is predicted to climb by around 40% by 2026 and by more than 50% by 2051 (Bera *et al.* 2021). Most urban settlements, particularly smaller urban settlements, are marked by haphazard and unplanned growth, incoherent land use, mushrooming unauthorized colonies, and land transformation from agriculture to urban areas leading to environmental degradation and low living quality. Proper administration of the urban regions requires regular information that is reliable and essential (Joy *et al.* 2021). The master plan/development plan is the primary tool for managing urban areas, which provides a precise allocation of urban/urban sustainable development. Most master/development plans are drawn up for 20 years and periodic review and change in 5-year intervals. Master plan preparation starts with the base map, current land use surveys, and the collecting of socio-economic data needed to assess the current situation and submit future land use plans (Kanga *et al.* 2021). With the advances made in remote sensing and geographic information systems, the plan-making process is accelerated by integrating the spatial and attribute-related data, which allow for a thorough assessment of the spatial growth of cities, land utilization status, physical infrastructure facilities, etc. (Adesipo *et al.* 2020). The most critical information for the construction of the master plan is an updated, precise basis map of the planning area, displaying roads and layouts, spatial expansion, and use the information for each parcel of land, etc. (Kumar *et al.* 2021). Preparing the basis maps for very high-resolution satellite (VHRS) images with geographic information system (GIS) technology can be both time and cost-efficient. Our development strategy uses a top-down approach at the national level, and satellites give macro data for influencing policy decisions (Meraj *et al.* 2021). On these maps, the villages are smaller than the thumbtacks that denote their positions. The maps do not provide local geography details, and hence decision-makers are obliged to call for and start great efforts with little or no data (Tomar *et al.* 2021). As a nation, our technical advance is incomplete until it reaches the villages. Our team was keen to deliver technology breakthroughs to India in the villages of Bharat. We visited the villages, carried out anthropological studies, and took the time to grasp the administrative framework involved in decision-making. We realized that precise data on localized locations was needed to support planning and development at the grassroots level.

The soul of India lives in its villages - Mahatma Gandhi. From Mahatma Gandhi's vision of the ideal village, "smart village" is an idea adopted by India's national, state and local governments as an initiative focusing on holistic rural development. Under the energetic, committed, and innovative leadership of Prime Minister Narendra Modi, the Government of India, is working on the Smart Cities and Smart Villages program, which is a good initiative (Kanga *et al.* 2021). The concept was created in 2010 by describing a village ecosystem, and mapping integrated design methods for building a smart village. In the present world era, the smart village has become another development concept of rural areas. Since "the smart village" concept was created, it is a solution to improve the quality of life of people in rural areas. The smart village ecosystem covers four aspects; (i) organization, (ii) resources, (iii) service chains, and (iv) service delivery technologies and processes. There were areas of focus in smart villages, including economy, people, information and communication technology (ICT), administration, environment, livelihood, and energy (Zavratnik *et al.* 2020). The inspiration behind "smart village" was to develop technology, improve education, create local business opportunities, improve health and wellness, increase democratic involvement and work to improve the living standards of villagers. According to statisticians, most India's population lives in villages, so smart villages are as crucial as smart cities. The growing population and rapid urbanization are one of India's massive challenges (Saurabh *et al.* 2021). A smart village-like smart city should be interactive and multifaceted; people need to participate in various activities. In the present context, the concept of a smart village seems more reasonable as there is a limit to the growth of cities resulting in the urban jungle, where the proportion of population per kilometer of land is much higher than the desired rule (Stojanova *et al.* 2021). Initially, it requires leading to a campaign at the national level as baby steps, then once this effort starts bearing fruits, there surely would be activities. Just like smart cities, a smart village should have active participation of people in various activities. Smart village is one that will automatically combine with local production and local distribution. A smart village will also have energy, knowledge, healthcare, technology, entrepreneurship, and internet connectivity (Adamowicz and Zwolińska-Ligaj, 2020).

Smart villages aim to have homes with toilets, safe drinking water, and regular electricity access. A smart village is about its citizens, available resources, and knows all the information about the services and schemes (Zhang and Zhang 2020). Each household has different livelihood opportunities or micro-enterprises, maintains its identity, culture, and heritage, and tracks development plans and their progress based on the person, resource, and service-centric information (Komorowski and Stanny 2020). It works towards revenue generation and has a functional solid/liquid waste management system to end preventable maternal deaths and infant deaths, which means providing good basic health facilities in the health-care center, 100% institutional deliveries, and communicates for the needs of government, NGOs, social entrepreneurs, and experts (Adesipo *et al.* 2020); the functional toilet, potable water electricity available in schools, health center; awareness of new technologies that can be implemented in villages, farms, and nearby places. *i.e.*, solar panel light systems like drip irrigation, and streetlights (Zavratnik and Stojmenova 2018).

The primary data extracted from these maps are identifying the orthomosaic village boundary and using photogrammetry to calculate the length of the highways, the land area, and the volume of pools. Furthermore, this allows different officials who may not be present on the site to share and cooperate (Mehta *et al.* 2021). UAVs or drones are referred to as unmanned planes or flying devices able to fly into the sky with a camera on them and be controlled independently, semi-autonomously, or manually (Spanò *et al.* 2018). With a camera installed on drones, high-resolution pictures of the floor can be obtained, and accurate, high-resolution maps can be analyzed. This map contains all the information obtained by a typical soil survey. The total time needed to map drones is also lower than traditional aerial methods. Many additional topographical and cadastral mapping surveys beyond Nepal have been used with UAVs. UAVs are used widely to build accurate maps cost-effectively and quickly. In addition to the intangible cultural complex linked with local resources, the harmonious development of an area and the exploitation of natural, environmental, and landscaping resources can become a modern engine of environmentally sustainable development (Cvar *et al.* 2020). With this in mind, landscaped terraces that constitute a pointed line in a much broader, virtuous design may likely be subject to reflections and tests supporting their promotion and reactivation. In recent years, the usage of UAV platforms for detection has been developing suddenly and increasingly in many various fields of application. Synthesis contributions are accessible to emphasize the possibility for the mapping, monitoring, and construction of 3D models, reporting multiple platform solutions and data from various sensors with which the vector is equipped. The literature that recommends the implementation of UAV systems in environmental sciences is agriculture and forestry, geomorphology, and multisensory vegetation surveillance. This chapter aims to create a base layer of existing land-use and landcover of the Dungarpur Reelka village. The study analyzes the current rural development scenario through the existing site survey and investigation (gap analysis) of Dungarpur Reelka village and finally creates an existing proposed smart village plan or land-use and landcover map to make Dungarpur Reelka village smart.

8.2 Study Area

Dungarpur Reelka is a village located in Gautam Buddha Nagar district in Uttar Pradesh, India, as shown in Figure 8.1. The Dungarpur Reelka village is located at 26.55° north latitude and 80.69° east longitude. It is situated 19 km away from the sub-district headquarter, Gautam Buddha Nagar. The total geographical region of the village is 357.52 hectares. There are about 433 houses in the village. It has an average elevation of 202 m. Figure 8.1 illustrates the geographical location of Dungarpur Reelka village. The rainfall distribution from June to September is 600 mm, followed by July when average rainfall is around 194.4 mm. The total number of houses is 433, with a literacy rate of 63.59 %, and the entire population is 2494.

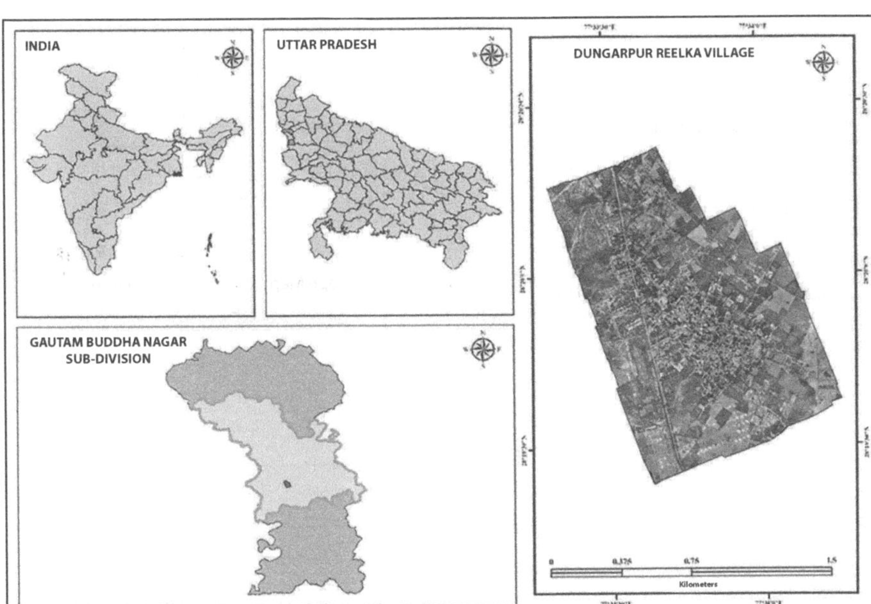

Figure 8.1 Geographical location of Dungarpur Reelka village.

8.3 Materials and Methods

Agisoft Viewer's software for drone image processing and GIS for base map and land use/land cover class classification. A pilot study was conducted at Dungarpur Reelka in Gautam Buddha Nagar tehsil of Uttar Pradesh to explore the geospatial technology in information and social research. The purpose was to understand the technological feasibility, scalability and gauge challenges arising in the process. Drones were used to create detailed images with geographic location altitude profiles, with a family-level survey that was conducted along with granular-level data from land references collected on land. The drone's name is Phantom 4 Pro, by which objects are captured by moving at high speed, which has continually been a challenge for flying cameras with an electronic shutter. The DJI Phantom 4 Pro drone flies at the height of 100 m, with a spatial resolution of 3 cm (Figure 8.2).

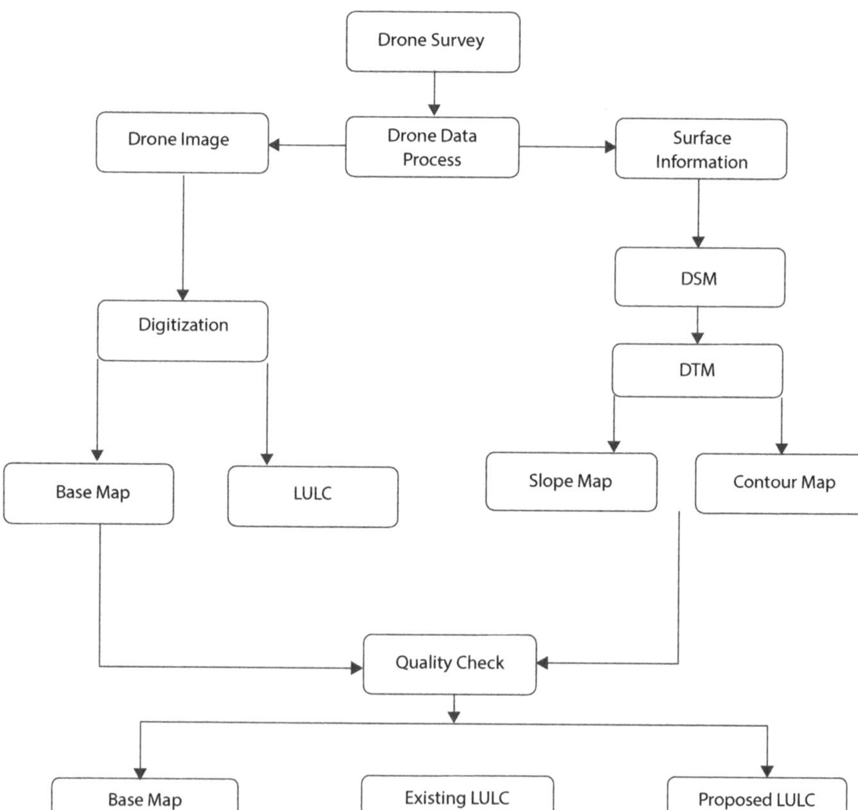

Figure 8.2 Methodology workflow.

8.3.1 Existing Site Survey and Investigation – Gap Analysis

8.3.1.1 Physical Infrastructure

The villages under investigation have limited physical infrastructure components such as water supply, sewage, sanitation, drainage, solid waste management, parks and playgrounds, and street lighting. The villagers are dependent on groundwater for all their daily needs. Unchecked use of bore wells in the villages is leading to depletion of the water table.

8.3.1.2 Sewage and Drainage

There is no effective system for drainage. Figures below show the status of drainage as observed on site. Since cattle breeding still abounds in the village, one can see cattle dung on most streets and no practical carriage by traditional methods. The villagers are dependent on septic tanks for the disposal of sewage as there are no sewage lines. Untreated wastewater is disposed of in the open spaces that would infiltrate and contaminate the groundwater.

8.3.1.3 Sanitation

There are insignificant cases of open defecation, as villagers seem to be aware of basic hygiene. There is the absence of community toilets in all the villages studied, as depicted in Figures 8.3, 8.4 and 8.5. Figure 8.3 shows the condition of water infrastructure in the villages. Figure 8.4 illustrates the sewage and drainage conditions observed on site, while Figure 8.5 highlights the sanitation system in the villages.

Figure 8.3 Condition of water infrastructure in villages.

Figure 8.4 Sewage and drainage conditions in villages.

Figure 8.5 The sanitation system in villages.

8.3.1.4 Power Supply

Villages receive irregular and low voltage supply of electricity through the network of open cables. Most of the installed solar lights are dysfunctional, leading to poorly lit streets.

8.3.1.5 Solid Waste Management

High population density poses problems for the collection and transport of municipal waste due to narrow streets, as illustrated in Figures 8.6 and 8.7. Figure 8.6 shows electric meters, solar street lights, and transformers, while Figure 8.7 depicts garbage dumped on vacant land. There has been no effort to manage solid waste thrown or dumped in the corners, roads, and other vacant sites. Dumping in the open creates unhygienic conditions

Figure 8.6 Electric meter, solar street light, and transformer.

Figure 8.7 Garbage dumped on vacant land.

of living and degradation of groundwater quality due to improper leachate management, increasing heaps of waste at disposal sites, and dumping garbage in drains.

8.3.1.6 Public Health

In this village where drainage and sanitation are inadequate, water drips over the ground during rains, encouraging and polluting the water. It contributes significantly to the spread of diseases such as typhoid and cholera, and the onset of insect infections.

8.3.1.7 Road Infrastructure

Approximately 85% of all roads in these villages are either paved or made of concrete, as shown in Figure 8.8. Most major roads are in good condition.

Internal roads will need to be almost completely reworked since the laying of water and sewage systems will require extensive digging to lay pipes. The area is connected through a primary network of roads accessed through the Yamuna Expressway. The interior streets of the villages are laid using cement concrete interlocking pavers flanked by open drains.

8.3.1.8 Housing Infrastructure

It has the potential to improve the living standard of the people. The house in this village is in good condition, and most of the homes are made of concrete.

8.3.1.9 Health Services

There are no Government primary health centers operational in the villages. Private clinics offering basic first aid were visible in some of the villages.

8.3.1.10 Education Facilities

The villages have operational Government primary and middle schools. The toilets in most schools are dysfunctional, often vandalized due to the absence of the school boundary walls.

Approximately 85% of all roads in these villages are either paved or made of concrete, as shown in Figure 8.8. The utility map of Dungarpur Reelka village is depicted in Figure 8.9. Figure 8.10 shows a private health center, while Figure 8.11 illustrates the primary education infrastructure in the villages.

Figure 8.8 Road infrastructure in villages.

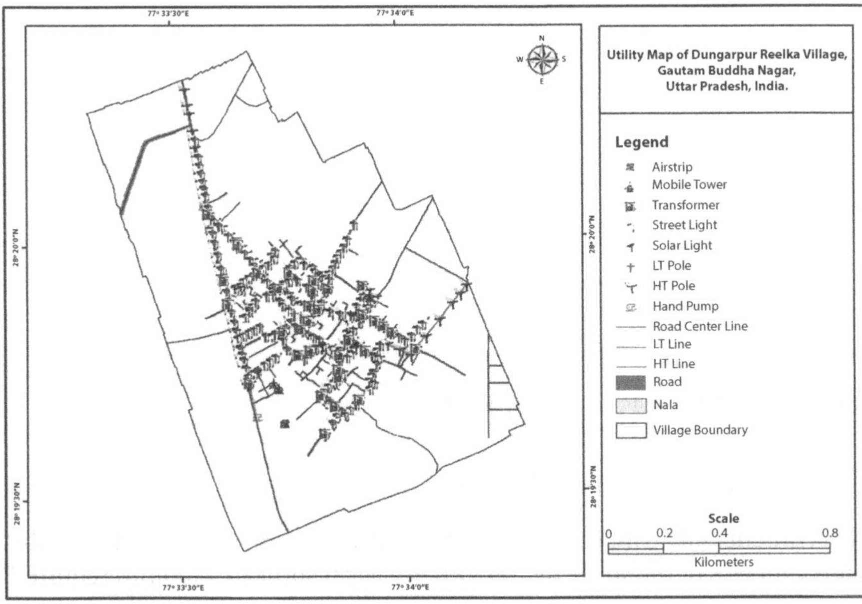

Figure 8.9 Utility map of Dungarpur Reelka village.

Figure 8.10 Private health center.

Figure 8.11 Primary education infrastructure of villages.

8.3.1.11 Maternity and Childcare

Government child care centers called Anganwadis are operational in all the visited villages of the study region, other basic healthcare facilities required for maternity services were not found in the villages.

8.3.1.12 Dungarpur Reelka Infrastructure Matrix

In Table 8.1, all this existing site survey gap analysis is structured and ranked by colors, red means poor—not existing, light green means average, and dark green means good.

Figure 8.12 shows an Anganwadi facility in villages Figures 8.13–8.15 represent the existing land use and land cover, base map, and relief map of Dungarpur Reelka village.

Table 8.1 Infrastructure matrix of the village.

No	Infrastructure matrix		Dungarpur Reelka
1	Solid waste management		
2	Drainage		
3	Sewage		
4	Water facilities		

(Continued)

Table 8.1 Infrastructure matrix of the village. (*Continued*)

No	Infrastructure matrix		Dungarpur Reelka
5	Electricity	Conventional	
		Renewable	
6	Road network		
7	Open spaces		
8	Educational Institutions	Primary school	
		Secondary school	
		Industrial Training Institute (ITI's)	
		Angan Wadi	
9	Medical facilities	Primary health center	
10	Gathering space	Local	
		Community hall	
11	Telecommunication		
12	Crematorium/burial site		
13	Markets		
14	Commercial institutions	Bank	
		ATM	
		Post office	
		Police station	
15	Transportation facilities		
Grading			
Good			
Average			
Poor/not existing			

Figure 8.12 An Anganwadi facility in villages.

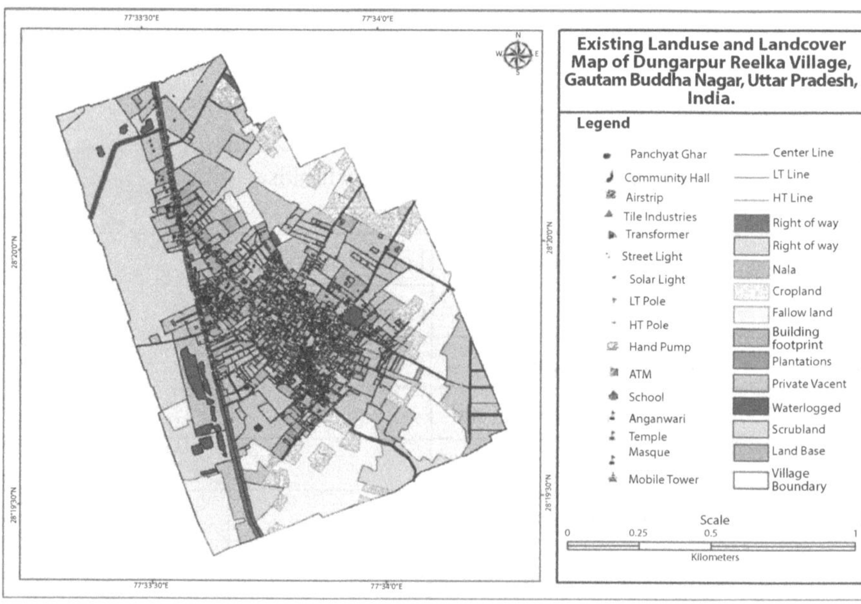

Figure 8.13 Existing land use and landcover of Dungarpur Reelka village.

8.4 Results and Discussion

Smart interventions will use digital technologies to provide villagers with access to many things such as medical and transport facilities, and skill development opportunities. I-CHAUPAL will be the center of information gathering and sharing. It will monitor various activities like solid waste management, water network, linkage, easy commuting, medical facility,

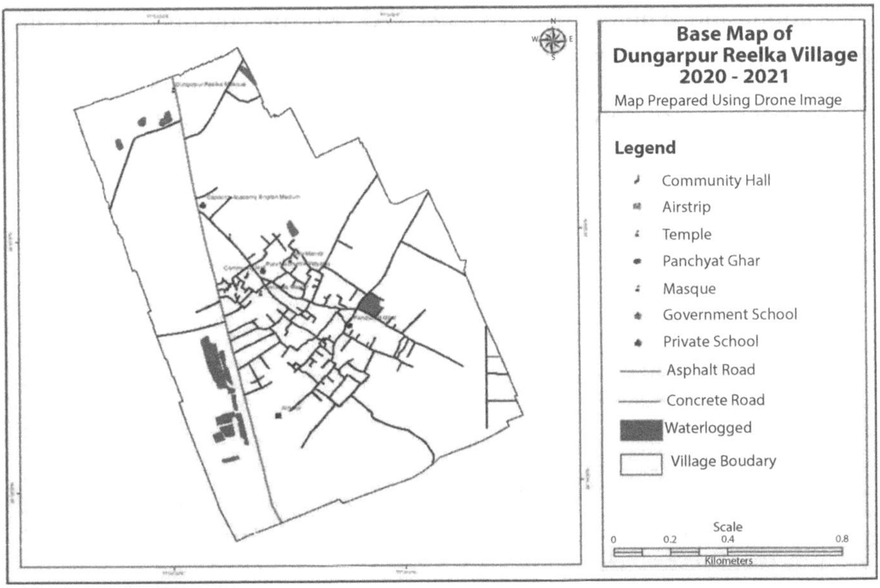

Figure 8.14 Base map of Dungarpur Reelka village.

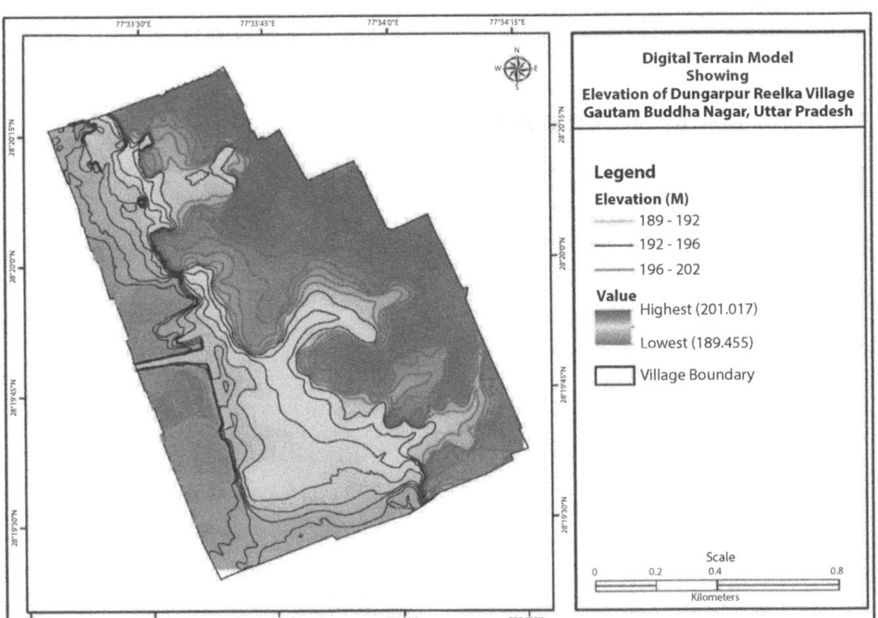

Figure 8.15 Relief map of Dungarpur Reelka village.

recreation, and employment. I-CHAUPAL will be equipped with computers, printers, internet, projector, UPS and will be manned by technical staff. I-CHAUPAL will also provide information about Wi-Fi and closed-circuit television (CCTV) coverage in the village. I-CHAUPAL will provide

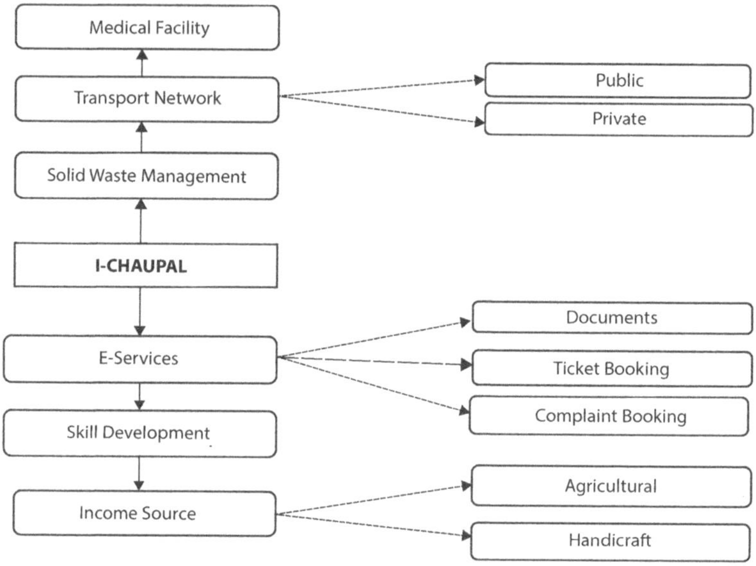

Figure 8.16 Flowchart of I-CHAUPAL.

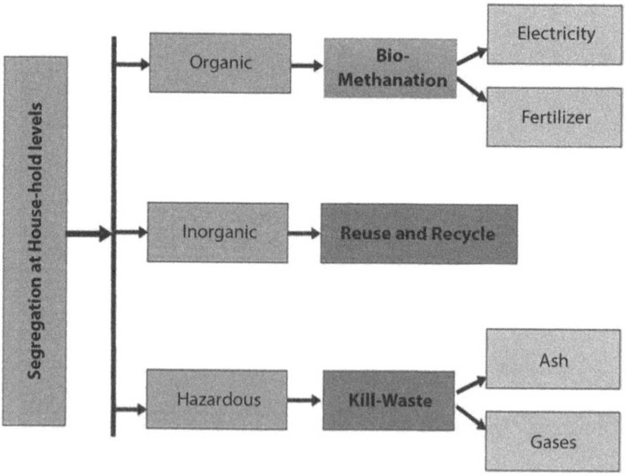

Figure 8.17 Solid waste management network.

information about mobile medical van facilities in village clusters. Figure 8.16 presents a flowchart of CHAUPAL. Villagers will use the I-CHAUPAL facility to connect with the digital interface of various government offices, banks, train ticket booking, etc. It will also enable the Skill Development Council and its affiliated organizations to reach out to the youth in the village for their skilling, training, and employment program. Micro-ATM may be installed, and space rented out to the service agency.

As per the policy and the desired objective, an agency will be set up to handle solid waste. The agency will do the following: The solid waste management network is illustrated in Figure 8.17.

8.4.1 Solid Waste Management

A Companion to the Facilitators of Swachh Bharat Mission (Gramin), by National Institute of Rural Development and Panchayati Raj. The Central Government of India gives this policy of solid waste management (SWM) under Swachh Bharat Abhiyan. It promotes waste segregation into three categories *i.e.*, biodegradable, non-biodegradable, and hazardous. Segregation of waste collected from various sources is done at two levels, primary level (at source segregation) and secondary segregation (at waste management center). The detailed process of waste segregation is elaborated in Figure 8.18. Further door-to-door collection of segregated waste will be performed, as well as segregation of waste into recyclable or saleable waste, and collection of user charges from each household, shops, schools, etc. Feeding the bio-degradable waste into the bio-methanation plant and maintaining the bio-methanation plant as per the installation and operation guidelines. Bio-methanation plant will be fitted with a generator to produce electricity from the gas. This electricity will be sold to the grid. I-CHAUPAL will monitor the entire process. The sizing and selection of bio-methanation and Incinerator technologies shall be decided by the agency managing the waste. When the bio-methanation's efficiency is reduced during the winter months, the generated waste can be stacked or fed into the kill-waste incinerator. Figure 8.19 demonstrate flow chart showing the process of bio-methanation.

8.4.2 Kill Waste

"Kill Waste" advanced fuel-free solid waste disposal is an eco-friendly and economical solution for solid waste disposal that operates on a controlled combustion policy through a controlled air supply and addresses major

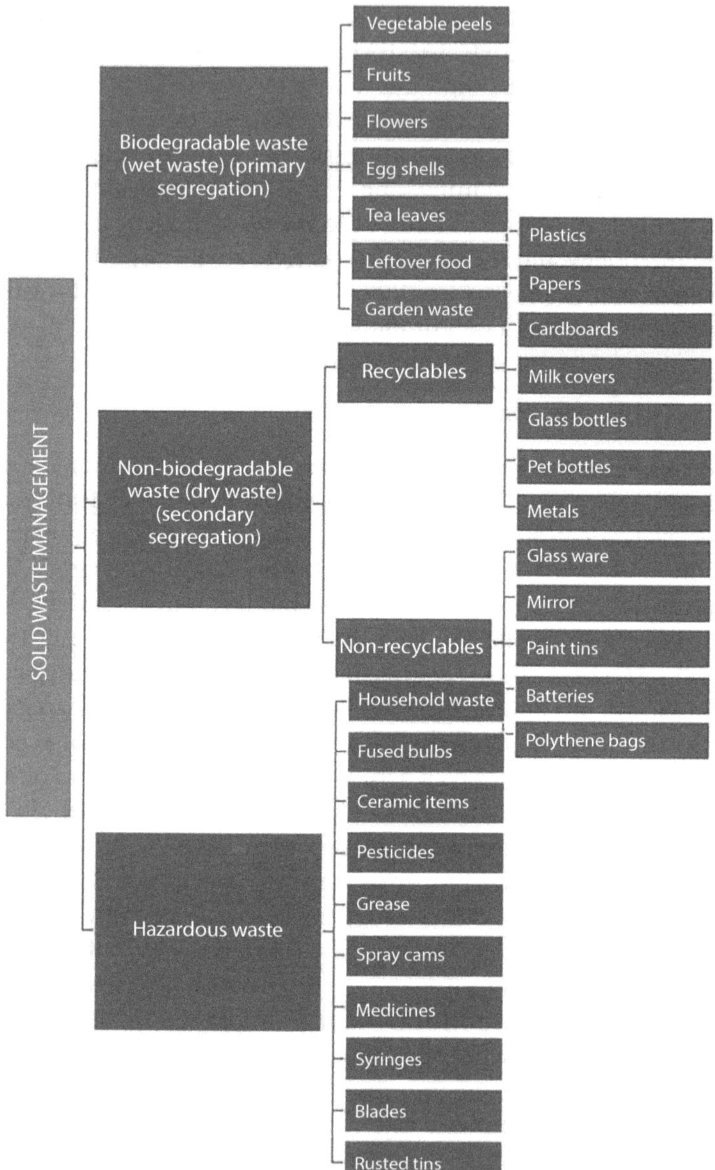

Figure 8.18 The schematic diagram for solid waste management.

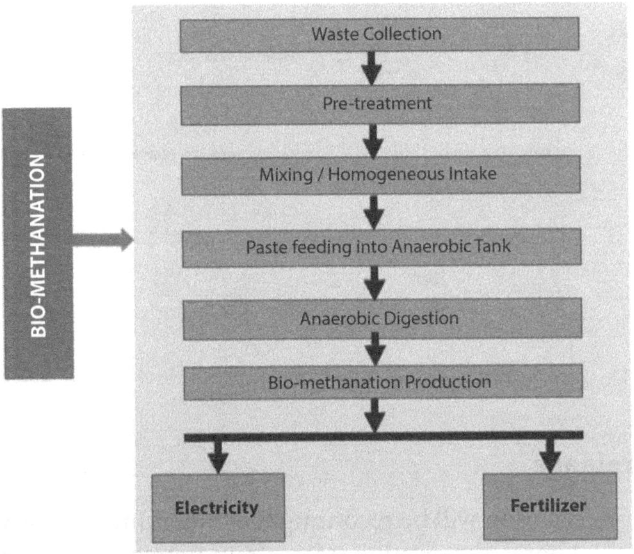

Figure 8.19 Flow chart showing the process of bio-methanation.

flaws in conventional combustion-based incinerator systems. Kill waste solid waste disposer works on air rotating technology that destroys general solid waste after simple segregation by controlled combustion using a scientifically regulated supply of atmospheric oxygen and will convert it into ash, heat, steam, and gas, thus causing air pollution. Relatively meager manufacturing costs, fuel costs avoid a recurrence-free operation, and minimum space for installation ensures compactness.

8.4.3 Drinking Water

An overhead tank will be constructed and fed from the water treatment plant. The supply would be monitored by an intelligent meter linked to the server at I-CHAUPAL. Each household will be fed from this water tank and will be billed as per the water meter reading. Each household will be encouraged to install rainwater harvesting systems in place of the existing bore wells. Net metering may be introduced for water to inculcate the habit of waste conservation and rainwater harvesting. This may be at the village level or household level. Figure 8.20 shows flow chart showing water supply network.

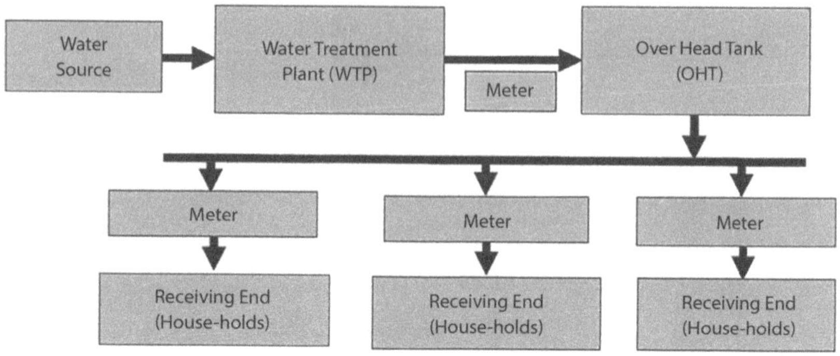

Figure 8.20 Flow chart showing water supply network.

8.4.4 Drainage

The smart intervention will be recording the total rainwater harvested by a village at its Pokhara. The existing street drain network will be redesigned to carry the rainwater to the current/proposed Pokhar(s). At the Pokhara a filter chamber will be built to clean the collected water. A meter will monitor the rainwater collected in the Pokhar and the amount of rainwater that is enough to meet the demand of the village. And surplus groundwater can be used to recharge. Figure 8.21 demonstrate flowchart showing stormwater drainage network.

8.4.5 Rainwater Harvesting

Rooftop rainwater harvesting is the most typical household rainwater harvesting (RWH) technique in rural areas; it is usually done on a small

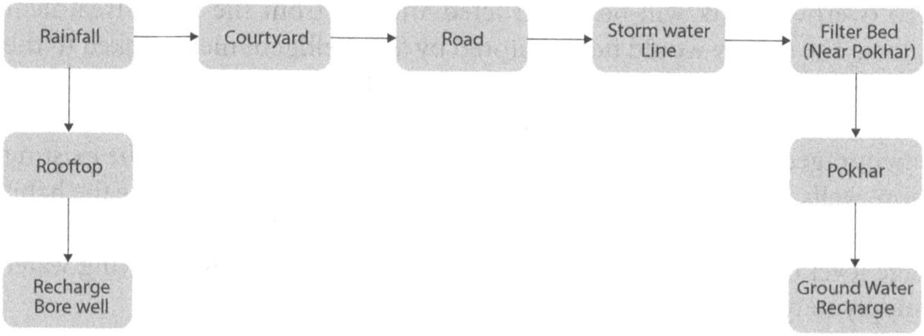

Figure 8.21 Flowchart showing stormwater drainage network.

scale. It is an easy, low-cost strategy that requires minimal specific skills or knowledge and offers several advantages. Rainwater is used to supplying water during use or to recharge any well or stream. After they become scarce or calibrated, harvesting will refill water sources, such as groundwater full of rain or contaminated surface water. However, the quality of rain may be laid low with pollution, animal or bird division, insects, dirt, and organic matter. Therefore regular maintenance (cleaning, repairs, etc.) as a treatment before the water is incredibly vital. For drinking water in rural areas, the most common strategy is to collect rainwater on small-sized roofs: rainwater is serene on the top and transferred to a reservoir with water, where it supplies water during use (for rainwater harvesting for agriculture). Water will supplement different water sources once they become scarce or area units of low quality like groundwater or contaminated surface water that is jolted throughout the monsoon. The sand wells dry up. This map Will benefit from presenting where the rainwater harvesting accurate location.

8.4.6 Sewage

The original sewage system is transformed as some of the sewage in the village is discharged through the ditches, which seriously affects the village's appearance at present. Each household is using a septic tank for sewage disposal. Defective and ill-maintained maintenance septic tanks can lead to groundwater contamination. Figure 8.22 shows process of a drinking water RWH. To avoid this, every household will be connected to a sewage network. This network will be connected to the sewage trunk lines of the sector, which will feed into the sewage treatment plant. Figure 8.23 demonstrate flow chart of sewage network.

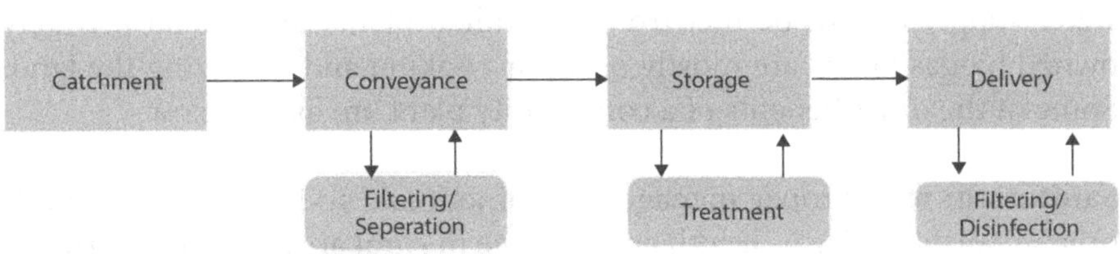

Figure 8.22 Process of a drinking water RWH.

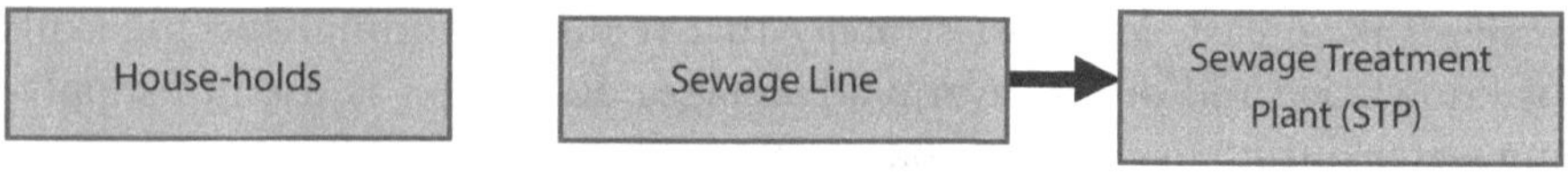

Figure 8.23 Flow chart of sewage network.

8.4.7 Solar Street Light

To increase sense of security and provide comfort of movement outside the house from dusk to dawn, street lights may be installed on every street. These may be at an interval of 24 to 30 m each. Energy for this is obtained either from batteries charged by solar panels or from electricity generated from waste. This will help to save the money required to pay the electricity bill for street lights regularly.

8.4.8 Solar Energy

Energy management is preventing wastage of electricity, monitoring the efficient use of energy, and conserving energy using green energy systems such as solar power. Efficient use of energy using wireless sensors, IoT (Internet of things) devices, solar panels, energy storage grid houses, and energy optimization strategies are vital parameters of intelligent energy management systems. Encouraging the installation of solar panels in individual homes can benefit both the user and the government. Solar panels store solar energy, and one can use it for daily use and sell unused or excess power to the government. Sensors mounted on solar panels can detect weather conditions and control electricity accordingly to prevent unnecessary energy.

8.4.9 Biogas Plant

General needs in rural areas such as abundant organic manure, light, and water supply cannot be met from individual plants. Individual privately owned biogas plants are mostly used for cooking and fertilizing the land. Some of the other benefits of a community plant are listed below:

Sanitation: with proper management of animals and other agricultural/organic wastes/villages, health and hygiene in rural areas will be improved, clean.
Energy conservation: converting organic waste to methane and using it as a fuel leads to energy conservation because fossil fuels do not last forever.

Pollution control: organic waste is usually emitted by greenhouse gases such as carbon monoxide or carbon dioxide. The methanation process reduces greenhouse gas emissions and helps reduce ozone depletion, which can achieve carbon credits.
Generation employment generation: such plants can be quickly established and managed in rural areas and can be controlled by local entrepreneurs with women self-help groups or per capita investment. Since the product has a captive market, the plant is economically viable, and it is bound to create employment opportunities for a vast number of people.

8.4.10 Smart Agriculture

For any village, its agroecosystem is a highly inter-ethnic identity directly related to its social, environmental, and economic structure. Skills and productivity in agriculture are directly related to farming adopted by the community. Fortunately, the consumer has opened many suppliers' opportunities with the technology and the intersection of agriculture farmers. These are some of the parameters depending on which agriculture depends. Most importantly, soil fertility needs to be calculated for the type of crop being planted. So, to understand these parameters, IoT is a new emerging technology that is helping in governing or determining the exact value or impact of these parameters on a particular land for healthy production. Like crop monitoring, cattle monitoring, and management, IoT-based agricultural sensors can be embedded in collar tags that can provide aggregate information about each cow's body temperature, health status, activity information, nutritional insights, and health status of the whole animal. Improving quality and quantity in agriculture using new technologies such as "sensors" and IoT will make a big difference soon through "smart farming."

8.4.11 Organic Farming

Organic farming is an environmentally friendly farm soil, environment, and health policy related to human beings, including animals. Products produced under organic farming are nutritious, safe, and free from chemical residues. On the other hand, chemical-based agriculture adversely affects soil and animal, and human health and ecosystems. Organic farming is the way to maintain and improve soil fertility. The production rate will be uniform and will remain constant for a long time due to the excellent health of the soil. So it will help farmers to double their production after proper crop management. Organic production, if certified, will have better demand in the market with premium prices than ordinary products, giving the producer an extra return to the farmers. These areas can easily be converted into organic farming for the production of various horticultural crops. On the other hand, it becomes difficult for farmers to change their minds to adopt fully organic methods in the pockets of commercial production.

8.4.12 Farming

These include agriculture, animal husbandry, and crops produced for raw materials and food. Different factors determine the type of farmer a farmer can adopt. Physical factors such as the distance of human resources to markets and labor and soil and climate determine the optimal farming for a given region. Farmers for their unique local physical environment, you need to choose the appropriate farms. They must make sure that their products are being sold in the local market.

8.4.13 Poultry Farming

Poultry farming is a fast-growing agricultural business. Poultry farming involves raising turkeys and chickens for meat and eggs. The government is already taking steps to triple egg production to improve children's health and benefit poultry farmers. The Government of India is already conducting a National Livestock Mission to promote a poultry farm business. Even the government gives below poverty line (BPL) families excellent investment and financial support to start a poultry farming business in India.

8.4.14 Fish Farming

Fish farming is also known as aquaculture fishery. This is the practice of raising fish in tanks or large fish ponds. According to Grist, despite the bad reputation of the fish farms, fish farms managed properly clean, durable, and high-quality protein supply. The National Horticultural Mission (NHM), which provides subsidies for excavated and lined farm ponds, has played a significant role in reducing popular fish. To encourage them to collect rainwater, the Indian government has provided a subsidy to poor farmers who want to build farm ponds on their land.

8.4.15 Ayurvedic Farming

There are many medicinal plants that farmers can cultivate. Since Ayurvedic medicine is popular in India, the growth of the growing herb tree should be increased. Tulsi, Aola, Chandan, Kalmegh, Satwari, Ashvahandha, Chirata, Katki, Shankapushpi, Ashoka, Gilo, Kokum, Safed Musuli, Atis, Google, Karth, Senna, Baibarang, Indian Barbiri, Licoris, Bayel, Isbagolish, and Lumb Popular have their demands.

8.4.16 Smart Dairy

Dairy raising cattle for dairy products is the second occupation of a large number of farmers. Dairy farmers in the era of perfect farming, which is considered more important for providing information and capturing a competitive market, need a variety of information sources containing dynamic and stable cow data about feeding, selection, nutrition, fertilization, and milk production process. The IoT began to affect milk production. The use of sensors and cameras in the barn helps farmers to manage their work better. Any changes can be notified immediately through alert messages, and necessary action can be taken. Optimal temperatures for cattle can be maintained using smart devices.

8.4.17 Food Processing Center

Food processing is converting agricultural products into food or converting one type of food into another. Food processing involves various processed foods, ranging from home-cooked raw flour to convenient industrial methods used for timely food processing and improves food security.

8.4.18 Handicraft

The rural industries of the neighboring countries include the three main sectors: the conversion industry, handicrafts, and small-scale industries. Neighboring countries have no conversion industry in rural areas. Handicrafts like the second-class neighboring countries have a long history, especially in rural areas, and many villagers still make a living through it. This industry is in the opposite situation. Rural handicrafts are influenced by geography and nature as this environment provides raw materials to the industries.

8.4.19 Grain Storage

They try to provide food security to the rural population and allow farmers to sell their surplus while getting better prices. With a capacity of 10 to 50, these stores often have construction features that do not qualify for low-loss storage. The same principles apply to the construction of small serial banks for large stores.

8.4.20 Plantation

Planting creates an instant forest; we do this by growing tall tree saplings in a short time, for which we provide fast-growing fruit trees, almond trees, etc. In arid tropical regions where rainfall is low, grasses for seed animals are scary in season, and quality feeder tree gardens contain imported feed items (nutrients) that are sometimes not available in the grass. In the village, trees are planted on the side of the road, and where the space is vacant, we can plant trees that help create an eco-friendly environment.

8.4.21 Smart Healthcare

ICT-enabled m-Health initiatives can enable mobile health diagnostic solutions, provide relatively low levels of local medical expertise, and access specialist healthcare services on an urban community basis where needed. Epidemiological data can be collected to provide more effective interventions and early warning capabilities for health-related challenges such as malnutrition, low birth weight, and anemic mothers. Smart health services are needed to improve living standards in rural areas. In advanced clinics and hospitals in the village, the devices are connected to the doctors.

8.4.22 Animal Clinic

At the rural area animal clinic, we know from the outset that pet families play an important role as part of the family. That is why we always take care of every pet that we care for with the same level of empathy and personal attention as if they were our own. We would like to be a faithful partner in all of their pet care needs and we look forward to welcoming them. Today's veterinarians are the only physicians trained to maintain the health of both animals and humans. They work hard to address the health and well-being needs of each species. Veterinarians also play a vital role in environmental protection, research, food security, and public health.

8.4.23 Smart Education

An innovative, healthy, and secure development is essential for the development of education. In a technology-friendly world, people need to be technically educated. The proposed land use and land cover map of Dungarpur Reelka village, illustrating various planned facilities, is shown in Figure 8.24. In a technology-friendly world, people need to be technically educated. Thus, smart education is necessary in every educational medium. An IoT-based education system focuses on interactive and attractive learning mediums. Presented in an exciting way, such as video games and lectures, it attracts students to learn interactively from textbooks to traditional thematic classroom teaching. Video conferencing and video gaming are two examples of smart education in the village. Smart villages aim to increase the time available for students to study and address the conventional factors that negatively affect students' acquisition of knowledge and skills needed to achieve economic goals and improve labor productivity. Holding on to access to the Internet and consequently the world's knowledge base, ending the experience isolation of many rural communities. New opportunities can be created for distance and adaptive learning.

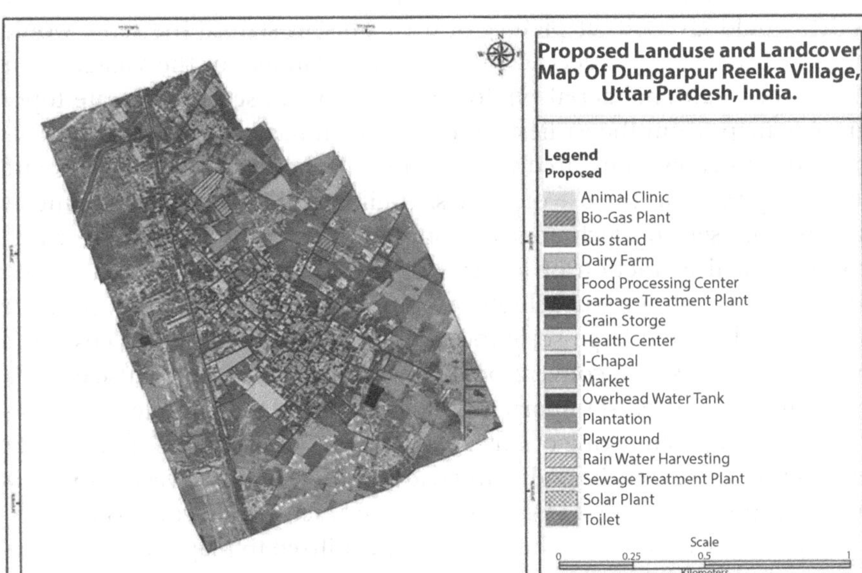

Figure 8.24 Proposed map of Dungarpur Reelka village.

8.4.24 Village Market

A village market is a place where villagers collect their daily necessities to buy and sell. It is an important place for the villagers. Various everyday conditions are sold here. Vegetables, fish, oil, milk, fruits, rice, sugar, cloth etc., are sold in the village market. A village market plays a vital role in the rural economic field. In the village market, villagers can sell their hand-made food and products. As a result, they can improve their financial situation. On the other hand, they can easily buy daily necessities from the village market. This saves them time and money. A village market has some social value. Villagers can get a chance to exchange their greetings. They can meet their relatives and kin. So that plays an essential role in the life of the villagers.

8.5 Conclusion

The present study reveals that agriculture production would be helpful for the farmers to achieve optimum crop production with the backing of modern technology. The preparation of various amenities like I-Chaupal, kill waste, clinic, health center, drinking water, rainwater harvesting, sewage treatment plant, solar light, biogas plant, organic farming, smart dairy, and toilet, would give a clear picture of their present status. The same would be very useful for the decision makers and planners of the villages. The notion of traditional aerial photogrammetry was used to generate topographic maps from the earliest times to date. It has nonetheless been outmoded, somehow, compared to other recent photogrammetric tools and technology such as the use of drones. Unlike traditional photogrammetry and remote sensing techniques, drones such as ortho correction, cloud coverage, and extraction of features have improved map characteristics. Therefore, it can produce high-resolution orthophotographs, superior altimetric data, and large-scale topographical maps, with fewer costs and human resources than conventional photogrammetry with the use of UAV photogrammetry. Most importantly, it is a practical tool for tolerable accuracy and budget surveying of tiny areas. We encourage UAV flights in clear weather for better results and altimetric accuracy. In dense vegetation, it is not recommended to fly a UAV with an optical sensor. However, other sensor types such as LiDAR technology can be utilized to map these locations accurately. In addition, the manual cloud point classification must be done correctly to provide a more accurate result.

References

ADAMOWICZ M., ZWOLIŃSKA-LIGAJ, M. (2020), The "Smart Village" as a Way to Achieve Sustainable Development in Rural Areas of Poland, *Sustainability* 12(16), 6503. https://doi.org/10.3390/su12166503

ADESIPO A., FADEYI O., KUCA K., KREJCAR O., MARESOVA P., SELAMAT A., ADENOLA M. (2020), Smart and Climate-Smart Agricultural Trends as Core Aspects of Smart Village Functions, *Sensors*, 20(21), 5977. https://doi.org/10.3390/s20215977

ADESIPO A., FADEYI O., KUCA K., KREJCAR O., MARESOVA P., SELAMAT A., ADENOLA M. (2020), Smart and Climate-Smart Agricultural Trends as Core Aspects of Smart Village Functions, *Sensors* 20(21), 5977. https://doi.org/10.3390/s20215977

BERA A., TALOOR A.K., MERAJ G., KANGA S., SINGH S.K., ĐURIN B., ANAND, S. (2021), Climate vulnerability and economic determinants: Linkages and risk reduction in Sagar Island, India; A geospatial approach, *Quat. Sci. Adv.* 4, 100038. https://doi.org/10.1016/j.qsa.2021.100038.

CVAR N., TRILAR J., KOS A., VOLK M., STOJMENOVA DUH E. (2020), The Use of IoT Technology in Smart Cities and Smart Villages: Similarities, Differences, and Future Prospects, *Sensors* 20(14), 3897. https://doi.org/10.3390/s20143897

JOY J., KANGA S., SINGH S.K., SUDHANSHU. (2021) *Cadastral level* Soil and Water conservation Priority Zonation using Geospatial technology, *Int. J. Agric. Syst.* 9 (1), 10-26.

KANGA S., MERAJ G., DAS B., FAROOQ M., CHAUDHURI S., SINGH S.K. (2021), Modeling the spatial pattern of sediment flow in lower Hugli estuary, West Bengal, India by quantifying suspended sediment concentration (SSC) and depth conditions using Geoinformatics, *Appl. Comput. Geosci.* 8, 10043.

KANGA S., RATHER M.A., FAROOQ M., SINGH, S.K. (2021), GIS Based Forest Fire Vulnerability Assessment and its Validation using field and MODIS Data: A Case Study of Bhaderwah Forest Division, Jammu and Kashmir (India), *Indian For.* 147 (2), 120-136.

KOMOROWSKI Ł., STANNY M. (2020), Smart Villages: Where Can They Happen? *Land* 9(5),151. https://doi.org/10.3390/land9050151

KUMAR A., KANGA S., TALOOR A.K., SINGH S.K., ĐURIN, B. (2021), Surface runoff estimation of Sind river basin using integrated SCS-CN and GIS techniques, HydroResearch, https://doi.org/10.1016/j.hydres.2021.08.001.

MEHTA PL., *KALRA R.* PRASAD, R. A.(2021), Backdrop Case Study of AI-Drones in Indian Demographic Characteristics Emphasizing the Role of AI in Global Cities Digitalization, *Wirel. Pers. Commun.* 118, 301–321. https://doi.org/10.1007/s11277-020-08014-6

MERAJ G., SINGH S.K., KANGA S., MD. NAZRUL, I. (2021), Modeling on comparison of ecosystem services concepts, tools, methods and their ecological-economic implications: a review, *Model. Earth Syst. Environ.* https://doi.org/10.1007/s40808-021-01131-6.

SAURABH M., SUDHANSHU, S., SINGH S.K., MATHUR S. (2021), Qualitative Assessment of Geoheritage for Geotourism promotion: A case study from Mehrangarh Ridge in Jodhpur city, Western Rajasthan, India. *Geoheritage*, (Accepted), Article in press

SPANÒ A., SAMMARTANO G., CALCAGNO TUNIN F., CERISE S., POSSI G. (2018), GIS-based detection of terraced landscape heritage: comparative tests using regional DEMs and UAV data, *Appl. Geomatics* 10, 77–97. https://doi.org/10.1007/s12518-018-0205-7.

STOJANOVA S., LENTINI G., NIEDERER P., EGGER T., CVAR N, KOS A., STOJMENOVA DUH E. (2021), Smart Villages Policies: Past, Present and Future, *Sustainability* 13(4), 1663. https://doi.org/10.3390/su13041663

TOMAR J.S., KRANJČIĆ N., ĐURIN B., KANGA S., SINGH, S.K. (2021), Forest Fire Hazards Vulnerability and Risk Assessment in Sirmaur District Forest of Himachal Pradesh (India): A Geospatial Approach, *ISPRS Int. J. Geo-Inf.* 10(7), pp. 447. https://doi.org/10.3390/ijgi10070447.

ZAVRATNIK V., KOS A., STOJMENOVA DUH E. (2018), Smart Villages: Comprehensive Review of Initiatives and Practices, *Sustainability* 10(7), 2559. https://doi.org/10.3390/su10072559

ZAVRATNIK V., PODJED D., TRILAR J., HLEBEC N., KOS A., STOJMENOVA DUH E. (2020), Sustainable and Community-Centred Development of Smart Cities and Villages, *Sustainability* 12(10), 3961. https://doi.org/10.3390/su12103961

ZHANG X., ZHANG Z. (2020), How Do Smart Villages Become a Way to Achieve Sustainable Development in Rural Areas? Smart Village Planning and Practices in China, *Sustainability* 12(24), 10510. https://doi.org/10.3390/su122410510

9

Varanasi's Land Mosaic: A Geospatial Analysis for Peri-Urban Area Planning and Management

Nitish Kumar Singh*, Vijay Kumar Baraik and Mahendra Singh Nathawat

Discipline of Geography, Indira Gandhi National Open University, New Delhi, India

Abstract
As one of the fastest-growing cities in India, Varanasi City has experienced unplanned and rapid land use/land cover changes in recent years. This study quantifies and characterizes the spatiotemporal land use/land cover patterns in Varanasi City and its peri-urban areas using Remote Sensing, GIS, and Landscape Metrics. The results of this study demonstrate a continuous increase in built-up and vacant lands replacing agricultural land, vegetation cover, and water bodies. This situation substantially hinders land use/land cover-related planning capabilities and targets in Varanasi City and its peri-urban areas. The dramatic urban expansion of the city of Varanasi has resulted in rapid changes in the landscape composition of peri-urban areas and increased landscape heterogeneity, leading to structural complexities at both class and landscape levels. This type of land use/land cover dynamics and related complexities are becoming increasingly challenging for local planners and policymakers. In addition, it has put pressure on local bodies to properly manage and utilize limited land resources. The thematic and landscape-based spatial information from this study facilitates an understanding of the land use/land cover dynamics of Varanasi City and its peri-urban areas. This study can further aid in the best possible decisions regarding land use planning and restoration of deteriorating urban ecosystems.

***Keywords*:** Land use/land cover, peri-urban area, landscape metrics, GIS, Varanasi City

**Corresponding author*: knack.008@gmail.com

Shruti Kanga, Gowhar Meraj, Suraj Kumar Singh, Majid Farooq and M. S. Nathawat (eds.) Geospatial Technology for Natural Resource Management, (237–258) © 2024 Scrivener Publishing LLC

9.1 Introduction

Over the last few decades, it has been observed that many large and medium-sized cities in developing countries such as India have rapidly attracted the rural population (Singh *et al.*, 2022). In rural areas, fewer employment opportunities (Singh *et al.*, 2023), lack of infrastructure facilities (Rani *et al.*, 2023), and poor living conditions (Ahmad *et al.*, 2023) have acted as push factors (Ahmad *et al.*, 2023). Similarly, in cities, more employment opportunities (Sharma *et al.*, 2023), availability of infrastructure facilities (Chaudhary *et al.*, 2023; Saqib *et al.*, 2023), and higher standard of living (Kumar *et al.*, 2023; Rai *et al.*, 2023) have contributed significantly to pull factors (Rathore *et al.*, 2023; Singh *et al.*, 2023). Under the influence of these two factors, all cities have expanded in an uncontrolled manner (Singh *et al.*, 2023; Tripathi *et al.*, 2023; Sud *et al.*, 2023). Consequently, land use and land cover have been altered significantly over time in peri-urban areas. It has been found that the expansion of any city has its own specific socioeconomic and environmental impacts (Upadhyay *et al.*, 2023; Lavane *et al.*, 2022). These impacts are currently a prime issue in the debate on sustainable urban development (Nath *et al.*, 2023; Kanga *et al.*, 2023; Gulati *et al.*, 2023). Therefore, spatiotemporal study of land use land cover in the city as well as in peri-urban areas is necessary for the planning and management of the city (Tomar *et al.*, 2023). The development of remote sensing has played an important role in these studies (Singh *et al.*, 2023). The evolution of remote sensing technology over time has made it easier to monitor, analyze, and study land use and land cover using various parameters. Remote sensing-based monitoring and the study of land use and land cover are very helpful in making long-term and sustainable development projects (Farooq *et al.*, 2023).

Varanasi is a famous religious city in northern India (Shruti *et al.*, 2020). Due to increasing socio-economic influence, the city has expanded rapidly in an uncontrolled manner (Kumar *et al.*, 2010). This has resulted in rapid LULC changes in the peri-urban regions. Achieving urban planning goals—balanced socio-economic development (Pandey and Bhardwaj, 2010; Raj *et al.*, 2014) and environmental quality (Mishra *et al.*, 2019; Srivastava *et al.*, 2020; Shruti *et al.*, 2020; Shruti *et al.*, 2022)—remains a major challenge. In this context, during the development work being done under the Varanasi Master Plan, it has been observed that most of the city's problems are closely related to the unplanned changes in its peri-urban areas. If unplanned changes in peri-urban areas are not included in the planning process, then the development goals of the Master Plan will not be successfully achieved. Therefore, class and patch-level analysis of the spatiotemporal pattern of LULC in peri-urban areas is necessary for purposes related to the decentralization of population density, environmental quality, traffic management, parking, decentralization of commercial activities, water supply, and sewerage. If these objectives are achieved in peri-urban areas, urban development and planning-related work can be made more sustainable. It has been found during the analysis of studies on land use and land cover in India related to planning works that the integrated study of class- and patch-level analysis of different land use and land cover is yet to be emphasized (Singh *et al.*, 2020). The class- and patch-level patterns of land use and land cover are closely influenced by the physical and socio-economic settings of the area concerned. Therefore, along with the traditional analysis of land use and land cover, landscape-level studies should be included in the planning works so that, along with the human aspects in the planning process, the nature and characteristics of the landscape of the area can also be understood in a comprehensive mode and the planning works can be made more sustainable and successful.

9.2 Description of Study Area, Data Source, and Methods

9.2.1 Study Area

The city of Varanasi is one of the oldest cities in the world and has continued its existence since ancient times. Geographically, this city (approximately 25° 15' N to 25° 22' N latitude and 82° 57' E to 83° 01' E longitude) is situated on the left crescent-shaped bank of the Ganges River in northern India (Figure 9.1). Since ancient times, the city has been an important religious center with traditions of Hindu devotion, pilgrimage, religious mysticism, death, and salvation, and represents the religious traditions of India. The land here is extremely fertile and is located on the mid-Gangetic plain. Therefore, most of the population of Varanasi City and its peri-urban areas is directly or indirectly engaged in agro-based economic activities. Climatologically, the city is influenced by a humid sub-tropical climate. Therefore, there is a great variation in the temperatures during the summer and winter. "*Loo*" in summer and "*Fog*" in winter are a common weather phenomenon in the city and its peri-urban areas.

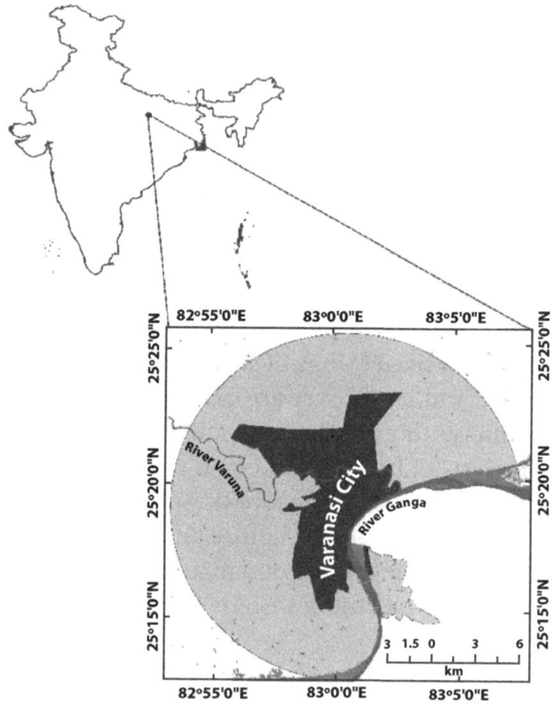

Figure 9.1 Location of Varanasi City.

9.2.2 Data Source

The main spatial and temporal data of the present study are Landsat satellite data from the USGS, which were acquired from the Open Access Landsat Imagery Service of the USGS Earth Explorer (https://earthexplorer.usgs.gov/). All Landsat images were from January to March of the respective year and downloaded on cloud-free parameters.

Similarly, the administrative boundaries were delineated with the help of the Survey of India Toposheet to create authentic reference data for the study area. After digitization, considering the CBD of Varanasi City as the center, a buffer zone was created up to a distance of 20 km from the city center. Subsequently, the final study area was created by clipping the shapefile of the digitized data using this buffer zone.

9.2.3 Method

In the first stage, a multiband composite image was created by a layer-stacking process using Erdas Imagine v.2015 software. The multiband composite images were mosaicked after the stacking process. These mosaicked images were then subset using the shapefile of the study area to obtain the final satellite images of the area of interest. Then, classification of this subset of data was performed using a supervised image classification method to generate land use/land cover data. After supervised image classification, an attempt was made to remove the classification inaccuracies present in the land use/land cover classes of each year (1972, 1980, 1990, 2000, 2010, and 2020) using the recoding process, and the accuracy assessment of the land use/land cover data generated was performed after the recoding process. For accuracy assessment, 240 sample pixels (40 samples for each land use/land cover class) were selected based on random sampling. Subsequently, the accuracy of these sample pixels was determined using toposheet and Google Earth historical data. After determining the pixel-level accuracy, the Producer and User accuracies for each land use/land cover class were calculated using Equations (9.1) and (9.2), respectively:

$$\text{Producer's Accuracy} = \frac{A}{B} \times 100 \tag{9.1}$$

Where A is number of correctly classified pixels in each class
B is total number of classified pixels in that class

$$\text{User's Accuracy} = \frac{C}{D} \times 100 \tag{9.2}$$

Where C is number of correctly classified pixels in each class
D is total number of classified pixels in that class

After calculating the Producer's and User's accuracies, the overall accuracy and Kappa values of the land use/land cover data for each year were calculated using Equations (9.3) and (9.4):

$$\text{Overall Accuracy} = \frac{E}{F} \times 100 \tag{9.3}$$

Where E is total number of correctly classified pixels (Diagonal)
F is total number of reference pixels

Table 9.1 Used landscape metrics.

Landscape-metrics	Metrics name	Description
CA	Class Area	measures area (ha) of each patch type
NP	Number of Patches	measures the number of patches of the corresponding patch type
MPS	Mean Patch Size	measures average patch size of the corresponding patch type
PLAND	Percentage of Landscape	measure the percentage of landscape comprised of a particular patch type

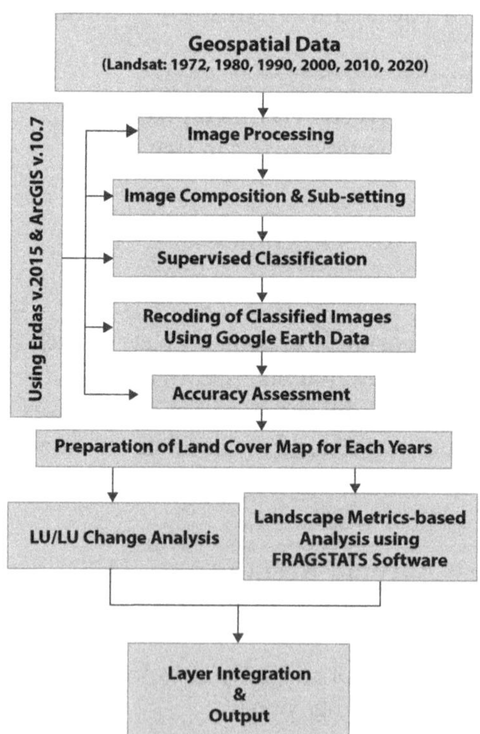

Figure 9.2 Methodology flow chart.

$$\text{Kappa value} = \frac{(TS - TCS) - \Sigma(Column\ Total \times Row\ Total)}{(TS)^2 - \Sigma(Column\ Total \times Row\ Total)} \times 100 \tag{9.4}$$

Where TS represent total number of sample, and

TCS represent total number of corrected samples.

After accuracy assessment, the final land use/land cover data for each year were obtained. Subsequently, these land use/land cover data has been used to generate a land use/land cover change matrix and landscape metrics-based structural data using ArcGIS and FRAGSTATS software, respectively. To demonstrate the landscape-based spatial configuration of land use/land cover classes, four metrics were selected in this study (Table 9.1):

Finally, land use/land cover change data and landscape metrics-based data were integrated and analyzed (Figure 9.2).

9.3 Results and Discussion

9.3.1 Accuracy Assessment

The analysis showed that the accuracy of the land use/land cover data for the years 1972 and 1980 was 87% and 87.60%, respectively (Table 9.2). The kappa values of the data for these two years were 84.4% and 85.20%, respectively. These kappa values show that the level of agreement of the land use/land cover data is of a strong category, due to which the classified land use/land cover data for the years 1972 and 1980 are reliable up to 64%–80%. Subsequently, the overall accuracy of the land use/land cover data for the years 1990, 2000, 2010, and 2020 was between 93% and 95% (Table 9.2). The kappa values of the land use/land cover data for these years were between 84% and 94%. The level of agreement of these data is of almost perfect category because these land use/land cover data are reliable up to 82–100%. Therefore, the higher the reliability percentage of land use/land cover data, the more accurate and realistic the study will be.

9.3.2 Dynamics of Agricultural Land

Unplanned development continues throughout Varanasi City, which is reflected by spatiotemporal changes in land use/land cover (Figure 9.3). The peri-urban area of Varanasi City is dominated by agricultural land. This

Table 9.2 Accuracy assessment matrix of land use/land cover.

LULC category	Year 1972		Year 1980		Year 1990		Year 2000		Year 2010		Year 2020	
	Producer's	*User's*	*Producer's*	*User's*	*Producer's*	*User's*	*Producer's*	*User's*	*Producer's*	*User's*	*Producer's*	*User's*
Agricultural Land	79	84	68	86	79	98	85	96	88	94	90	90
Built-up Land	88	90	93	90	95	90	100	100	90	94	96	96
Sandy Area (Riverine)	87	86	100	92	100	100	100	100	98	98	100	98
Vacant Land	97	86	97	88	97	88	97	84	95	86	87	84
Vegetation Cover	87	86	86	80	97	88	93	88	90	96	90	98
Water Body	83	90	86	90	98	100	92	100	100	94	95	94
Overall Accuracy	***87***		***87.6***		***94***		***94.6***		***93.6***		***93.3***	
Kappa Value	***84.4***		***85.2***		***92.8***		***93.6***		***92.4***		***92***	

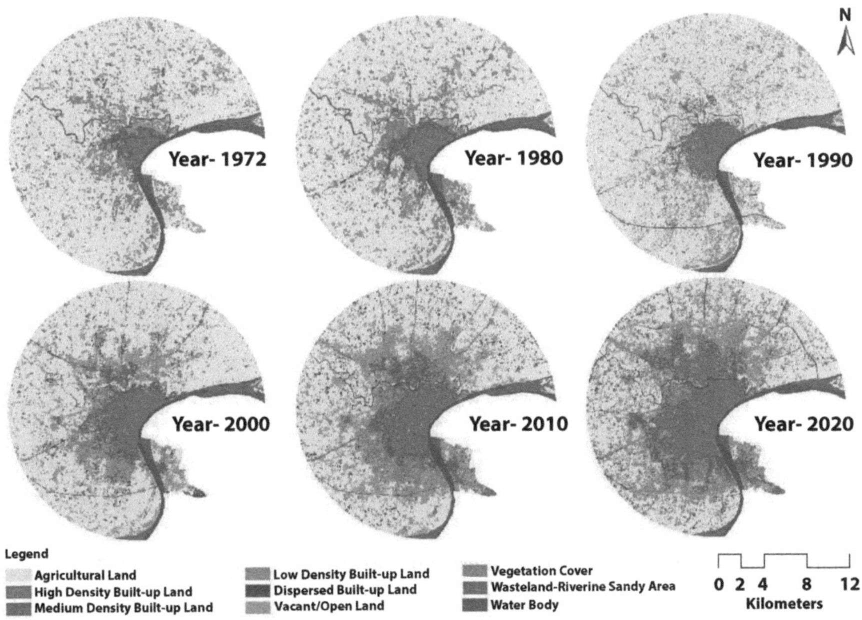

Figure 9.3 Spatiotemporal land use land cover pattern in the study area (from 1973 to 2020).

study found that the agricultural land in the city and its peri-urban area has been continuously decreasing since 1972 (Figure 9.4), due to which the agricultural land has decreased from 25,919.24 ha (year-1972) to 13,557.23 ha (year-2020). Agricultural land decreased at an average rate of 257.5 ha per year from 1972 to 2020. Decadal-level analysis showed that agricultural land increased only between 1980 and 1990. In the past decade, agricultural land has increased at a rate of 189.81 ha per year. The maximum decrease in agricultural land occurred during 1990–2000; in this decade, agricultural land decreased at a rate of 761.13 ha per year (Figure 9.5). Structural analysis of agricultural land at the landscape level found that the PLAND value of agricultural land has been decreasing, indicating that the dominance of agricultural land at the landscape level has decreased over time (Figure 9.6). The decrease in the mean patch size (MPS) and increase in the number of patches (NP) are the main causes of fragmentation in agricultural lands; as a result, agricultural land has been continuously decreasing (Figure 9.7). Table 9.3 shows that 55% of the area of agricultural land has been converted into other land use/land cover types due to various socio-economic development activities. In these changes, most of

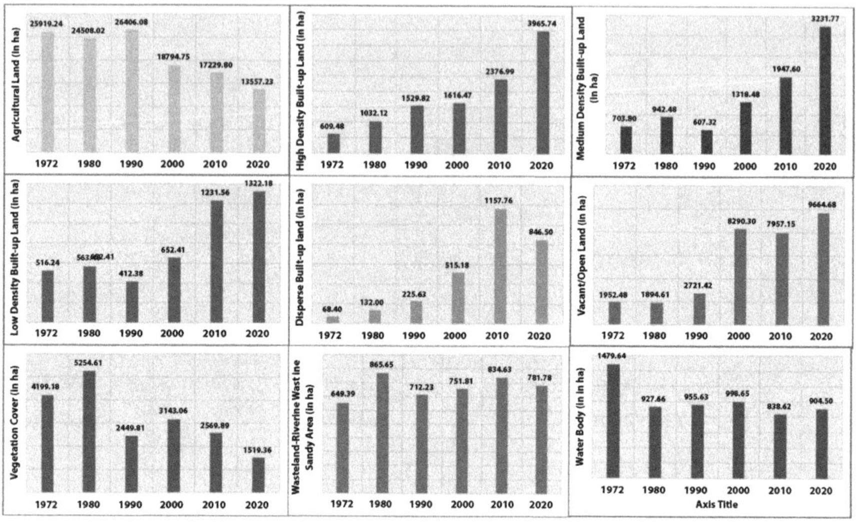

Figure 9.4 Class areas of different land use/land cover in the study area (1972–2020).

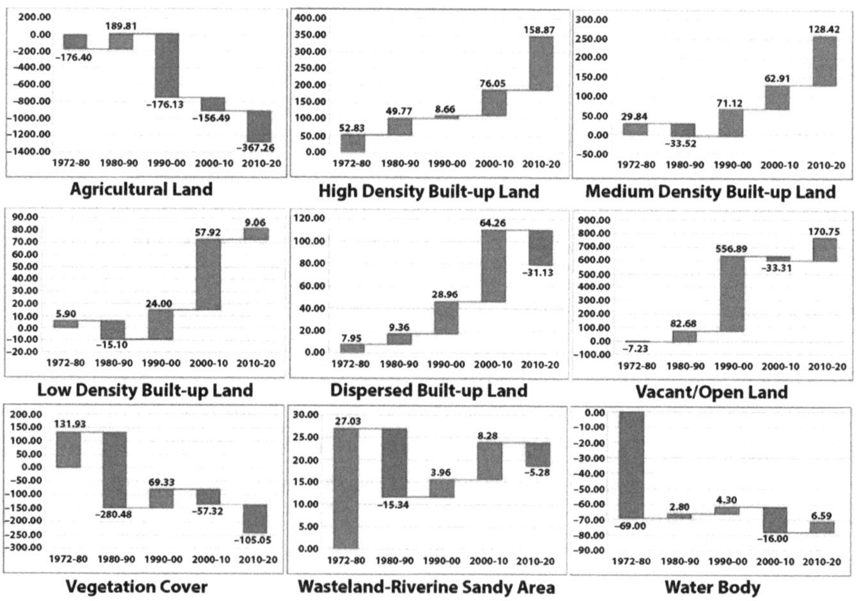

Figure 9.5 Gain and loss in areas of different land use/land cover classes in ha/year (1972–2020).

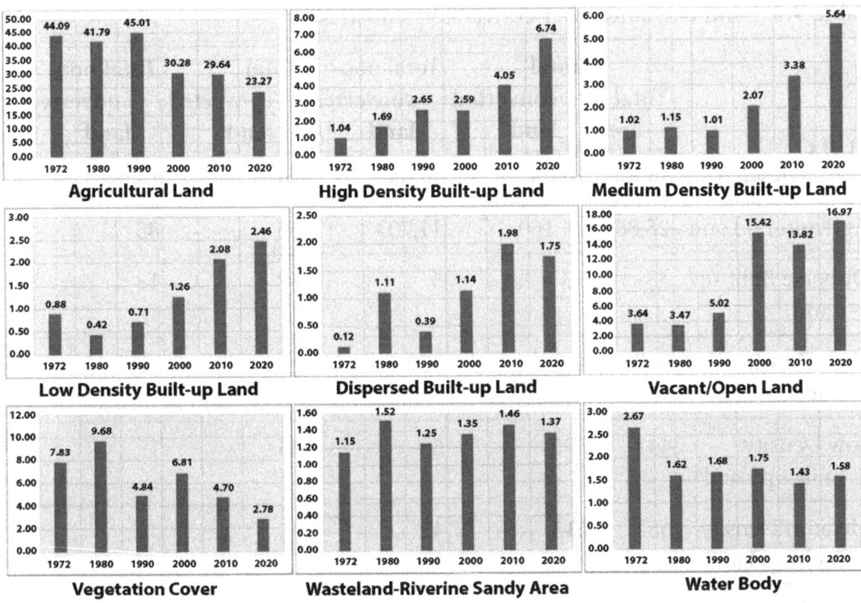

Figure 9.6 Percentage of landscape (PLAND) pattern of different land use/land cover (1972–2020).

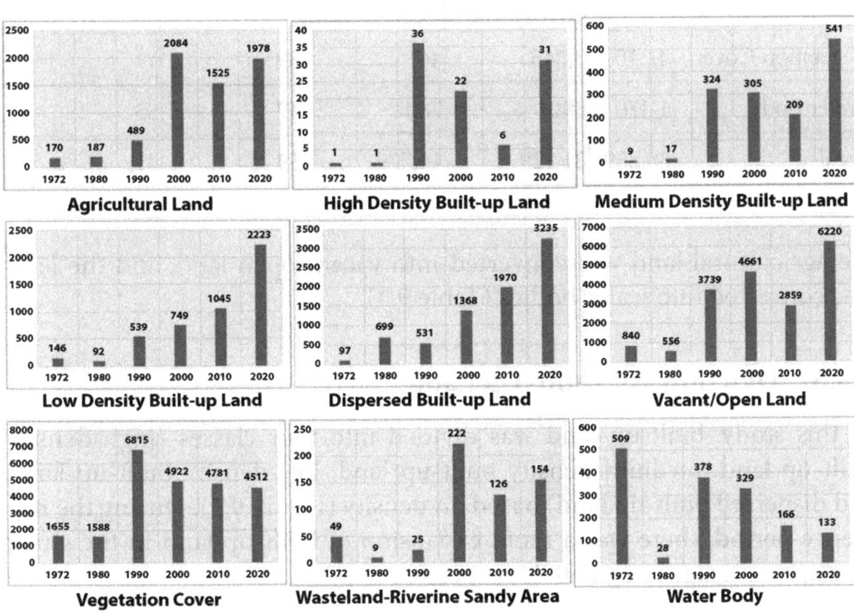

Figure 9.7 Number of patches (NP) in different land use/land cover (1972–2020).

Table 9.3 Land use/land cover conversion statistics.

LU/LC transformation	Total area	Total converted land	Total non-converted land	Total converted land	Total non-converted land
	In ha	In ha	In ha	In %	In %
Agricultural Land	25,863	14,160	11,703	55	45
Disperse Built-up Land	38	33	5	87	13
High Density Built-up Land	580	0	580	0	100
Low Density Built-up Land	441	375	66	85	15
Medium Density Built-up Land	662	615	47	93	7
Waste Land-Riverine Sandy Area	648	141	508	22	78
Vacant/Open Land	1,947	1,371	576	70	30
Vegetation Cover	4,210	3,745	465	89	11
Water Body	1,470	690	780	47	53
Total	35,859	21,129	14,729	59	41

the agricultural land was converted into vacant/open land, and the least was converted into water bodies (Table 9.4).

9.3.3 Dynamics of Built-Up Land

In this study, built-up land was divided into four classes (high-density built-up land, medium density built-up land, low-density built-up land, and dispersed built-up land) based on density (Figure 9.3). During the reference period, there was a rapid expansion of built-up land in the study area at an average rate of 69.92 ha per year.

Table 9.4 Statistics of converted land use/land cover classes (1972–2020).

LU/LC transformation		Unit	Year- 2020								
			AGR	BDL	HDBL	LDBL	MDBL	WRSA	V/O L	VEG	WB
Year- 1972	AGR	in ha		**821**	**1,298**	**1,130**	**2,112**	**58**	**7,663**	**1,040**	**38**
		in %		*5.8*	*9.2*	*8.0*	*14.9*	*0.4*	*54.1*	*7.3*	*0.3*
	DBL	in ha			**6**	**19**	**7**				
		in %			*19.6*	*57.6*	*22.6*				
	HDBL	in ha									
		in %									
	LDBL	in ha			**277**		**98**				
		in %			*73.9*		*26.3*				
	MDBL	in ha			**615**						
		in %			*100*						
	WRSA	in ha	**47**	**3**		**1**	**2**		**27**	**2**	**58**
		in %	*33.4*	*2.2*		*1.0*	*1.6*		*19.3*	*1.5*	*40.9*
	V/O L	in ha	**635**	**47**	**378**	**77**	**164**	**5**		**59**	**5**
		in %	*46.3*	*3.5*	*27.6*	*5.6*	*12.0*	*0.4*		*4.3*	*0.3*
	VEG	in ha	**794**	**137**	**699**	**204**	**652**	**13**	**1,237**		**9**
		in %	*21.2*	*3.7*	*18.7*	*5.4*	*17.4*	*0.4*	*33.0*		*0.2*
	WB	in ha	**171**	**17**	**46**	**37**	**57**	**196**	**161**	**5**	
		in %	*24.8*	*2.5*	*6.7*	*5.3*	*8.3*	*28.4*	*23.4*	*0.7*	

Legend: AGR, Agricultural Land; DBL, Dispersed Built-up Land; HDBL, High-Density Built-up Land; MDBL, Medium-Density Built-up Land; LDBL, Low-Density Built-up Land; WRSA, Waste Land-Riverine Sandy Area; V/O L, Vacant/Open Land; VEG, Vegetation Cover; WB, Water Body.

9.3.3.1 *High-Density Built-Up Land*

The highest pace of increase in high-density built-up land occurred during 2010–2020, when high-density built-up land increased at an average rate of 185.87 ha per year (Figure 9.5). High-density built-up areas have absorbed 3,320 ha of other land use/land cover during their expansion. Agricultural land and vegetation cover were the categories most absorbed by high-density built-up lands (Table 9.4). It was found that high-density built-up land absorbed most of the other land use/land cover within it because these were the places where new residential and commercial buildings came up. The increasing patch number (Figure 9.7) and decreasing mean patch size (Figure 9.8) indicate that high-density built-up land has expanded as smaller patches. The dominance of high-density built-up land increased almost six-fold compared to that in 1972 (Figure 9.6).

9.3.3.2 *Medium-Density Built-Up Land*

The area of medium-density built-up land (MDBL) also increased by approximately 2,528 ha in 1972–2020 at an average rate of 52.67 ha per year (Figure 9.4). The highest rate of increase in MDBL was in 2010–2020,

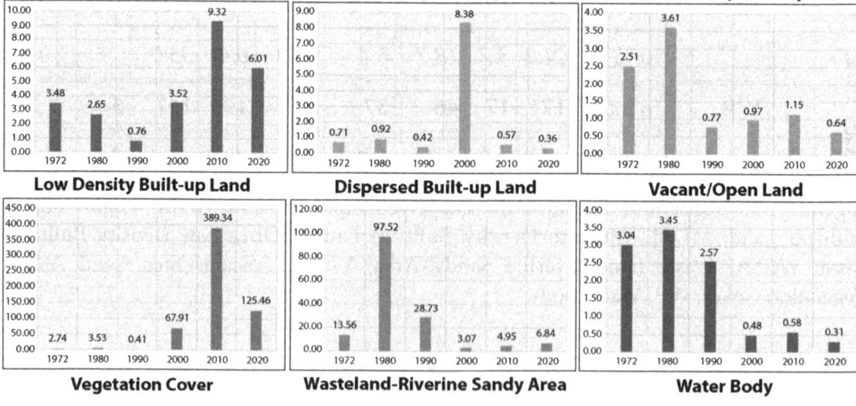

Figure 9.8 Mean patch sizes (MPSs) of different land use/land cover classes (1972–2020).

when MDBL increased at a rate of 128.42 ha per year (Figure 9.5). This study shows that the main reason for the increase in the area of MDBL is associated with the diffusion of commercial and residential facilities from HDBL. Due to the diffusion of these services, a rapid increase in MDBL has been observed. The increasing value of PLAND indicates the increasing dominance of MDBL in the study area (Figure 9.6). With increasing dominance, the NP of MDBL increased from 9 to 541 (Figure 9.7), and the MPS decreased from 65.56 ha to 5.12 ha (Figure 9.8). In the reference period, 93% of MDBL changed to HDBL, while 7% remained unchanged (Table 9.3). MDBL absorbed an area of 3,093 ha of other land use/land cover around it, with the maximum absorbed land being agricultural land and vegetation cover (Table 9.4).

9.3.3.3 *Low-Density Built-Up Land*

The area of low-density built-up land (LDBL) increased by approximately 806 ha from 1972 to 2020 (Figure 9.4), at an average rate of 16.80 ha per year in the study area. LDBL decreased at a rate of 15.10 ha per year in 1980–1990 but increased progressively after 1990 (Figure 9.5). The highest rate of increase in LDBL occurred in 2000–2010; in this decade, LDBL increased at a rate of 57.92 ha per year. Landscape analysis showed that the PLAND value of LDBL has been continuously increasing since 1980; that is, the preponderance of LDBL has been continuously increasing in the study area (Figure 9.6). With the increasing dominance of LDBL, the Mean Patch Size of LDBL increased (Figure 9.8), and along with the Mean Patch Size, the number of patches also increased rapidly (Figure 9.7). The overall gist of the above analysis is that LDBL expanded rapidly over time in the form of incremental patches. Increased means of transportation outside the MDBL and relatively low land prices have increasingly attracted growing populations to areas in the LDBL. This has led to a rapid increase in the size and area of LDBL. From 1972 to 2020, 85% of LDBL was converted to other land use/land cover types, while 15% remained unchanged (Table 9.3). A total of 277 ha of LDBL was converted into HDBL and 98 ha into MDBL due to the infrastructural development. In contrast, LDBL has absorbed an area of 1,468 ha of other land use/land cover in the vicinity due to its territorial expansion. In this absorption process, agricultural land and vegetation cover were the land most absorbed by the LDBL (Table 9.4).

9.3.3.4 *Dispersed Built-Up Land*

The built-up land in the border part of the peri-urban areas of Varanasi City was included in the dispersed built-up land (DBL). Figure 9.4 shows that the area of DBL increased by 778.10 ha in 1972–2020 at an average rate of 16.21 ha per year. Decadal analysis showed that the most rapid increase in DBL occurred during 2000–2010, when DBL increased at an average rate of 64.26 ha per year (Figure 9.5). The landscape-level PLAND value of DBL increased from 0.12 to 1.75, indicating that the dominance of DBL increased in the outer parts of peri-urban areas (Figure 9.6). The increasing value of NP (Figure 9.7) and decreasing value of MPS (Figure 9.8) show that DBL is expanding in the form of small patches. The main reason for the increase in DBL is the relatively low cost of residential and commercial land. The establishment of educational institutions and small-scale industries has increased rapidly in these areas, promoting DBL's growth. From 1972 to 2020, 87% of DBL was converted to other land use/land cover, while 13% remained unchanged (Table 9.3). Through its expansion, DBL has absorbed approximately 1,026 ha of other land use/land cover, the largest part being agricultural land and vegetation cover (Table 9.4).

9.3.4 Dynamics of Vacant/Open Land

Vacant/open land is the land on which no construction (residential or commercial) has taken place and is not being used. The vacant/open land in the study area mainly appears in the form of plots that have been deliberately left empty so that they can be used for residential or commercial purposes in the future. Figure 9.4 shows that vacant/open land was 1,952.48 ha of the entire study area in 1972 and increased to 9,664.68 ha in 2020 at the average rate of 160.67 ha per year. Vacant/open land had the highest average growth rate (556.89 ha/year) during 1990–2000. In contrast, vacant/open land decreased by 7.23 ha per year and 33.31 ha per year in 1970–1980 and 2000–2010, respectively (Figure 9.5). Landscape-level analysis showed a sharp increase in the PLAND value of vacant/open land, indicating that vacant/open land dominance has increased rapidly over time (Figure 9.6). The increasing value of NP (Figure 9.7) and decreasing value of MPS (Figure 9.8) show that the vacant/open land in the study area has increased in fragment form and expanded to new areas in the form of new patches. Table 9.3 shows that 70% of the vacant land was converted to other land use/land cover, while 30% remained unchanged. Vacant/open land has absorbed approximately 9,088 ha of other land use/land cover due to its expansion. Vacant/open land has expanded mostly on terms of agricultural land and vegetation cover. This transformation pattern shows that agricultural and plantation lands are being rapidly converted into non- agricultural and non-plantation lands in the study area (Table 9.4).

9.3.5 Dynamics of Vegetation Cover

Vegetation cover is defined as the area covered by green vegetation such as gardens, plantations, and forests. The present study observed that vegetation cover has been declining at an average rate of 55.83 ha per year in 1972–2020. At decadal level, the most rapid decrease in vegetation cover occurred in 1980–1990, when vegetation cover decreased at an average rate of 280.48 ha per year. On the other hand, vegetation cover increased at the average rate of 131.93 ha per year and 69.33 ha per year in 1972–1980 and 1990–2000, respectively (Figure 9.5). The decreasing value of PLAND shows the decreasing dominance of vegetation cover at the landscape level (Figure 9.6). This decreasing dominance has resulted in the rapid fragmentation of large patches of vegetation cover, which has caused the area of vegetation cover to shrink over time. It has been observed that patches of vegetative cover have been rapidly destroyed in the form of gardens and naturally grown vegetation outside city limits. Therefore, only vegetation covered on government land could survive in the study area with less interference. Efforts have not been made to increase vegetation cover. Due to some positive efforts, an increase in vegetation cover of 1,107 ha has been observed. However, there was also a destruction of vegetation cover of 3,745 ha in the same period (Table 9.3). Most of the vegetation cover has been transformed into vacant/open land. Therefore, due to the loss being greater than the increase, there has been a decrease in vegetation cover over time.

9.3.6 Dynamics of Wasteland-Riverine Sandy Area

Due to the presence of the Ganga and Varuna rivers, the expansion of the sandy area is found in the eastern and central parts of the study area, which remain almost in a constant state or sometimes exhibit minor spatial changes. From 1972 to 2020, the sandy area (riverine) increased by 132.39 ha at an average rate of 2.76 ha per year. At the decadal level, the most rapid growth rates in the sandy area (riverine) occurred during 1972–1980; because of this, the sandy area (riverine) increased at an average rate of 27.03 per year. In contrast, it declined rapidly (15.34 ha per year) during 1980–1990 (Figure 9.5). Figure 9.6 shows that the dominance of the sandy area (riverine) increased at the landscape level. The decrease in the MPS of the sandy area (Figure 9.8) and the increase in the number of patches show that the sandy area (riverine) is increasing (Figure 9.7). According to Table 9.4, the sandy area (riverine) absorbed 273 ha of other land use/land cover. In contrast, other land use/land cover has expanded to 141 ha of sandy area (Table 9.3).

9.3.7 Dynamics of Water Body

In the study area, rivers, ponds, and small drains (Nalas) come under the water body. As shown in Figure 9.4, the largest decrease in the water body area occurred during 1972–1980. The water body decreased to 927.66 ha during this decade at an average rate of 69 ha per year. After 2010, due to the renovation of old ponds and water sources, there has been an increase in the area of the water body at an average rate of 6.59 ha per year (Figure 9.5). Despite this increase, the average pattern of PLAND showed that the dominance of the water body decreased from 1972 to 2020 (Figure 9.6). There was a significant decline in the NP of the water body, indicating that small patches (mainly small ponds and lowland wet areas) of the water body were destroyed (Figure 9.7). Similarly, the size of the MPS has also decreased significantly, which shows that the water bodies in the study area have shrunk and their size has decreased significantly (Figure 9.8). The water body is spread over approximately 109 ha of other land use and land cover. Therefore, the maximum loss was caused by sandy areas (riverine) and agricultural land (Table 9.4). In contrast, the area under other land use/land cover spread over a 690 ha area of the water body, which decreased the area of the water body (Table 9.3). The maximum parts of the water body were transformed into sandy areas, agricultural land, and vacant land (Table 9.4).

9.4 Conclusion

The unplanned and rapid expansion of built-up land in Varanasi City and its peri-urban areas has resulted in a rapid land use/land cover transformation. The development of national and state highways and related link roads has also accelerated the expansion of built-up land in the northwest, southwest, and southern parts of the Varanasi city. This unplanned built-up land expansion has resulted in the rapid loss of other land use/land cover in these areas. Over time, the expansion of community-level facilities, transportation facilities, and infrastructure (water, sewerage, education, etc.)

facilities has moved from Varanasi City to peri-urban areas, which has greatly promoted land use/land cover changes in peri-urban areas. These reasons have led to the fragmentation of patches of land use/land cover classes in the study area. This has led to a rapid increase in landscape-level spatial heterogeneity in land use/land cover classes. Increasing spatial heterogeneity has reduced continuity in the mosaic of land use/land cover classes.

Land use, land governance, and land ownership are interdependent factors that have a significant impact on land use planning in the area. Due to this effect, planning and sustainable development are stressful tasks. This stress is compounded when peri-urban areas are excluded from the planning process. City planning cannot successfully achieve its objectives if peri-urban areas are left unplanned. Therefore, when formulating policies for urban growth and redevelopment, special attention should be given to the city along with its peri-urban areas so that it becomes easier to design plans and programs related to the spatial arrangement of land use/land cover. To obtain a balanced insight into the existing spatial arrangement of land use/land cover in the urban planning process, it is necessary to analyze the spatiotemporal pattern of the landscape structure of land use/land cover along with the traditional analysis of land use/land cover. Hence, the output of this study may help local planners, stakeholders, and concerned scholars better understand land use/land cover dynamics at the local level with landscape-based insight.

References

Singh, S., Singh, S. K., Kumar, R., Shrama, A., & Kanga, S. (2022). *Finding Alternative to River Sand in Building Construction.*

Singh, S., Meraj, G., Kumar, P., Singh, S. K., Kanga, S., Johnson, (2023). Decoding Chambal River Shoreline Transformations: A Comprehensive Analysis Using Remote Sensing, GIS, and DSAS. *Water*, 15(9), 1793.

Singh SK, Kanga S, Gulati B, Raič M, Sajan B, Đurin B, Singh S (2023) Spatial and Temporal Analysis of Hydrological Modelling in the Beas Basin Using SWAT+ Model. *Water* 15, 3338. https://doi.org/10.3390/w1519333

Rani A, Gupta SK, Singh SK, Meraj G, Kumar P, Kanga S, Đurin B, Dogančić D (2023) Predicting Future Land Use Utilizing Economic and Land Surface Parameters with ANN and Markov Chain Models. *Earth* 4(3):728-751.

Ahmad T, Gupta SK, Singh SK, Meraj G, Kumar P, Kanga S (2023) Unveiling Nature's Resilience: Exploring Vegetation Dynamics during the COVID-19 Era in Jharkhand, India, with the Google Earth Engine. *Climate* 11(9):187.

Sharma M, Upadhyay RK, Tripathi G, Kishore N, Shakya A, Meraj G, Kanga S, Singh SK, Kumar P, Johnson BA, Thakur SN (2023) Assessing Landslide Susceptibility along India's National Highway 58: A Comprehensive Approach Integrating Remote Sensing, GIS, and Logistic Regression Analysis. *Conservation* 3(3):444-459.

Saqib N, Rai PK, Kanga S, Kumar D, Đurin B, Singh SK (2023) Assessment of Ground Water Quality of Lucknow City under GIS Framework Using Water Quality Index (WQI). *Water* 15(17):3048.

Choudhury U, Singh SK, Kumar A, Meraj G, Kumar P, Kanga S (2023) Assessing land use/land cover changes and urban heat island intensification: A case study of Kamrup Metropolitan District, Northeast India (2000–2032). *Earth* 4(3):503-521.

Rai NK, Mathur S, Singh SK, Tiwari M, Singh VK, Haque R, Tiwari S, Sharma LK (2023) Differential regulation of mitochondrial complex I and oxidative stress based on metastatic potential of colorectal cancer cells. *Oncol. Lett.* 26(1):1-1.

Debnath J, Sahariah D, Lahon D, Nath N, Chand K, Meraj G, Kumar P, Singh SK, Kanga S, Farooq M (2023) Assessing the impacts of current and future changes of the planforms of river Brahmaputra on its land use-land cover. *Geosci. Front.* 14(4):101557. Elsevier.

Kumar P, Singh SK, Gupta SK, Kanga S, Meraj G (2023) Uncovering the hydro-meteorological drivers responsible for forest fires utilizing geospatial techniques. *Theor. Appl. Climatol.* https://doi.org/10.1007/s00704-023-04497. Springer Nature.

Singh SK, Kotecha MJ, Kanga S, Pankhaniya SK, Agrawal S, Meraj G (2023) Urban Heat Island (UHI) Resilience Plan in Varying Climatic Conditions Using Geospatial Approach: A Case Study of Rajkot City. In: *Climate Change, Agriculture and Society*, vol 1. pp 307-338.

Singh S, Singh SK, Prajapat DK, Pandey V, Kanga S, Kumar P, Meraj G (2023) Assessing the Impact of the 2004 Indian Ocean Tsunami on South Andaman's Coastal Shoreline: A Geospatial Analysis of Erosion and Accretion Patterns. *J. Mar. Sci. Eng.* 11(6):1134.

Rathore SS, Mathur SC, Mathur S, Sharma S, Singh SK (2023) Late Cretaceous–Early Paleocene Geodiversity of Fatehgarh Formation of Petroliferous Barmer Basin, Western Rajasthan, India: A Potential Geopark Site. *Geoheritage* 15:54.

Singh S, Meraj G, Kumar P, Singh SK, Kanga S, Johnson BA, Prajapat DK, Debnath J, Sahariah D (2023) Decoding Chambal River Shoreline Transformations: A Comprehensive Analysis Using Remote Sensing, GIS, and DSAS. *Water* 15(9):1793.

Tripathi G, Shakya A, Upadhyay RK, Singh SK, Kanga S, Pandey SK (2023) Landslide Susceptibility Mapping of Tehri Reservoir Region Using Geospatial Approach. In: *Climate Change Adaptation, Risk Management and Sustainable Practices in the Himalaya*, pp 135-156.

Sud A, Kanga R, Singh SK, Meraj G, Kanga S, Kumar P, Ramanathan AL, Bhardwaj V (2023) Simulating Groundwater Potential Zones in Mountainous Indian Himalayas—A Case Study of Himachal Pradesh. *Hydrology* 10(3):65.

Upadhyay RK, Tripathi G, Đurin B, Šamanović S, Cetl V, Kishore N, Sharma M, Singh SK, Kanga S, Wasim Md, Rai PK, Bhardwaj V (2023) Groundwater Potential Zone Mapping in the Ghaggar River Basin, North-West India, Using Integrated Remote Sensing and GIS Techniques. *Water* 15(5):961.

Lavane K, Kumar P, Meraj G, Han TG, Ngan LHB, Lien BTB, Ty TV, Thanh NT, Downes NK, Nam NGD, Minh HVT, Singh SK, Kanga S (2023) Assessing the effects of drought on rice yields in the Mekong Delta. *Climate* 11(1):13.

Nath N, Sahariah D, Meraj G, Debnath J, Kumar P, Lahon D, Chand K, Farooq M, Chandan P, Singh SK, Kanga S (2023) Land use and land cover change monitoring and prediction of a UNESCO world heritage site: Kaziranga eco-sensitive zone using cellular automata-Markov model. *Land* 12(1):151.

Kanga S, Upadhyay RK, Sharma M, Sharma S, Tripathi G, Singh SK (2023) Geotechnical Stability Assessment of Road-cut Slopes: A Case Study of Srinagar, Garhwal Himalaya, India. *Biolog. Forum* 15(5):1071-1080.

Gulati B, Sharma R, Kanga S, Singh SK, Sajan B, Meraj G, Kumar P, Ramanathan AL (2023) Unraveling the Relationship Between Stubble Burning and Air Quality Degradation in Punjab: A Temporal and Spatial Analysis (2019-2022). *J. Clim. Change* 9(2):43-53, IOS Press.

Tomar P, Singh SK, Kanga S, Pattanaik A (2023) Water Bodies mapping and monitoring using high-resolution satellite images. *Sustain. Agri Food Environ. Res.* 11.

Singh H, Meraj G, Singh S, Shrivastava V, Sharma V, Farooq M, Kanga S, Singh SK, Kumar P (2022) Status of Air Pollution during COVID-19-Induced Lockdown in Delhi, India. *Atmosphere* 13(12):2090.

Farooq M, Mushtaq F, Meraj G, Singh SK, Kanga S, Gupta A, Kumar P, Singh D, Avtar R (2022) Strategic Slum Upgrading and Redevelopment Action Plan for Jammu City. Resources 11(12):120.[11] Singh, S., Singh, S. K., Prajapat, D. K., Pandey, V., Kanga, S., Kumar, P., & Meraj, G. (2023). Assessing the Impact of the 2004 Indian Ocean Tsunami on South Andaman's Coastal Shoreline: A Geospatial Analysis of Erosion and Accretion Patterns. *J. Mar. Sci. Eng.*, 11(6), 1134.

Shruti, S., Singh, P. K., & Ohri, A. (2020). Evaluating the environmental sustainability of smart cities in India: The design and application of the Indian smart city environmental sustainability index. *Sustainability*, 13(1), 327.

Shruti, S., Singh, P. K., Ohri, A., & Singh, R. S. (2022). Development of Environmental Decision Support System for Sustainable Smart Cities in India. *Environ. Prog. Sustain. Energy*, e13817.

Singh, R. P., Kumar, S., & Rana, P. S. (2020). UN SDGs and context of Holy-Heritage cities in India: a study of Ayodhya and Varanasi. In *Geoecology of Landscape Dynamics* (pp. 187-206). Springer, Singapore.

10

Human Resource Influences on Online Shopping Behavior: Factors, Preferences, and Satisfaction Among Consumers in North Bengal, India

Subham Dey Sarkar[1]*, Sushmita Singh[1], Anuusua Poddar[1], Tuhin Dey Roy[2], Pranoy Dey[1,3] and Arindam Basak[1]

[1]*Department of Geography & Applied Geography, University of North Bengal, Dist. Darjeeling, West Bengal, India*
[2]*Department of Geography, Siliguri Mahila Mahabidyalaya, Siliguri, Dist. Darjeeling, West Bengal, India*
[3]*Department of Geography, Birpara College, Dist. Alipurduar, West Bengal, India*

Abstract
Technology and media have propelled global online shopping. Through customer service strategies, marketing efforts, and the cultivation of brand loyalty, human resource significantly influences online shopping behavior. Comfort, product variety, cost savings, and information accessibility are driving online purchasing. Age, income, and Internet use impact online shopping. Website, product, and security quality are key to online shopping, impacting consumer satisfaction. This study used questionnaires to collect data from 700 respondents in Siliguri, Jalpaiguri, Alipurduar, Koch Bihar, Raiganj, Balurghat, and Malda. Between April and July 2023, data was collected through survey on respondents' demographics, Internet usage, and purchasing preferences. Analyses included descriptive statistics, chi-square testing, factor analysis, and importance-performance. Demographic data showed that most North Bengal Internet buyers were 26–35 years old, had greater monthly earnings, and spent more time online daily. Consumers favored electronics, apparel, and cosmetics. Based on factor analysis, convenience, online shopping obstacles, product comparison and variety, and pricing awareness influence online shopping. According to the Importance-Performance Analysis, online shopping

**Corresponding author*: rs_subhamd@nbu.ac.in

Shruti Kanga, Gowhar Meraj, Suraj Kumar Singh, Majid Farooq and M. S. Nathawat (eds.) Geospatial Technology for Natural Resource Management, (259–286) © 2024 Scrivener Publishing LLC

satisfied customers in terms of account management, user-friendly navigation and return and refund policies. However, customer support and security needed improvement. Regional development and customer satisfaction will depend on addressing performance gaps and ensuring a secure and seamless online shopping experience.

Keywords: Online shopping, consumer, human resource, factor analysis, importance-performance analysis

Introduction

The act of purchasing and selling products and services through the Internet has taken over the worldwide market. Customers cannot touch, smell, or feel things when they purchase online, which is analogous to browsing in a paper catalogue. Customers may explore the company's products and services online by examining the photographs and images, product details, features, and prices specified therein and making real-time purchases (Jahwari *et al.*, 2018). In recent years, the development of mass media and technical advances has made online shopping the most popular type of shopping (Lim *et al.*, 2016; Wu *et al.*, 2011). For more than a decade, the Internet has been fostering consumer empowerment (Pires *et al.*, 2006), resulting in a constant growth in the number of online shoppers as well as an increase in the revenue earned from online purchasing (Ozen & Engizek, 2014). Customers nowadays choose to purchase online rather than physically visiting market places for the purpose of buying goods and services. Nonetheless, e-commerce has benefited physical firms as they are now attempting to obtain a competitive edge by interacting with their clients through e-commerce (Lee and Lin, 2005). One of the main reasons behind increase in online shopping can be attributed to the fact that online shopping saves time and effort since customers may purchase from the comfort of their homes at any time, which is especially beneficial for individuals who work long hours and do not have much spare time (Wolfinbarger & Gilly, 2001). Aside from being more comfortable and less time consuming, Internet shopping provides shoppers with a greater product variety, reduced costs, unique and customized service, and simple access to information (Ahn, Ryu, & Han, 2004). Despite the numerous benefits of Internet buying, not all online customers view it the same way. Purchasing in general and Internet purchasing in particular, is exceedingly personal, and hence highly subjective in nature. Individuals' motivations and goals for purchasing online differ. Similarly, online consumers' and purchasers' perceptions of online purchasing and e-services differ. External variables such as consumer attributes, situational circumstances, product characteristics, past online buying experiences, and trust in online shopping impact people's attitudes regarding online shopping and their intents to shop online (Monsuwe, Delleart, & Ruyter, 2004).

Consumer behavior in the context of online buying is also heavily influenced by their demographic traits. People from various social and economic backgrounds respond differently to Internet buying. Age, gender, income, marital status, level of education, and religion can all have a significant influence on their purchasing habits. Buyers' income or purchasing power impacts whether a firm may gain a market share among a target population (Kotler and Armstrong, 2005). The more the disposable income, greater is the demand for goods and services. Consumer behavior is a complicated, dynamic, multi-dimensional process, and all marketing choices are founded on consumer behavior assumptions (Bhinde & Joshi, 2016). Donthu and Garcia (1999) discovered a favorable link between income and Internet buying intentions. Higher-income customers are likewise recognized to profit more from online services (Lambrecht and Seim 2006). Consumers who are "income rich but time poor" may adopt a time-saving purchasing orientation, whereas those who are "time rich but income poor" may adopt a money-saving buying orientation. Consumer characteristics such as income, education, and generational age moderate the relationship between consumer beliefs and online purchase behavior, as do consumption values such as the proclivity to consider many alternatives before making a decision, the enjoyment of shopping, and the proclivity to research products before making a purchase (Punj, 2011). According to the findings of the study, different demographic parameters such as age, gender, marital status, family size, and income have a major impact on online purchasing in India. Gender, as one of the demographics factors, influences consumer Internet possession and frequency of online purchases (G. Nagra and R. Gopal, 2013).

Yet another equally important aspect of online shopping is customer satisfaction. Oliver (1980) has opined that "customer satisfaction is a summary of cognitive and affective reaction to a service incident (or sometime to a long-term service relationship). Satisfaction (or dissatisfaction) results from experiencing a service quality encounter and comparing that encounter with what was expected." The most difficult task for online buying is providing and maintaining consumer happiness. A service-focused approach is a critical success element in surviving in a fiercely competitive e-environment. A corporation must provide outstanding service experiences to its consumers in order for them to repurchase and remain loyal to the company (Gounaris *et al.*, 2010). Consumers' intent to buy is influenced by a variety of elements, including the look of the website, quality content, and video clipped narration of the items, rather than actual experience (Kolesar and Galbraith, 2000). All of this has an effect on the degree of client satisfaction. According to Lin, Wu, and Chang (2010), the most significant criteria influencing online consumer satisfaction are delivery quality, product quality, service quality, system quality, and a user-friendly interface. Zamri (2014) revealed that three aspects impact online shopping customer satisfaction: product quality, brand variety, and shopping experience. According to Al Karim (2013), the information provided with diverse items, affordable prices, a simple purchasing procedure, and damage-free delivery are the catalysts for online consumer happiness. Client loyalty and the successful adoption of online commerce are driven by the privacy and security of client information, as well as the speed of service (Zatalini and Pamungkas, 2016). The dangers connected with online purchases can influence customer behavior and pleasure. A customer examines the numerous hazards involved with online transactions before acquiring a product, which are referred to as perceived or expected risks (Javadi *et al.*, 2012). The kind and degree of risk perceived by a consumer while considering a certain purchasing choice is referred to as perceived risk (Cox and Rich, 1964). Similarly, security is a fundamental issue for online buyers; improved system security consequently improves client happiness (Suh and Han, 2003). Attracting and retaining customers is important to the success of online merchants. The researchers focus more on customer demographics and experience. According to Lakshmanan (2016), the factors that influence consumer satisfaction with online purchasing include product preferences, customer issues, a greater number of items with discounts, risk perception, and after-sales services. Financial transaction security, convenience, and site design all have a substantial influence on client satisfaction (Evanschitzky *et al*, 2004). According to Szymanski and Hise (2000), the dominant elements influencing e-satisfaction ratings include consumer views of online ease, merchandising (product offers and product information), site design, and financial security. Similarly, entertainment, convenience, dependability, information quality, and speed all play essential roles in online consumer satisfaction (Sang Yong Kim & Young Jun Lim, 2001). Human factor (number of visits to site), amusement, instructive, irritability, usefulness, attitude towards website, and intention to return are all significant aspects of a successful website that influence consumers' online purchasing behavior (Hausman & Siekpe, 2008). Website quality, according to Bai *et al.* (2008), has a direct and positive influence on customer satisfaction, and customer satisfaction has a direct and positive impact on purchase intentions. For any business to sustain itself in the long run, customer loyalty is of utmost importance. It is therefore prerequisite for the providers of online goods and services and entrepreneurs to have perfect knowledge about the taste and preferences of the customers and be aware of the factors that affect online customers' level of satisfaction.

Influences of Human Resource on Online Shopping Behavior

Human resources influence online shopping behavior in several significant ways, particularly through the development and implementation of customer service strategies, marketing efforts, and the cultivation of brand loyalty. By training customer service representatives to provide exceptional support, companies can enhance the overall shopping experience, thereby increasing consumer satisfaction and repeat purchases. Marketing teams, often managed and coordinated by human resource departments, play a crucial role in understanding consumer preferences and tailoring advertisements to meet the specific needs of the target demographic. Additionally, HR practices that focus on employee engagement and satisfaction can lead to a more motivated workforce, which in turn improves the quality of interactions consumers have with the brand, whether through social media, live chat, or other communication channels. This holistic approach to managing human resources ensures that every touchpoint between the consumer and the company contributes positively to the online shopping experience, ultimately driving higher sales and stronger customer loyalty.

Database and Methodology

In general, the study's findings are descriptive. Questionnaires were used to assemble the data set. Survey was carried out in seven cities of North Bengal, India, *viz.*, Siliguri, Jalpaiguri, Alipurduar, Koch Bihar, Raiganj, Balurghat, and Malda, with an equal representative sample of urban residents from these cities. Seven hundred people participated in the survey. The information was gathered between April 2023 and July 2023 by direct contact (including e-mail). Clarity and accuracy issues with the questionnaire were anticipated through pilot testing. As a result of the comments and input, numerous adjustments were made to enhance the representation of the contents. There are three distinct sections that make up the questionnaire. The first section is comprised of introductory questions that provide specifics regarding the socio-economic profile of the consumers as well as their behavior with regard to the utilization and the amount of time spent on the Internet, and the shopping websites that are preferred while making the purchase. The second section is comprised of questions that are related to the factors that influence online shopping. A questionnaire with a total of 14 items was used to investigate the factors that are relevant for consumers' preferences regarding online shopping. The questionnaire used a five-point Likert scale with response ranges from "strongly agree" to "strongly disagree," with the following equivalencies: "strongly disagree" = 1, "disagree" = 2, "neutral" = 3, "agree" = 4, and "strongly agree" = 5. The third section is comprised of questions to highlight consumer satisfaction on 20 online shopping attributes. In this section, a five-point Likert scale was also used to find out the feedback of respondents on the importance and performance of these 20 online shopping attributes. Following the completion of the data collection, it was entered manually into the Excel sheet. Descriptive statistics and a chi-square test were used to signify the results of the analysis of demographic profile of shoppers and most visited category of items. For the analysis of reasons for adoption or non-adoption of online shopping, factors analysis has been performed on Statistical Package for the Social Science (SPSS) for the purpose of factor reduction. To study the impact of demographic characteristics of the respondents on their shopping behavior, data related age, sex, education, monthly income and usage of Internet have been collected during the course of the primary

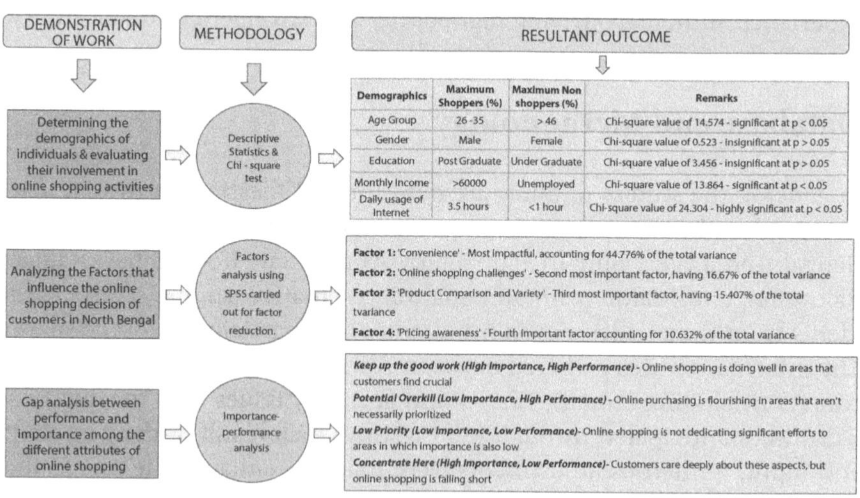

Figure 10.1 Methodological framework.

survey. To study the customer satisfaction level, importance-performance analysis has been used to better understand customers' perceptions of the significance of elements impacting their online shopping behavior and their actual experience with it (Figure 10.1).

Factor Analysis

Factor analysis began in the early 1900s with Charles Spearman's interest in human capacity and the creation of the Two-Factor Theory, which led to a flurry of work on the ideas and mathematical concepts of component analysis (Harman, 1976). As a result of computer technical improvements, factor analysis is now employed in various domains such as behavioral and social sciences, medicine, economics, and geography (Rummel, 1970). Factor analysis is based on the idea that measurable and observable variables may be reduced to fewer latent variables that have a similar variance but are unobservable, a process known as dimensionality reduction (Bartholomew, Knott, & Moustaki, 2011). A researcher would often use a survey to ask multiple questions about a product while researching client satisfaction. These questions will cover topics like product features, simplicity of purchasing, usability, cost, visual appeal, and so on. These are usually measured on a numerical scale. A researcher, on the other hand, seeks the underlying dimensions or "factors" of consumer pleasure. These are usually psychological or emotional reactions to the product that cannot be assessed directly. The variables from the survey are used in factor analysis to determine them indirectly. These survey variables are condensed into one or more factors when a researcher assumes them. These factors were discovered using principal component analysis. The first factor is obtained by calculating the largest variance. The variance is then subtracted and replaced by the factor. The second factor is then chosen by the next biggest variance, and the procedure is repeated until no further variances exist. Once the factors have been retrieved, the questionnaire's score is believed to be linearly connected to the components. Along with all of the factors in the equation, the margin of error is considered.

There are essentially two types of factor analysis. A) **Exploratory factor analysis (EFA)**: in exploratory factor analysis, the researcher does not make any assumptions about prior relationships between factors. In this method, any variable can be related to any factor. This helps identify complex relationships among variables and group them based on common factors. The EFA is used when a researcher wants to discover the number of factors influencing variables and to analyze which variables 'go together' (De Coster,

1998). B) **Confirmatory factor analysis (CFA):** the confirmatory factor analysis, on the other hand, believes that variables are associated to certain factors and use pre-established theory to validate the model's assumptions. The CFA is used to test hypotheses and represents variables and components using route analysis diagrams (Child, 2006). This approach has been frequently used by academics to identify the characteristics that influence customer behavior while purchasing online. To empirically determine the key underlying dimensions in the motivations data, principal component factor analysis is a commonly used method. Convenience, product variety, purchasing environment, information depth, and brand are emergent motivational factors that affect online shopping behavior of customers (Bagdonienel & Zemblyte, 2009). Similarly, convenience, accessibility, scope, attraction, reliability, experience, and clarity were important factors considered by the online shoppers (Rishi, 2008). Exploratory factor analysis was undertaken for the buyers group to analyze the underlying structure of obstacles to buying online from the perspective of Internet users who do shop online. An EFA for the buyer categories may be undertaken to analyze the underlying structure of obstacles to buying online from the perspective of Internet users who do purchase online (Kumar & Dange, 2014). To find the characteristics that influenced consumers' propensity to purchase products from an online shop, a principle components analysis was used to minimize the number of criteria to a smaller number of factors. Confirmatory factor analysis was used to ensure the accuracy of the factors (Bucko *et al.*, 2018).

Importance-Performance Analysis

Importance-performance analysis (IPA) is a business research technique that was originally developed as a market research tool to analyze and recommend management strategies. The IPA combines customer perception and importance metrics into a two-dimensional (2D) plot to make it easier to interpret data (Martilla & James, 1977). Importance-performance analysis employs a simple graphical approach to compare the mean score for 'perceived importance' of various attributes with the corresponding 'satisfaction rating' using a two-dimensional grid. This grid classifies mean scores into four categories to aid in data interpretation and assessing management priorities: 'concentrate here,' 'keep up the good work,' 'low priority,' and 'possible overkill,' allowing management to identify the areas of highest concern that warrant the use of available resources (Bagri & Kala, 2015). Its primary goal is to identify the performance of various product or service characteristics, while making it easier to interpret data and generate actionable recommendations for management (Dwyer *et al.*, 2012). The IPA identifies the most important characteristics, strengths and weaknesses. This gives insight into which areas of product or service providers should focus their efforts on. It prioritizes managerial initiatives to propose the most effective use of scarce resources that should enhance and maintain customer satisfaction. Importance can be defined as the perceived value or importance of the attributes of a product or service to an individual (Abalo *et al.*, 2007; Chu & Choi, 2000). The IPA is widely regarded as an effective tool for assessing customer satisfaction and managing strategies. According to Huang, 2015, IPA is an effective way to identify those aspects of a product or service that need the most improvement or are candidates for cost-reduction conditions without significantly impacting the overall quality of the product or service. Magal *et al.*, (2005) emphasizes the value of IPA, which was first developed by Martilla (1977) and James (2001). It provides managerial services with useful information to measure satisfaction and allocate resources effectively. Importance-performance analysis is a widely accepted analytical approach that provides guidance for the management of customer satisfaction. It is characterized by its 2D structure, which is based on the perceived significance of service characteristics and attribute performance, and thus provides a useful tool for making strategic decisions in customer satisfaction management (Matzler *et al.*, (2003). The IPA, according to the article, facilitates a more effective allocation of resources in order to improve tourist satisfaction and enhance the attractiveness of a destination (Sever, 2015). Wyryd & Biesok (2017) highlight the visual representation of the data through IPA as a valuable tool for identifying areas of concern, allowing you to make informed decisions and make targeted changes to your marketing strategies. Importance-performance analysis is a tool of marketing research that is very versatile and practical. It is easy to use, especially when you want to highlight the key factors and attributes of your product or service that need urgent attention and improvement. Susanty *et al.* (2019) contribute to the understanding of IPA by highlighting its utility in mapping problems into four quadrants, enabling more targeted decision-making. The authors stress that each quadrant serves a distinct function, allowing for nuanced and specific actions based on the identified issues. The IPA, as described by Susanty *et al.*, unveils the relationship between the importance of an indicator and the perceived satisfaction or performance, providing a structured approach to addressing concerns and improving strategic decision-making.

The present study has therefore made use of importance-performance gap analysis to better understand the perception of customers about the importance of the factors affecting their online purchase behavior and the performance or the actual experience of the same while doing online shopping. It also highlights their satisfaction level and helps the in formulation of better marketing strategies. For this, the online customers and buyers were made to rate the importance they assign of different parameters of online shopping before actually shopping on the scale of 1 to 5, where 1 = very bad, 2 = bad, 3 = average, 4 = good, and 5 = very good. Similarly, the respondents were again made to rate their actual experience of the same parameters of online shopping on the scale of 1 to 5, where 1 = very bad, 2 = bad, 3 = average, 4 = good, and 5 = very good. Further, the difference between importance and performance for each variable was calculated wherein, positive difference is indicative of greater importance but lesser performance and vice-versa.

Result

The SPSS version 20 was utilized in order to carry out a comprehensive quantification and analysis of the primary data that was obtained from the questionnaires. This statistical tool makes it possible to be accurate while also making it very simple to comprehend data.

Demographic Profile of Online Shopping Adopters and Non-Adopters

Table 10.1 represents the results of demographic status of the respondents along with question related to usage of Internet. According to the results, the maximum percentage of shoppers belong to 26–35 age group and on the other hand, the maximum percentage of non-shoppers belong to >46 age group. The chi-square value for age as a determining factor of online shoppers and non-shoppers is depicting that it has a significant effect. In the next category, the maximum number of shoppers belongs to males and the maximum number of non-shoppers belongs to females. In this case, chi-square value is insignificant so it can be said that gender has less influence on shoppers and non-shoppers. Whereas the educationally maximum shoppers belong to post graduate group and the maximum non-shoppers belong to undergraduate group, here, the chi-square value is also insignificant to infer anything. Next, the monthly income has a significant effect in determining the shoppers and non-shoppers as the chi-square value

Table 10.1 Online shoppers vs non-shoppers: demographics.

Demographics no.		Shoppers		Non-shoppers		Chi-square
		%	No.	%		
Age	15–25	66	55.46	53	44.54	df = 3 Calculated chi-square = 14.574 Critical value= 7.815 p-Value = 0.002 Significance level = 0.05
	26–35	151	63.18	88	36.82	
	36–45	143	52.00	132	48.00	
	>46	26	38.81	41	61.19	
Gender	Male	195	56.52	150	43.48	df= 1 calculated chi-square= 0.523 critical value= 3.841 p-value= 0.470 significance level= 0.05
	Female	191	53.80	164	46.20	
Education	Undergraduate	106	50.96	102	49.04	df = 3 Calculated chi-square = 3.456 Critical value = 7.815 p-Value = 0.326 Significance level= 0.05
	Graduate	133	55.19	108	44.81	
	Post graduate	104	60.47	68	39.53	
	Any other	43	54.43	36	45.57	
Monthly income	Unemployed	69	42.86	92	57.14	df = 3 Calculated chi-square = 13.864 Critical value = 7.815 p-Value = 0.003 Significance level = 0.05
	< Rs.15,000	58	56.86	44	43.14	
	Rs.15,000–Rs.30,000	113	56.78	86	43.22	
	> Rs.30,000	146	61.34	92	38.66	
Daily usage of Internet	< 1 hour	53	38.69	84	61.31	df = 3 Calculated chi-square = 24.304 Critical value = 7.815 p-Value = 0.000 Significance level= 0.05
	1–3 hours	104	55.03	85	44.97	
	3–5 hours	112	56.57	86	43.43	
	> 5 hours	117	66.48	59	33.52	

indicates. The income picture suggests that maximum shoppers belong to >60,000 income group whereas maximum non-shoppers belong to unemployed group. The income level is very much positively correlated with shoppers as more income suggests more positive changes in lifestyle. The

Table 10.2 Consumer preferences on the category of items in online shopping.

Category of items	Mean	Rank
Electronics, computer hardware and gadgets	3.59	1
Apparel/clothing and footwear	3.53	2
Cosmetics, jewelry, and accessories	3.37	3
Home and kitchen appliance	3.35	4
Food and grocery	2.96	5
Gifts, greetings, and flowers	2.63	6
Movies, music, and video games	2.30	7
Books	2.29	8
Toys and baby products	1.80	9
Stationary items	1.79	10
Sports, fitness, bags, and luggage	1.51	11
Health and personal care	1.41	12

chi-square value of daily Internet usage also suggests that it has a significance effect on determination of shoppers and non-shoppers. The maximum percentage of shoppers belongs to that group which have a usage of daily 3–5 hours and on the other hand the maximum non-shoppers belong to the lowest Internet usage group.

The responses to the questionnaires on the items that are most frequently purchased through online shopping sites are presented in Table 10.2. The replies from customers placed "electronics, computer hardware, and gadgets" at the top of the list, followed by "apparel/clothing and footwear" and "cosmetics, jewelry, and accessories" in that order. Consumers in North Bengal put "health and personal care" products last on their list of priorities.

Factors of Using Online Shopping

Factor analysis was used to determine the various motivations for online shopping that were expressed by consumers. The responses that were

Table 10.3 KMO and Bartlett's test.

Kaiser-Meyer-Olkin measure of sampling adequacy		**0.744**
Bartlett's test of sphericity	Approx. chi-square	14058.576
	df	91
	Sig.	0.000

obtained were analyzed using this method, and the data that resulted were subjected to the Kaiser-Meyer-Olkin (KMO) measure of sampling adequacy and the Bartlett's test of sphericity. According to Table 10.3, the KMO value for this study was 0.744, which is more than 0.60; this demonstrates that the values are adequate, therefore factor analysis can proceed. The results of the Bartlett's test of sphericity are supported by the findings presented in Table 10.3, which demonstrate that the chi-square value of 14058.576 with 91 degrees of freedom (df) is statistically significant at the 0.000 level.

According to Table 10.4, the first four factors account for 85.489% of the total variation. Therefore, only the four factors with eigen values larger than 1.0 are kept, and the rest are eliminated. In light of these eigen values, shown in Table 10.4, it is possible to reduce the original set of 14 variables down to only four factors.

Factor loadings represent direct associations between the variables and the factors. In this instance, the Varimax rotation method has been employed (Table 10.5). To improve the factors' interpretability, this orthogonal rotation technique reduces the number of variables with high loadings on a given factor. The prevalence of the four interrelated factors is suggested by the rotated component matrix.

Principal component analysis using the rotation technique (Varimax with Kaiser normalization) converged after five iterations. Four components can be extracted, as shown in Table 10.6:

Factor 1: Component 2, 4, 6, 11 (convenience)
Factor 2: Component 7, 9, 10, 12 (online shopping challenges)
Factor 3: Component 3, 13, 14 (product comparison and variety)
Factor 4: Component 1, 5, 8 (pricing awareness)

Factor 1: among the factors considered, "convenience" was shown to be the most impactful, accounting for 44.776% of the total variance and having a higher eigenvalue (6.269). Very strong correlations exist between the four

Table 10.4 Total variance explained.

Component	Initial eigenvalues			Extraction sums of squared loadings			Rotation sums of squared loadings		
	Total	% of Variance	Cumulative %	Total	% of Variance	Cumulative %	Total	% of Variance	Cumulative %
1	6.269	44.776	44.776	6.269	44.776	44.776	3.419	24.420	24.420
2	2.334	16.673	61.449	2.334	16.673	61.449	3.246	23.186	47.606
3	2.157	15.407	76.857	2.157	15.407	76.857	3.000	21.425	69.032
4	1.489	10.632	87.489	1.489	10.632	87.489	2.584	18.457	87.489
5	0.975	6.963	94.452						
6	0.201	1.437	95.889						
7	0.142	1.016	96.906						
8	0.124	0.887	97.793						
9	0.089	0.636	98.429						
10	0.075	0.536	98.965						
11	0.051	0.366	99.331						
12	0.039	0.275	99.606						
13	0.033	0.235	99.842						
14	0.022	0.158	100.000						

Extraction method: principal component analysis.

Table 10.5 Rotated component matrix.

Statements	Component			
	1	2	3	4
I choose online shopping because of the substantial discounts provided.	0.108	−0.097	0.096	**0.948**
It is more convenient to shop online instead than in-store.	**0.824**	−0.386	0.014	0.058
Online shopping provides more alternatives to select from.	0.085	−0.184	**0.914**	0.076
It is more convenient to pay online using a credit card or bank transfer.	**0.850**	−0.377	−0.050	−0.022
The items purchased online from reputable retailers are of superior quality and are priced affordably.	0.008	−0.352	−0.087	**0.770**
Online shopping saves me time and money since I can avoid heavy traffic and save on fuel.	**0.900**	0.005	0.240	0.146
I have some reservations regarding the security and privacy of my online payments.	−0.237	**0.911**	−0.205	−0.186
The discount doesn't apply to all of the items, so I can only buy a few.	0.137	−0.064	0.100	**0.945**
There are occasional delays in the product's delivering.	−0.292	**0.810**	−0.320	−0.217
I don't use credit cards, which limits my access to online shopping.	−0.347	**0.570**	−0.429	−0.238
I choose online shopping owing to the convenience of Internet purchasing methods.	**0.932**	−0.147	0.144	0.134
I can't buy something online without first seeing it in person.	−0.162	**0.951**	−0.095	−0.105
When I shop online, I always make sure to check the reviews first so that I can get the best product at the best price.	0.079	−0.138	**0.943**	0.018
Online shopping makes it simple to acquire things from a specific brand.	0.091	−0.165	**0.911**	0.008

Extraction method: principal component analysis.
Rotation method: Varimax with Kaiser normalization.
Rotation converged in five iterations.

Table 10.6 Formation of factors.

Factor no.	Name of factors	Item no.	Variables	Factor loading
1	Convenience	2	It is more convenient to shop online instead than in-store.	**0.824**
		4	It is more convenient to pay online using a credit card or bank transfer.	**0.850**
		6	Online shopping saves me time and money since I can avoid heavy traffic and save on fuel.	**0.900**
		11	I choose online shopping owing to the convenience of Internet purchasing methods.	**0.932**
2	Online shopping challenges	7	I have some reservations regarding the security and privacy of my online payments.	**0.911**
		9	There are occasional delays in the product's delivering.	**0.810**
		10	I don't use credit cards, which limits my access to online shopping.	**0.570**
		12	I can't buy something online without first seeing it in person.	**0.951**
3	Product comparison and variety	3	Online shopping provides more alternatives to select from.	**0.914**
		13	When I shop online, I always make sure to check the reviews first so that I can get the best product at the best price.	**0.943**
		14	Online shopping makes it simple to acquire things from a specific brand.	**0.911**
4	Pricing awareness	1	I choose online shopping because of the substantial discounts provided.	**0.948**
		5	The items purchased online from reputable retailers are of superior quality and are priced affordably.	**0.770**
		8	The discount doesn't apply to all of the items, so I can only buy a few.	**0.945**

statements loaded into Factor 1. According to the study, shoppers save time and energy by purchasing goods online rather than braving the crowds and gridlock of traditional brick-and-mortar stores. Online payment options like credit cards and electronic payments are seen as more convenient by some consumers.

Factor 2: "online shopping challenges" has been identified as the second most important factor, explaining 16.67% of the total variation. There are four statements that define the difficulties of online shopping. Given that they must provide their credit card information to purchase online, customers' main concern is the legitimacy of some websites. Some customers experience issues with sporadic delivery delays. Some consumers are hesitant to purchase online since they cannot physically touch the merchandise.

Factor 3: "product comparison and variety" appeared as the third most important factor after further investigation. It was discovered that 15.407% of the factor's variants fit the description. There were three statements about this factor, and all of them were linked to each other. Consumers believe they have more alternatives to select from when they purchase online. It also indicates that online shoppers are seeking name-brand products. Some buyers rely on ratings and comments to help them choose the best product at the lowest price.

Factor 4: "pricing awareness" was the fourth important factor that came out of the factor analysis. It accounted for 10.632% of the total variance and was made up of three statements that were closely related. When compared to in-store purchases, the substantial savings offered by shopping online for certain items makes online shopping the preferred method for the vast majority of customers. Also, consumers think that there are good websites that they can trust for both price and quality.

Consumer Satisfaction on Online Shopping Attributes

From Table 10.7 it has been observed that among all the given attributes, the "return and refund policy" A8, has a high priority rating from customers, with an importance rating of 4.14. With a performance score of 3.58, social media has performed admirably in this regard. The performance rating of "customer support - code A7 (live chat, phone, e-mail)", is noticeably lowest at 1.21 among all the attributes, in spite of it having a comparatively high importance rating of 3.19, which indicates that consumers value this attribute highly but online retail sites are not living up to customers' expectations in this area. The performance rating, at 4.39, is considerably greater for "social media integration (code A20)", however, it received a

Table 10.7 Gap analysis between importance and performance among the different attributes of online shopping.

Attributes	Codes	Importance (I)	Performance (P)	(I-P)	Pearson correlation	t Stat	p Value
Account management (profile, order history)	A1	3.21	3.16	0.05	0.827	3.848	0.000
User-friendly navigation	A2	3.59	3.42	0.17	0.754	6.110	0.000
Product reviews and ratings	A3	3.90	2.41	1.49	0.165	29.990	0.000
Pricing transparency	A4	3.62	2.66	0.96	−0.021	23.080	0.000
Product descriptions and images	A5	3.97	2.28	1.69	0.211	31.082	0.000
Customer feedback and surveys	A6	3.80	2.36	1.44	0.263	25.329	0.000
Customer support (live chat, phone, e-mail)	A7	3.19	1.21	1.98	0.152	37.326	0.000
Return and refund policy	A8	4.14	3.58	0.56	0.496	13.800	0.000
Website loading speed	A9	2.21	3.51	−1.29	0.275	−22.198	0.000
Responsiveness (mobile and desktop)	A10	1.95	3.75	−1.81	0.065	−25.094	0.000
Product search functionality	A11	2.38	2.29	0.09	0.940	4.892	0.000
Wishlist and cart saving	A12	3.32	3.82	−0.49	0.732	−12.719	0.000
Loyalty programs and discounts	A13	3.95	2.83	1.12	0.292	22.739	0.000
Payment options	A14	3.93	4.11	−0.18	0.811	−8.912	0.000
Product reviews and ratings	A15	2.43	4.38	−1.95	−0.334	−44.272	0.000
Security and privacy	A16	3.91	2.37	1.54	0.392	27.935	0.000
Order tracking	A17	3.50	3.56	−0.06	0.891	−2.977	0.002
Mobile-friendly design	A18	2.37	3.66	−1.29	0.223	−21.655	0.000
Shipping options (speed and cost)	A19	3.89	3.34	0.55	0.471	13.386	0.000
Social media integration	A20	1.72	4.39	−2.68	−0.102	−40.019	0.000

lower importance rating from consumers (1.72). In IPA, the gap refers to the discrepancy between the value customers attach to a specific attribute and their perception of its performance. As illustrated in Figure 10.2, positive gaps refer to areas where performance surpasses the importance of the attribute, indicating strengths that customers may not fully appreciate. Conversely, negative gaps indicate areas where improvement is required, as performance does not match the perceived importance of the attribute. Therefore, it is important to focus on addressing those with negative gaps in order to align performance to customer expectations and enhance customer satisfaction. Conversely, positive gaps highlight areas of strength, which can be used in marketing or improved for a competitive advantage. From the perspective of highest and lowest gaps, it can be interpreted that, in the given figure, "product search functionality" has the lowest gap with an attribute score of 0.09, having an importance of 2.38 and a performance rating of 2.29 which means that online shopping has performed considerably well in this attribute in reference to the Importance mentioned by customers having a very low gap. On the other hand, "social media

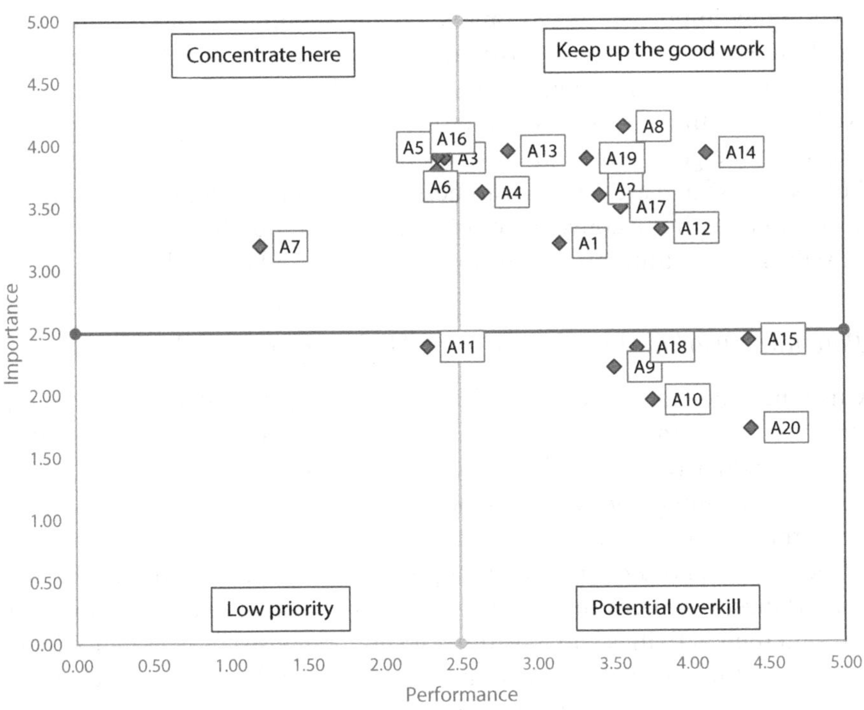

Figure 10.2 Importance-performance analysis of online shopping attributes.

integration" has the highest gap with an attribute score of –2.68, having an importance of 1.72 and a performance rating of 4.39 which means that online shopping exceeds its performance in this attribute in reference to the Importance, however the importance is quite low. All the given data are statistically significant.

With the help of the mean scores of each attribute, importance-performance matrix has been prepared wherein the mean scores of each attribute have been plotted against the customers' perceived importance of the service attributes and the performance of the attributes along Y- axis and X-axis, respectively. The following section briefly discusses the four quadrants of the IPA matrix.

Keep Up the Good Work (High Importance, High Performance)

Attributes in this quadrant [namely, A1 - account management (profile, order history), A2 - user-friendly navigation, A4 - pricing transparency, A8 - return and refund policy, A12 - wishlist and cart saving, A13 - loyalty programs and discounts, A14 - payment options, A17 - order tracking, A19 - shipping options (speed and cost)] as depicted in Figure 10.2 are considered highly important by customers. Online shopping is performing well on these attributes, meeting customer expectations. Since both importance and performance are high, it suggests that the online shopping is doing well in areas that customers find crucial. The IPA diagram's attributes for "keep up the good work" point to a situation where online shopping has been successful in areas that customers value highly. Online shopping provides users with the convenience of being able to purchase from a variety of sellers, at any time, and from any location around the globe.

Potential Overkill (Low Importance, High Performance)

Customers do not place a significant value on these attributes [namely, A9 - website loading speed, A10 - responsiveness (mobile and desktop), A15 - product reviews and ratings, A18 - mobile-friendly design, A20 - social media integration]. Online purchasing, meanwhile, is flourishing in areas that are not necessarily prioritized. It could be wise to reevaluate how resources are allocated. If online shopping is overinvesting in areas of low importance, resources might be better used elsewhere. For firms looking to align their activities with customer goals and consistently enhance their overall performance, it is imperative to comprehend and proactively handle each quadrant in an IPA analysis.

Low Priority (Low Importance, Low Performance)

The attribute A11- product search functionality is not considered highly important by customers. Also, online shopping is not dedicating significant efforts to these areas, and performance is low. Due to their low relevance and performance, these characteristics could not have a big effect on consumer satisfaction.

Concentrate Here (High Importance, Low Performance)

The attributes in this quadrant [namely, A3 - product reviews and ratings, A5 - product descriptions and images, A6 - customer feedback and surveys, A7 - customer support (live chat, phone, e-mail), A16 - security and privacy] are of the utmost importance to customers. However, online shopping is not meeting expectations in these areas; there is a performance gap. Customers care deeply about these aspects, but online shopping is falling short.

Discussion

The analysis of the dynamics of Internet buying demonstrates the complex interaction of various factors. Age turns out to be a key factor, as seen by a significant chi-square value. Interestingly, males predominate among shoppers while women predominate among non-shoppers. Women are less likely than males to realize the benefits of Internet purchasing (Bassam, 2010). Surprisingly, the gender-related chi-square value is negligible in this situation, indicating that both the shopper and non-shopper profiles are less affected. Higher education is another important factor, with postgraduates making up most of the shoppers and undergraduates making up the bulk of the non-shoppers. The chi-square value in this educational setting, however, fails to produce decisive insights, advising caution in making firm conclusions regarding the link between education and online buying behavior. According to our data, there is a positive relationship between purchasing habits and income, proving that greater income levels are associated with better lifestyle changes that are achieved through increasing online shopping. Several studies also mention that income has an influence on how consumers shop online (Li *et al.*, 1999; Swinyard & Smith, 2003; Sonwaney & Chincholkar, 2019). The chi-square value emphasizes the importance of daily Internet use as a crucial predictor for classifying people as consumers or non-consumers. Notably, individuals who spend

3–5 hours online each day had the largest proportion of shoppers, while those who use the Internet the least frequently have the highest proportion of non-shoppers. People and researchers are drawn to the Internet as a shopping medium. Online consumers are catching the researcher's eye as more and more individuals do their buying there. More study is being done by researchers on how people purchase online. Internet informed individuals may easily search for and find information online. Most individuals do not have the time to visit stores and instead prefer to get their necessities online. For them, the convenience and speed of online buying are important benefits. (Teo, 2006; Jarvelainen, 2007).

Factor analysis reveals key characteristics that have a big impact on online purchasing behavior. Factor 1 – 'convenience,' is revealed to be the most important consideration for consumers. The analysis demonstrates strong relationships between claims made about the simplicity and effectiveness of Internet buying. This is consistent with published research that highlights the advantages of online purchasing over traditional brick-and-mortar retailers in terms of saving time and energy. Additionally, customers favorably evaluate online payment methods like credit cards or electronic payments, which are perceived as convenient (Sonwaney & Chincholkar, 2019). The second factor titled "online shopping challenges," is of secondary relevance. The difficulties mentioned, such as not being able to physically touch things, make consumers reluctant to make online purchases, as the touch and feel factor often demoralizes consumers to buy online (Desai, 2012). The third most significant factor was found to be "product comparison and variety." Product type affects consumer spending because it influences the perception of satisfaction and trust. Moreover, having a wide range of products gives the buyer a wide range of options that can meet their needs. Furthermore, women prefer to buy multiple items that fit together, like a dress, a handbag, and heels. This is why women are attracted to product diversity. Therefore, customers want to find all these products in one online store (Al Hamli, & Sobaih, 2023). Online consumers rely heavily on virtual interactions rather than in-person meetings. That is why consumer reviews on specific products determine whether sales will be high or low. It is also a highly subjective factor. Consumers use online reviews to form preconceived notions about the quality of a product based on other online consumers' reviews (Alotaibi, 2021). The fourth crucial component is identified as "pricing awareness." Consumers see online buying favorably because they consider it to be more cost-effective than in-store purchasing, saving them a substantial amount of money. One of the main benefits of online shopping is that customers are price sensitive and often look for value products. Another key benefit of e-commerce is the ease of transactions and the wide variety of products and services available. Customers also prefer shopping online because they can do it from anywhere, anytime, and avoid the hassle of rush hour traffic, parking issues, and long lines (Bellenger, 1980; Brassington & Pettitt, 2000; Jayawardhena *et al.*, 2007; Lim & Dubinsky, 2004; Prasad &Aryasri, 2009; Childers *et al.*, 2001).

Critical facets of customer priorities and the effectiveness of online shopping are revealed by the examination of qualities inside the defined quadrants. Customers place a high value on features in Quadrant 1 [keep up the good work (high importance, high performance)] that include account management, user-friendly navigation, pricing transparency, and others, and online shopping is admirably living up to these expectations. Through the use of practical window shopping, all users are able to shop at the same time and on a variety of marketplaces. This type of shopping is time-efficient due to the availability of a wide range of items on the Internet, allowing users to purchase anything at any time and from any location (Sharma, 2018; Yang, Zhao, & Wan, 2010). In contrast, Quadrant 2 [*potential overkill (low importance, high performance)*] emphasizes features that customers do not place much importance in, such as website loading speed, responsiveness, and social media integration. Interestingly, while being given less importance, Internet shopping outperforms in these categories. According to Quadrant 3, [*low priority (low importance, low performance)*] product search functionality, which is given less attention by online shopping platforms and is hence less effective, is viewed as being less significant by customers. According to Quadrant 4, [*concentrate here (high importance, low performance)],* buyers place a high value on features like product reviews, descriptions, security and privacy, and customer support. However, online shopping falls short of these expectations, demonstrating a performance gap. The practice of online shopping involves the sharing of personal information, which can lead to the theft of personal data and its subsequent misuse. The more personal information provided online, including credit card numbers, names, and addresses, the more likely it is that the information will be stolen (Gupta *et al.*, 2013). In addition, to promote online shopping, marketers must create more effective return policies, enhance product quality and after-sales services, and charge competitively (Rahman *et al.*, 2018). The importance of strategic alignment is emphasized by this research, which places a focus on high-priority characteristics in customer-centric development plans for online marketplaces.

Conclusion

This study examines the various factors that influence consumer behavior in North Bengal, such as age, income, and Internet usage. It reveals that the products chosen for online shopping are not only purchases, but also a reflection of an individual's lifestyle and social aspirations. The results of the analysis, which includes descriptive statistics, chi-square tests and factor analysis, reveal that electronic gadgets, clothing, and jewelry are the most popular categories for online shopping, with a strong correlation between them and the prevalence of online shopping in the region. In a fast-paced modern life where time is a valuable commodity, convenience has become a driving force. Home shopping and doorstep delivery has become an attractive proposition for consumers. However, challenges remain, such as online fraud, product delay, and misleading information. The study highlights the importance of discounts and the variety of options in influencing online shopping decisions. In short, North Bengal is on the path to a higher standard of living, driven by rising income levels and the increasing trend of online shopping, especially in urban centers. This trend offers convenience and access to endless choices, but it also requires addressing challenges to provide a safe and smooth online shopping experience. As technology continues to influence consumer behavior, the landscape of online shopping in North Bengal is undergoing a transformative transformation.

References

1. Abalo, J., Varela, J., & Manzano, V. (2007). Importance values for importance performance analysis: A formula for spreading out values derived from preference rankings. *J. Bus. Res., 60,* 115-121.
2. Al Karim, R. (2013). Customer Satisfaction in Online Shopping: a study into the reasons for motivations and inhibitions. *IOSR J. Bus. Manage.*, 11(6), 13-20, retrieved from http://www.iosrjournals.org/iosrjbm/papers/Vol11-issue6/B01161320.pdf
3. Ahn, T., Ryu, S., & Han, I. (2007). The impact of Web quality and playfulness on user acceptance of online retailing. *Inf. Manage., 44(3), 263–275.* doi:10.1016/j.im.2006.12.008
4. Al Hamli, S. S., &Sobaih, A. E. E. (2023). Factors Influencing Consumer Behavior towards Online Shopping in Saudi Arabia Amid COVID-19: Implications for E-Businesses Post Pandemic. *J. Risk Fin. Manag., 16*(1), 36. MDPI AG. Retrieved from http://dx.doi.org/10.3390/jrfm16010036.

5. Alotaibi, R. S. (2021). Understanding customer loyalty of M-commerce applications in Saudi Arabia. *Int. Trans. J. Eng. Manage. Appl. Sci. Technol., 12*, 1-12.
6. Bagdoniene. L & Zemblyte. J, (2009) Online Shopping Motivation Factors And Their Effect On Lithuanian Consumers. *Ekon. Vadyb. Econ. Manage.*
7. Bagri, S. C., & Devkant, K. A. L. A. (2015). TOURISTS'SATISFACTION AT TRIJUGINARAYAN, INDIA: AN IMPORTANCE-PERFORMANCE ANALYSIS. *Adv. Hosp. Tour. Res. (AHTR), 3*(2), 89-115.
8. Bai Billy, Law Rob, Wen Ivan (2008). The impact of website quality on customer satisfaction and purchase intentions: Evidence from Chinese online visitors. *Int. J. Hosp. Manage.* 27, pp. 391–402.
9. Bartholomew, D., Knotts, M., & Moustaki, I. (2011). *Latent variable models and factor analysis: A unified approach.* (3rd ed.). West Sussex, UK: John Wiley & Sons.
10. Bellenger, D. N. (1980). Profiling the recreational shopper. *J. Retail., 56*(3), 77-92.
11. Bhinde, Hetal and Bhinde, Hetal and Joshi, Shrinivas V., Impact of Gender, Age and Income on Consumers' Purchasing Responsiveness to White Goods (December 31, 2016). Available at SSRN: https://ssrn.com/abstract=3308290 http://dx.doi.org/10.2139/ssrn.3308290
12. Brassington, F., & Pettitt, S. (2000). *Principles of Marketing.* 2002. Financial Times/Prentice Hall, ISBN, 273644440.
13. Bucko, J., Kakalejčík, L., & Ferencová, M. (2018). Online shopping: Factors that affect consumer purchasing behaviour. *Cogent Bus. Manage., 5*(1), 1535751.
14. Child, D. (2006). *The essentials of factor analysis.* (3rd ed.). New York, NY: Continuum International Publishing Group.
15. Childers, Terry & Carr, Christopher & Peck, Joann & Carson, Stephen. (2001). Hedonic and Utilitarian Motivations for Online Retail Shopping Behavior. *J. Retail. 77. 511-535. 10.1016/S0022-4359(01)00056-2.*
16. Chu, R. K. S., & Choi, T. (2000). An importance-performance analysis of hotel selection factors in the Hong Kong hotel industry: a comparison of business and leisure travellers. *Tour. Manage.*, 21, 363-377.
17. Desai, H. V. (2012). E-commerce and the Rural Sector. *Int. J. Adv. Res. Comput. Eng. Technol. (IJARCET), 1*(9), pp-144.
18. Donthu, Naveen and Adriana Garcia (1999), "The Internet Shopper," *J. Advert. Res.*, 3, 9, 52–8.
19. Dwyer, L., Knezevic Celbar, L., Edwards, D., & Mihalic, T. (2012). Fashioning a destination tourism future: the case of Slovenia. *Tour. Manage.*, 33, 305-316.
20. Evanschitzky Heiner, Iyer Gopalkrishnan R, Hesse Josef, Ahlert Dieter (2004). E-satisfaction: a re-examination. *J. Retail.* Vol. 80 (2004) pp. 239–247.

21. Gupta, A., Bansal, R., & Bansal, A. (2013). Online Shopping: A Shining Future. *Int. J. Techno-Manag. Res., 01*, 1. https://www.researchgate.net/publication/318224640
22. Harman, H.H. (1976). *Modern factor analysis* (3rd ed. revised). Chicago, IL: University of Chicago Press.
23. Hasan, Bassam. (2010). Exploring gender difference in online shopping attitude. *Comput. Hum. Behav. 26*. 597-601. 10.1016/j.chb.2009.12.012.
24. Jahwari, *et al.*, (2018). Factors Influencing Customer Satisfaction Of Online Shopping In Oman – Youth Perspective. *Humanit. Soc. Sci. Rev.* 6(2), 64-73 https://doi.org/10.18510/hssr.2018.628
25. Järveläinen, J. (2007). Online Purchase Intentions: An Empirical Testing of a Multiple-Theory Model. *J. Organ. Comput. Electron. Commer., 17(1), 53–74.*
26. Jayawardhena, C., Tiu Wright, L., & Dennis, C. (2007). Consumers online: intentions, orientations and segmentation. *Int. J. Retail Distrib. Manage.*, 35(6), 515-526.
27. Javadi, M.H.M., Dolatabadi, H.R., Nourbakhsh, M., Poursaeedi, A. and Asadollahi, A. R. (2012). An Analysis of Factors Affecting on Online Shopping Behavior of Consumers. *Int. J. Mark. Stud.*, 4(5), 81-98. doi: https://doi.org/10.5539/ijms.v4n5p81
28. Rishi, B. J. (2008). An Empirical Study of Online Shopping Behaviour: A Factor Analysis Approach. *J. Mark. Commun., 3*(3).
29. Kotler, P. & Amstrong, G. (2013). *Principles of Marketing* (15th ed.) New Jersey: Pearson Education Inc.
30. Kolesar, M.B. & Galbraith, R.W. (2000). A Services-Marketing perspective on e-retailing: Implications of e-retailers and Directions for Further Research. *Internet Res.*, 10(5), 424-438. doi: https://doi.org/10.1108/10662240010349444
31. Kumar, D. V., & Dange, U. (2014). A study on perceived risk in online shopping of youth in Pune: A factor analysis. *Available at SSRN 2518293.*
32. Lakshmanan, A. (2016). Customers Satisfaction towards Online Shopping in Amazon.Com–A Study with Reference to Udumalpet Taluk. *Int. J. Recent Res. Appl. Stud.*, Vol. 3, 12(16), pp. 68-74.
33. Lambrecht, Anja and Katja Seim (2006), "Adoption and Usage of Online Services in the Presence of Complementary Offline Services," Retail Banking, NET Institute Working Paper
34. Li, Hairong& Kuo, Cheng & Russell, Martha. (1999). The Impact of Perceived Channel Utilities, Shopping Orientations, and Demographics on the Consumer's Online Buying Behavior. *J. Comput.-Mediat. Commun. 5.* 10.1111/j.1083-6101.1999.tb00336.x.
35. Lim, Y. J., Osman, A., Salahuddin, S. N., Romle, A. R., & Abdullah, S. (2016). Factors influencing online shopping behavior: the mediating role of purchase intention. *Proc. Econ. Financ.*, 35(5), 401-410. https://doi.org/10.1016/s2212-5671(16)00050-2

36. Lim, H., & Dubinsky, A. J. (2004). Consumers' perceptions of e-shopping characteristics: an expectancy-value approach. *J. Serv. Mark.*, *18*(7), 500-513.
37. Lin, C., Wu, H. & Chang, Y. (2010). The Critical Factors Impact on online Customer Satisfaction. *Proc. Comput. Sci.*, 3, 276-281. doi: https://doi.org/10.1016/j.procs.2010.12.047
38. Magal, S.R. and Levenburg, N.M. (2005) Using Importance-Performance Analysis to Evaluate E-Business Strategies among Small Businesses. *IEEE Proceedings of the 38th Hawaii International Conference on System Sciences,* Big Island, HI, 3-6 January 2005, 176a.
39. Martilla, J.A. and James, J.C. (1977) Importance-Performance Analysis. *J. Mark., 41*, 77-79. http://dx.doi.org/10.2307/1250495
40. Matzler, K., Sauerwein, E., &Heischmidt, K. A. (2003). Importance-performance analysis revisited: The role of the factor structure of customer satisfaction. *Serv. Ind. J., 23*(2), 112–129. https://doi.org/10.1080/02642060412331300912
41. Monsuwe, T. P. Y., Dellaert, B. G. C., & Ruyter, K. D. (2004). What Drives Consumers to Shop Online? A Literature Review. *Int. J. Serv. Ind. Manage.,15*(1), 102-121, DOI :10.1108/09564230410523358
42. Nagra, G., & Gopal, R. (2013). A study of factors affecting on online shopping behavior of consumers. *Int. J. Sci. Res. Publ.*, *3*(6), 1-4.
43. Ozen, H., & Engizek, N. (2014). Shopping online without thinking: being emotional or rational? *Asia Pac. J. Mark. Logist.*, 26(1), 78-93. https://doi.org/10.1108/apjml-06-2013-0066
44. Prasad, C. J., &Aryasri, A. R. (2009). Determinants of shopper behaviour in e-tailing: An empirical analysis. *Paradigm, 13*(1), 73-83.
45. Punj, G. (2011). Effect of consumer beliefs on online purchase behavior: The influence of demographic characteristics and consumption values. *J. Interact. Mark.*, *25*(3), 134-144.
46. Rahman, M. A., Islam, M. A., Esha, B. H., Sultana, N., & Chakravorty, S. (2018). Consumer buying behavior towards online shopping: An empirical study on Dhaka City, Bangladesh. *Cogent Bus. Manage.*, 0(0). doi:10.1080/23311975.2018.1514940
47. Rummel, R.J. (1970). *Applied factor analysis*. Evanston, IL: Northwestern University Press.
48. Sever, I. (2015). Importance-performance analysis: A valid management tool? *Tour. Manage.*, 48, 43–53. https://doi.org/10.1016/j.tourman.2014.10.022
49. Sharma, S. (2018). Trends Of Online Shopping, Their Pros and Cons: An Extensive Review. *J. Manage. Res. Anal. (JMRA), 878*(03), 72–77. http://jmraonline.comhttp://jmraonline.com.
50. Sonwaney, V &Chincholkar, S. (2019). Identifying The Factors Impacting Online Consumer Buying Behaviour. *Int. J. Sci. Technol. Res.* Volume 8, Issue 08. Pp. 445 – 456

51. Suh, B. & Han, I. (2003). The Impact of Customer Trust and Perception of Security Control on the Acceptance of Electronic Commerce. *Int. J. Electron. Commer.*, 7(3), 135-161, retrieved from http://www.jstor.org/stable/27751068
52. Susanty, M., Wahyu Ningsih, T., &Faizin Rissa, M. (2019). Analysis of Customer Satisfaction with Importance Performance Analysis (IPA) Method in One of the Banking Industries. *Int. J. Innov. Sci. Res. Technol.* (Vol. 4, Issue 7).
53. Swinyard, W. R., & Smith, S. M. (2003). Why people (don't) shop online: A lifestyle study of the internet consumer. *Psychol. Mark.*, *20*(7), 567-597.
54. Szymanski David M., Hise Richard T. (2000). e-Satisfaction: An Initial Examination. *J. Retail.*, Vol. 76(3) pp. 309–322.
55. Teo, T. S. H. (2006). To buy or not to buy online: adopters and non-adopters of online shopping in Singapore. *Behav. Inf. Technol., 25(6), 497–509.*
56. Wolfinbarger, M., & Gilly, M.C. (2001). Shopping online for freedom, control, and fun. *Calif. Manage. Rev.*, *43*(2), 42-55. http://dx.doi.org/10.2307/41166074
57. Wu, L., Cai, Y. & Liu, D. (2011). Online shopping among Chinese customers: an exploratory investigation of demographics and value orientation. *Int. J. Consum. Stud.*, 35, 458-469. https://doi.org/10.1111/j.1470-6431.2010.00982.x
58. Wyród-Wróbel, J., & Biesok, G. (2017). Decision Making on Various Approaches to Importance-Performance Analysis (IPA). *Eur. J. Bus. Sci. Technol.*, *3*(2), 123-131. doi: 10.11118/ejobsat.v3i2.82
59. Yang, J., Zhao, H., & Wan, J. (2010). *Research on the Advantages and Disadvantages of Online Shopping and Corresponding Strategies. 2010 International Conference on E-Product E-Service and E-Entertainment.* doi:10.1109/iceee.2010.5660278
60. Yong, A. G., & Pearce, S. (2013). A beginner's guide to factor analysis: Focusing on exploratory factor analysis. *Tutor. Quant. Methods Psychol.*, *9*(2), 79-94.
61. Zamri, N.N.A. (2014). Factors affecting Customer Satisfaction towards Online Shopping among University Students at UniversitiTeknikal Malaysia, Melaka. Bachelor Degree Research report, submitted to the Faculty of Technology Management and Technopreneurship, UniversitiTeknikal Malaysia, Melaka, Jun. 2014, 1-78, retrieved from http://eprints.utem.edu.my/16204/
62. Zatalini, M.A. &Pamungkas, T.N. (2017). Exploring the Success Factors of E-CRM Implementation On B2C E-Commerce: Satisfaction and Loyalty A Conceptual Framework, *J. Ekon. Bisnis*, 22(2), 94-106, retrieved from http://ejournal.gunadarma.ac.id/index.php/ekbis/article/view/1620

11

Tourism and Protected Areas in India—A Symbiotic Relationship

Sagar Sood[1]*, Kesar Chand[1], Aditi Kohli[2], Shruti Kanga[3], Suraj Kumar Singh[4] and Gowhar Meraj[5]

[1]*Govind Ballabh Pant National Institute of Himalayan Environment, Himachal Regional Centre, Mohal - Kullu, Himachal Pradesh, India*
[2]*Government College, Kullu, Himachal Pradesh, India*
[3]*Department of Geography, School of Environment and Earth Sciences, Central University of Punjab, Punjab, India*
[4]*Centre for Climate Change and Water Research, Suresh Gyan Vihar University, Jaipur, India*
[5]*Department of Ecosystem Studies, Graduate School of Agricultural and Life Sciences, The University of Tokyo, Tokyo, Japan*

Abstract

Protected areas (PAs) and tourism share a symbiotic relationship, with PAs offering unique and pristine environments for tourism, while tourism provides economic support for conservation efforts. This review explores the status and growth of PAs in India, which cover approximately 5% of the country's geographical area, and highlights the interrelationship between tourism and PAs. The study also examines the legislative framework, acts, programs, and bodies responsible for notifying and managing PAs, with an emphasis on tourism in India. The International Union for Conservation of Nature (IUCN) defines PAs as clearly defined geographical spaces dedicated to the long-term conservation of nature, associated ecosystem services, and cultural values. Protected areas play a crucial role in biodiversity conservation, maintaining key habitats, providing refugia, and ensuring the maintenance of natural processes across the landscape. They also contribute to human well-being by providing livelihoods, drinking water, and food security. Well-managed PAs with participatory and equitable governance mechanisms yield significant benefits beyond their boundaries, contributing to poverty reduction and sustainable development. As the impact of climate change threatens the planet, PAs offer a solution

**Corresponding author*: sagarsood87@gmail.com

Shruti Kanga, Gowhar Meraj, Suraj Kumar Singh, Majid Farooq and M. S. Nathawat (eds.) Geospatial Technology for Natural Resource Management, (287–302) © 2024 Scrivener Publishing LLC

for both mitigation and adaptation responses. The study emphasizes the importance of better managed, connected, governed, and financed PAs in addressing the challenges posed by climate change and achieving conservation goals.

Keywords: Protected areas, tourism, conservation, climate change, India

Introduction

Humans always have an urge to explore the unexplored. As tourism and environment are closely linked together (Green *et al.*, 1990), people are inclined to visit areas having beautiful landscapes, healthy wildlife, and authentic cultures. With the population growth and invention of better means of transport, the travelers have become more capable of moving over to far distances and developed an urge to visit new places, especially the areas having the rich biodiversity, indigenous cultures, scenic areas, etc., which are mostly present inside or in the vicinity of protected areas (PAs). Perhaps, PAs tourism has shown budding potential in the global economy. Being an important ecosystem service, it generates support for conservation (Karanth & DeFries, 2011).

The relationship between PAs and tourism is as old as the history of PAs (Eagles *et al.*, 2002). Though this relation in multifaceted and occasionally antagonistic, tourism is always a critical component to consider in the formation and management of PAs. Tourism and PAs share a long-lasting and mutually beneficial relationship (Bushell *et al.*, 2007; Reinius & Fredman, 2007; Strickland-Munro & Moore, 2014). The PAs offer an exclusive, clean and desirable environment for the tourism product, while tourism offers the park management an economic source that can directly assist in conservation efforts. Further, the engagement of local people in tourism activities in and outside the PAs provides them an alternate source of livelihood, thus making a direct impact on their social, cultural, educational and economic wellbeing (Cooper, 2008; Keller & Beiger, 2011; McCabe *et al.*, 2011; Swanson & Edgell, 2013; Pratheep, 2017).

Tourist attractions located in the PAs constitute approximately 20–40% of global tourism (WAP, 2019). Tourism in PAs is estimated to contribute 10% towards the global gross domestic product (GDP) (Weaver, 1999; Balmford *et al.*, 2009). Tourism activities, if done sustainably, can contribute to the conservation of PAs throughout the world. Perhaps, it can contribute for the Strategic Plan for Biodiversity 2011–2020 of the Convention on Biological Diversity (CBD), focused on promoting conservation, community development and public awareness (McCool & Moisey,

2008; UNWTO, 2010; Buckley, 2012a; Hvenegaard, *et al.*, 2012; CBD, 2015; UNEP-WCMC& IUCN, 2016). Similarly, the adoption of Guidelines on Biodiversity and Tourism Development (CBD, 2004) by CBD in 2004 has helped in the promotion of tourism in PAs.

Furthermore, tourism can also help in the achievement of the United Nation's (UNs) Sustainable Development Goals (SDGs) (set of 17 goals to be achieved by 2030). These SDGs are core components of UN's sustainable development agenda with tourism featuring in Goal 8 (sustainable economic growth), Goal 12 (sustainable consumption and production) which can directly or indirectly amalgam with Goal 14 (conservation and sustainable use of oceans, seas, and marine resources), and Goal 15 (protect, restore and promote sustainable use of terrestrial ecosystems, sustainably manage forests, combat desertification, and halt and reverse land degradation and halt biodiversity loss) to project tourism as a core component in conservation of PAs.

Tourism being such an integral part of the environment, especially in the PAs, as it conserves the environment and presents the virgin beauty of the planet to visitors. Thus, the present study focuses on tourism in the PAs of India, which covers approximately 5% of the geographical area of the country. The study highlights the status and growth of PAs in the country, throws light on the interrelationship between tourism and PAs and highlights the legislative framework, acts, programs and bodies for notifying and management of PAs with emphasis on tourism in the country.

Methodology

The present study is a review of tourism in PAs in India. The study has incorporated all forms of PAs which include national parks, wildlife sanctuaries, community reserves, and conservation reserves. Various sources have been used for the collection of data which includes research articles, books, government reports, and gazettes. The language was restricted to English only for the search of the literature. The keywords included protected areas, tourism, ecotourism, wildlife tourism, India, and conservation.

What is a Protected Area (PA)?

The IUCN (International Union for Conservation of Nature) defines PA as a 'clearly defined geographical space, recognized, dedicated and managed, through legal or other effective means, to achieve the long-term conservation of nature with associated ecosystem services and cultural values' (IUCN definition, 2008). Protected areas are a cultural artifact and have a long history. It is believed by many researchers that some areas were set aside by people of India for conservation and protection of natural resources over two millennia ago (Holdgate, 1999). Protected areas are the cornerstone of biodiversity conservation; they maintain key habitats, provide refugia, allow for species migration and movement, and ensure the maintenance of natural processes across the landscape. Not only do PAs secure biodiversity conservation, but they also secure the well-being of humanity itself. Protected areas provide livelihoods for nearly 1.1 billion people, are the primary source of drinking water for over a third of the world's largest cities and are a major factor in ensuring global food security. Well-managed PAs harboring participatory and equitable governance mechanisms yield significant benefits far beyond their boundaries, which can be translated into cumulative advantages across a national economy and contribute to poverty reduction and sustainable development including achievement of the Millennium Development Goals (MDGs) and Sustainable Development Goals (SDGs). As the detrimental impact of climate change threatens the planet, PAs provide a convenient solution to an inconvenient truth. Better managed, better connected, better governed and better-financed PAs are recognized as the key to both mitigation and adaptation responses to climate change.

Status of PAs in India

Globally, India is unique in the richness of biodiversity for its physical and climatic diversity (Mandal, 2003). It is considered one of the 17 mega diverse countries around the world (Mittermeier *et al.*, 1997; Mittermeier *et al.*, 2011; Puri, 2019). It houses more than 400 species of mammals, 1200 species of birds, and over 600 species of reptiles and amphibians in different habitats (Vattakaven *et al.*, 2016). Globally, approximately 15% of the land is under PAs, India officially protects 5% of its area (Dinerstein *et al.*, 2017). The PAs of India includes marine protected areas, national parks, wildlife sanctuaries, conservation reserves, and community reserves (Table 11.1). These PAs could further be divided into various categories which include tiger reserves, elephant reserves, biosphere reserves, RAMSAR wetland sites, natural world heritage sites, important coastal and marine biodiversity areas (ICMBAs), important bird areas (IBAs), key biodiversity areas, and biodiversity heritage sites.

Table 11.1 Protected areas in India.

	No.	Total area (km^2)	Coverage % of country
National parks (NPs)	104	40,501.13	1.23
Wildlife sanctuaries (WLSs)	551	119,775.8	3.64
Conservation reserves (CRs)	88	4356.49	0.13
Community reserves	127	525.22	0.02
Protected areas (PAs)	**870**	**165,158.54**	**5.02**

Source: *National Wildlife Database. Wildlife Institute of India.*

Growth of PAs in India

The introduction of India Wildlife Protection Act in 1972, provided special legal protection to wildlife and to the endangered species of fauna in particular, which resulted in the creation of new PAs throughout the country, expanding from 127 in 1970 to 870 today (Table 11.1) (Banerjee, 2012; NWD, 2019). The launch of various projects (Project Tiger, 1973; Crocodile Conservation Project, 1975; Project Elephant, 1992; Sea Turtle Project, 1999; Project Snow Leopard; 2014) further cemented the wildlife protection regime in the country. After 1970, the creation of PAs in the country acquired a great pace, which resulted in the displacement of people from their native land and laid restrictions on hunting, grazing, collection of timber, fuel, fodder, and non-timber forest produces (NTFPs). This gave rise to 'conflict of interest' between local communities and forest authorities. Furthermore, the modus operandi of forest authorities, dictated by the philosophy of 'fortress conservation,' frequently led to harassment of local people and caused substantial anguish in the absence of livelihood alternatives (Kothari *et al.*, 1995; Saberwal *et al.*, 2001; Banerjee, 2012). It resulted in the increased incidents of poaching, illegal tree felling and clashes of local communities with police and forest authorities. For minimizing these conflicts between the forest stakeholders, community reserves and conservation reserves were incorporated under the PA category through the Wildlife (Protection) Amendment Act, 2002—the amendment to the Wildlife Protection Act of 1972. According to Wildlife Institute of India, these reserves typically act as buffer zones and migration corridors between established national parks, wildlife sanctuaries and reserved forests (WII, 2019). These buffer zones provided an opportunity for alternative livelihoods for the people, along with being centric in the development and management of the regions near to PAs. On the positive side, the implementation of the Wildlife Protection Act, 1972, lead to the protection and preservation of fauna and flora species in the PAs (Shirili & Dhivya, 2018).

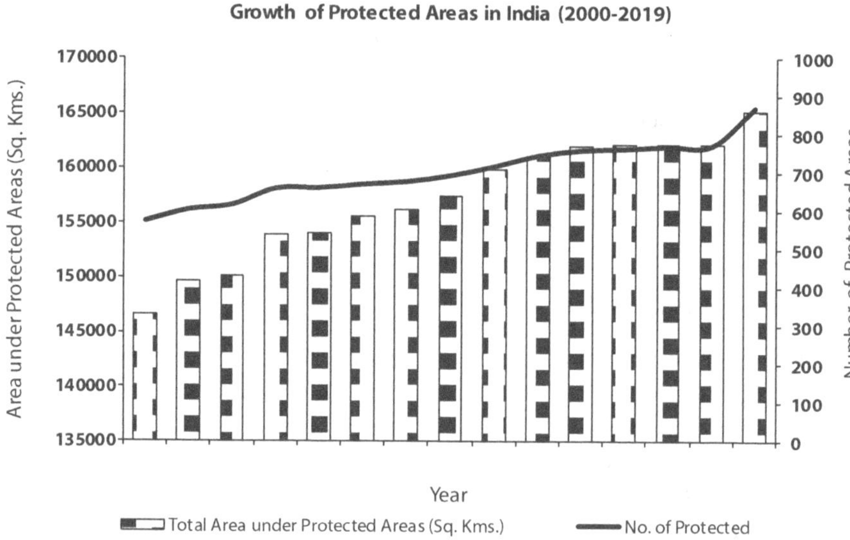

Source: Ministry of Environment, Forest and Climate Change.

Tourism and PAs in India

The wildlife tourism nurtured in the PAs of India has emerged as an important sector of tourism. With more than 800 PAs in India, covering 5.02% of the total geographical area of the country possesses a great potential of tourism in these PAs (Banerjee, 2012; Dinerstein *et al.*, 2017; NWD, 2019). The growth in the development of tourism in PAs started in the 1990s with the economic liberalization in the country (World Bank, 1991). Urbanization, globalization and swift economic growth have doubled the middle class from 1990 to 2005 (Beinhocker *et al.*, 2007; Gandhi & Orr, 2007; Das, 2009; Karanth & DeFries, 2011). This has encouraged the common middle-class people to travel, which assisted in the development of tourism, precisely wildlife tourism in the PAs, which has become a recent trend in India (Hannam, 2005). The 90% of PAs in India are open for tourism activities, out of which are tiger reserves, which accounts highest tourism influx during the tourist season (Kumar, 2002). Specifically, around tiger reserves, safari-based tourism is currently permitted in only 20% of the core or critical tiger habitats of a tiger reserve. The 28 tiger reserves alone receive more than 1 million visitors a year (TTF, 2005). Rules that govern tourism around other PAs are less stringent and provide opportunities for trekking, bird-watching, rafting, camping, etc. Due to India's rich biodiversity, tourism around the PAs has seen a growth of 6% in the last decade (Karanth, Jain, & Mariyam, 2017).

However, in India, the institutional framework to develop wildlife tourism is weak (Sekhar, 2003). Research on tourism trends in Indian PAs suggests that wildlife tourism is growing (the mean growth rate was 14.9% from 2002 to 2008) and is largely domestic (Karanth & DeFries, 2011). This growth in wildlife tourism potentially generates public interest in conservation but presents new challenges in a landscape where conservation efforts are already challenged by the growth of infrastructure, high densities and resource dependence of local people living around parks, and declining wildlife populations (Sekhar, 2003; Hannam, 2005; TTF, 2005; DeFries *et al.*, 2010; Karanth & DeFries, 2011; Karanth & Nepal, 2012). Recognizing the potential to earn revenues from tourism based around protected areas, the Central and State Governments, and private tourism enterprises (*i.e.*, resorts, camps, and tour operators) have promoted various forms of nature-based tourism (Puri, 2019).

The Tourists

The primary purpose of PAs in any region, is to protect the natural and cultural heritage, the secondary one is often tourism, focused on the economic development of the local people and providing tourists with recreation, spiritual refreshment, natural exquisiteness and learning experiences (Cochrane, 2006). The number of tourists visiting the PAs in the country is very small as compared to the total visits in the country. In 2015, 46,23,134 tourists visited the 118 wildlife reserves in India, which is phenomenally lower as compared to domestic and foreign tourists in the country (Karanth, *et al.*, 2017). Out of which, 32,33,665 people visited national parks, 18,51,280 visited wildlife sanctuaries, and 14,66,728 visited the tiger reserves (Karanth, *et al.*, 2017). The tourist data was not available for community reserves and conservation reserves.

Previous studies on tourism in PAs in India suggests that tourism in these PAs is rising (mean growth rate was 14.9% from 2002 to 2008) and is largely domestic (Karanth & DeFries, 2011), which has increased from 90% to 96% from 2005 to 2015. When fragmented, across all wildlife reserves, domestic visitors increased from 90% to 96% from 2005 to 2015. In national parks, domestic visitors increased from 85% to 95% in the same decade. In wildlife sanctuaries, domestic visitors also increased from 97% to 98% in 2014–2015. Among tiger reserves, domestic tourism increased from 83% to 89% (Karanth, *et al.*, 2017). This growth of tourism has generated stakeholders' consciousness in conservation and protection in and around PAs.

Local Communities

In 1992, during the United Nations Rio de Janeiro conference on Environment and Development, various efforts were made for linking PAs with sustainable livelihood opportunities for local (Naughton-Treves *et al.*, 2005).

The basic approach for management of PAs has been isolationist, based on the questionable assumption that certain areas are pristine or primary and that management must protect the park from people living in surrounding areas and shield wildlife and other natural resources from exploitation. This is achieved through the strict enforcement of legislation, patrols to prevent illegal activities and infrastructure maintenance. In this scenario, attempts to protect PAs from human intervention by coercion have often led to hostile attitudes of local people towards wildlife management and forestry staff, and sometimes to open conflict.

Constitutional and Legislative Framework with Respect to Tourism Management in PAs of India

Constitutional Framework for PAs in India

The Constitution of India has provided significant provisions for PAs of the country. Under the Directive Principle of State Policy, Article 48-A of the Constitution enjoins that the State shall endeavor to protect and improve the environment and to safeguard the forests and wildlife of the country, and Article 51-A (g) defines the fundamental duty of every citizen of India to protect and improve the natural environment including forests, lakes, rivers, and wildlife and to have compassion for living creatures (Upadhyay, 2019). The Ministry of Environment Forest and Climate Change (MoEFCC), Govt. of India, set up the PAs and as well as the improvement of the reserve forests in line with the Wildlife Protection Act,

1972 along with its later amendments (Dejouhanet, 2010). Furthermore, for the strengthening of PAs, in 1976, wildlife, the essence of PAs, was placed in the Concurrent List of the Indian constitution (Jariwala, 2015).

Legislative Framework

With the enactment of the Wildlife Protection Act (1972), Forest Conservation Act (1980), and Wildlife Protection Act (1986), the restrictions were imposed on local communities for use of resources of PAs. In the process, the local communities, the tribal, and the forest dwellers inside the PAs had to be displaced and resettled elsewhere from their conventional lifestyles and mode of living. Their fundamental livelihoods were the resources of the forest which they had to forego. These areas were later opened up for tourism to provide them with alternative livelihoods.

S. no.	Acts/rules/programs	Tourism component
1.	Wildlife Protection Act, 1972	Chapter IV (No. 28) empowers Chief Wildlife Warden to grant permission for: • Photography • Scientific research • Tourism • Transaction of lawful business (tourism)
2.	Forest Conservation Act, 1980	Chapter 12 Clause 14 accords ecotourism as a non-forestry activity and will be allowed in PAs if the said activities are part of the Management Plan/Tiger Conservation Plan
3.	Environment Protection Act, 1986	• Strict building and expansion regulations for pilgrim sites located inside PAs must be designated as sacred groves. • The first benefit of ecotourism must go to the local people. • A sustainable partnership between forest department, tourism professionals and local communities.

(Continued)

(*Continued*)

4.	The Noise Pollution (Regulation and Control) (Amendment) Rules, 2010	All tourism facilities located within 5 km of PAs must adhere to ambient noise rules for silence zone which stands at 50 dB and 40 dB at day time and night time, respectively.
5.	Integrated Development of Wildlife Habitats	• Providing local people living in the vicinity of PAs, an opportunity to find employment as guides, drivers, hospitality personnel and other ancillary tourism jobs. • Generating alternate economic resources for residents through tourists visit in and around PAs. • Promoting wildlife conservation along with sustaining life support systems as well as ensuring the food, water, and livelihood security.
6.	National Tiger Conservancy Authority	• Adopt low-impact wildlife tourism that protects the ecological integrity of forest and wildlife areas. • Develop mechanisms to generate revenues from wildlife tourism for the welfare and economic upliftment of local communities • Highlight the heritage value of India's wilderness and tiger reserves. • Build environmental, cultural awareness and respect. • Facilitate the sustainability of tourism enterprises and activities. • Provide livelihood opportunities to local communities. • Promote sustainable use of indigenous materials for tourism activities. • Promote processes for forest dwellers to control and maintain their resources, culture, and rights to minimize negative impacts.

Conclusion

Tourism and protected areas (PAs) have complementing relationship where both are mutually beneficial to each other and contribute towards each other's growth. But only sustainable tourism can ensure long lasting existence of their mutually beneficial relationship. Conservation of exclusive and rich biodiversity in certain special pockets of land is major goal which is accomplished by establishment of protected areas which is often designated by different names like national park, wildlife sanctuary, wilderness areas, and nature reserves. Protected areas often set in peaceful and scenic backgrounds are endowed with rich biological resources in their most pure form. With growing chaos and unrest in modern lifestyle people all around the globe are often in search of peaceful break close to nature. This kind of tourism demand can only be fulfilled inside or in vicinity of PAs. Protected areas in turn gets economic assistance for their better management and accomplishment of their prime goal of conservation. Goal 15 of SDGs also highlights tourism as a major driver in conservation of PAs. Protected areas tourism holds great potential of making a significant contribution in the global economy. With year on year increase in number of tourists (majorly domestic), the PAs have become an important part of the tourism industry in India. Further, the constitutional and legislative frameworks have regulated the tourism activities in the PAs and has empowered the stakeholders i.e. local communities by providing economic benefits and social upliftment and the tourists by providing natural areas to explore in the country. Thus, the tourism holds the key for sustainable development of PAs in India.

References

Al-Tokhais, A., & Thapa, B. (2019). Stakeholder Perspectives Towards National Parks and Protected Areas in Saudi Arabia. *Sustainability, 11*(8), 2323.

Bahuguna, V. K. (2000). Forests in the economy of the rural poor: an estimation of the dependency level. *AMBIO: A Journal of the Human Environment, 29*(3), 126-129.

Balmford, A., Beresford, J., Green, J., Naidoo, R., Walpole, M., & Manica, A. (2009). A global perspective on trends in nature-based tourism. *PLoS Biol., 7*(6), e1000144.

Banerjee, A. (2012). Is wildlife tourism benefiting Indian protected areas? A survey. *Curr. Issues Tourism, 15*(3), 211-227.

Beinhocker, E. D., Farrell, D., & Zainulbhai, A. S. (2007). Tracking the growth of India's middle class. *McKinsey Q.*, 3, 50.

Boyd, S. W., & Butler, R. W. (1993). *Review of the development of ecotourism with respect to identifying criteria for ecotourism for Northern Ontario*. Report. Sault Ste. Marie, Ontario: Department of Natural Resources/Forestry Ministry of Natural Resources.

Buckley, R. (2012). Tourism, conservation and the Aichi targets. *Parks, 18*(2), 12-19.

Bushell, R., Staiff, R. & Eagles, P.F.J. (2007). 'Tourism and protected areas: Benefits beyond boundaries', in R. Bushell & P.J.F. Eagles (eds.). *Tourism and protected areas: Benefits beyond boundaries*. 1–11. CAB International, Wallingford.

CBD (2015). *Tourism Supporting Biodiversity: A Manual on Applying the CBD Guidelines on Biodiversity and Tourism Development*. Montreal: Secretariat of the Convention on Biological Diversity. https://www.cbd.int/tourism/doc/tourism-manual-2015-en.pdf. Accessed 02 November 2019.

CBD (Secretariat of the Convention on Biological Diversity) (2004). *Guidelines on Biodiversity and Tourism Development*. Montreal: Secretariat of the Convention on Biological Diversity. http://www.cbd.int/tourism/guidelines.shtml. Accessed 03 November 2019.

Cochrane, J. (2006). A typology of tourists to protected areas. *Protected Areas Programme*, 10.

Cooper, C. (2008). *Tourism: Principles and practice*. Pearson education.

Das, D. K. (2009). Globalisation and an emerging global middle class. *Econ. Aff.*, *29*(3), 89-92.

DeFries, R., Karanth, K. K., & Pareeth, S. (2010). Interactions between protected areas and their surroundings in human-dominated tropical landscapes. *Biol. Conserv., 143*(12), 2870-2880.

Dejouhanet, L. (2010). Participatory eco-development in question: the case of the Parambikulam wildlife sanctuary in South India. *J. Alp. Res./ Rev. Géogr. Alp.*, (98-1).

Dinerstein, E., Olson, D., Joshi, A., Vynne, C., Burgess, N.D., Wikramanayake, E., Hahn,N., Palminteri, S., Hedao, P., Noss, R., Hansen, M., Locke, H., Ellis, E.C., Jones, B.,Barber, C.V., Hayes, R., Kormos, C., Martin, V., Crist, E., Sechrest, W., Price, L.,Baillie, J.E.M., Weeden, D., Suckling, K., Davis, C., Sizer, N., Moore, R., Thau, D.,Birch, T., Potapov, P., Turubanova, S., Tyukavina, A., de Souza, N., Pintea, L., Brito,J.C., Llewellyn, O.A., Miller, A.G., Patzelt, A., Ghazanfar, S.A., Timberlake, J., Klöser,H., Shennan-Farpón, Y., Kindt, R., Lillesø, J.-P.B., van Breugel, P., Graudal, L., Voge,M., Al-Shammari, K.F.& Saleem, M. (2017). An ecoregion-based approach to protectinghalf the terrestrial realm. *BioScience, 67*(6), 534–545.

Eagles, P. F., McCool, S. F., Haynes, C. D., & Phillips, A. (2002). *Sustainable tourism in protected areas: Guidelines for planning and management* (Vol. 8). Gland: IUCN.

Gadgil, M. (1990). India's deforestation: patterns and processes. *Soc. Nat. Resour.*, *3*(2), 131-143.

Gandhi, S. S., & Orr, R. J. (2007). Large scale urban development in India: past and present. Collaboratory for Research on Global Projects, DU.

Green, H., Hunter, C. & Moore, B. (1990). Assessing the environmental impact of tourism development: use of the Delphi technique. *Tour. Manag.*, 11(2), 111-120.

Hannam, K. (2005). Tourism management issues in India's national parks: an analysis of the Rajiv Gandhi (Nagarahole) national park. *Curr. Issues Tourism*, *8*(2-3), 165-180.

Holdgate, M. (2014). *The green web: a union for world conservation*. Routledge.

Hvenegaard, G. T., Halpenny, E. A., & McCool, S. F. (2012). Protected area tourism and the Aichi Targets. *Parks*, *18*(2), 6-11.

IUCN (2008). What is a protected area?. https://www.iucn.org/theme/protected-areas/about. Accessed 7 Jaunuary, 2020.

Jariwala, C. M. (2015). The Constitution 42nd Amendment Act and the Environment.

Karanth, K. K., & DeFries, R. (2011). Nature-based tourism in Indian protected areas: New challenges for park management. *Conserv. Lett.*, *4*(2), 137-149.

Karanth, K. K., & Nepal, S. K. (2012). Local residents perception of benefits and losses from protected areas in India and Nepal. *Environ. Manage.*, *49*(2), 372-386.

Karanth, K. K., Jain, S., & Mariyam, D. (2017). Emerging Trends in Wildlife and Tiger Tourism in India. *Nature tourism*, 159.

Keller, P., & Bieger, T. (Eds.). (2011). *Tourism development after the crises: global imbalances-poverty alleviation* (Vol. 5). International Progress Organization.

Kothari, A., Suri, S., & Singh, N. (1995). People and protected areas: Rethinking conservation in India. *Ecologist*, 25, 188–194.

Kumar, S. (2002). Wildlife tourism in India: Need to tread with care. In B.D. Sharma (Ed.), *Indian wildlife: Threats and preservation*. 72–94. New Delhi: Anmol Publications.

Kuvan, Y. & Akan, P. (2005). Residents' attitudes toward general and forest related impacts of tourism: the case of Belek, Antalya. *Tour. Manage.*, *26*(5), 691-706.

Mandal, M. (2003). Protected area management in India: A perspective. *XII World Forestry Congress at Quebec*, Canada, September 21-28, 2003.

McCabe, S., Minnaert, L., & Diekmann, A. (Eds.). (2011). *Social tourism in Europe: Theory and practice* (Vol. 52). Channel View Publications.

McCool, S. F., & Moisey, R. N. (2008). Introduction: Pathways and pitfalls in the search for sustainable tourism. *Tourism, recreation and sustainability*, 1-16.

Mittermeier, R. A., Gil, P. R., & Mittermeier, C. G. (1997). *Megadiversity: Earth's biologically wealthiest nations*. Washington, DC: Conservation International.

Mittermeier, R. A., Turner, W. R., Larsen, F. W., Brooks, T. M., & Gascon, C. (2011). Global biodiversity conservation: The critical role of hotspots. In F. Zachos, & J. Habel (Eds.), *Biodiversity hotspots* (pp. 3–22). Heidelberg: Springer.

National Wildlife Database (2019). ENVIS Centre on Wildlife & Protected Areas. http://www.wiienvis.nic.in/Database/Protected_Area_854.aspx. Accessed 20 November, 2019.

Naughton-Treves, L., Holland, M. B., & Brandon, K. (2005). The role of protected areas in conserving biodiversity and sustaining local livelihoods. *Annu. Rev. Environ. Resour.*, *30*, 219-252.

Pimm, S. L., Jenkins, C. N., Abell, R., Brooks, T. M., Gittleman, J. L., Joppa, L. N., ... & Sexton, J. O. (2014). The biodiversity of species and their rates of extinction, distribution, and protection. *Science, 344*(6187), 1246752.

Pratheep, P. S. (2017). The Impact of Tourism on Indian Culture. *KnE Soc. Sci.*, 1(3), 429-437.

Press Information Bureau (2012). *Guidelines for Tourism in and Around Tiger Reserves.* https://pib.gov.in/newsite/PrintRelease.aspx?relid=90542. Accessed 10 January, 2020.

Puri, M., Karanth, K. K., & Thapa, B. (2019). Trends and pathways for ecotourism research in India. *J. Ecotourism*, *18*(2), 122-141.

Rangarajan, M. (2001). *India's wildlife history*. New Delhi: Permanent Black.

Reinius, S. W., & Fredman, P. (2007). Protected areas as attractions. *Ann. Tour. Res.*, *34*(4), 839-854.

Saberwal, V., Rangarajan, M., & Kothari, A. (2001). People, parks and wildlife: *Towards co-existence*. New Delhi: Orient Longman.

Sekhar, N. U. (2003). Local people's attitudes towards conservation and wildlife tourism around Sariska Tiger Reserve, India. *J. Environ. Manage.*, *69*(4), 339-347.

Shirili, V. R. & Dhivya, R. (2018). A comparative study of wildlife protection act of 1972 and wildlife (protection) amendment bill, 2013 with regard to national parks, sanctuaries and close areas. *Int. J. Pure Appl. Math.. 120* (5), 2135-2146.

Strickland-Munro, J., & Moore, S. (2014). Exploring the impacts of protected area tourism on local communities using a resilience approach. *koedoe*, *56*(2), 1-10.

Swanson, J. R., & Edgell Sr., D. L. (2013). *Tourism policy and planning: Yesterday, today, and tomorrow*. Routledge. 394-400.

TTF (2005). *Joining the Dots*. Project Tiger, Ministry of Forests and Environment, Government of India, New Delhi, India.

UNEP-WCMC (UNEP-World Conservation Monitoring Centre) and IUCN (2016). Protected Planet Report 2016. Cambridge UK, and Gland, Switzerland: UNEP-WCMC and IUCN. https://www.protectedplanet.net/c/protected-planet-report-2016.Accessed 03November 2019.

UNWTO (2010). *Tourism and Biodiversity—Achieving Common Goals towards Sustainability*. Madrid: UNWTO.

Upadhyay, V. (2019). Session 9: Lecture Notes on Legal and Policy Frameworks related to Forest Conservation. http://awsassets.wwfindia.org/downloads/lecture_notes_session_9_1.pdf. Accessed December 28, 2019.

Vattakaven, T., George, R., Balasubramanian, D., Réjou-Méchain, M., Muthusankar, G., Ramesh, B., & Prabhakar, R. (2016). India biodiversity portal: An integrated, interactive and participatory biodiversity informatics platform. *Biodivers. Data J.*, 4, e10279.

Weaver, D. B. (1999). Magnitude of ecotourism in Costa Rica and Kenya. *Ann. Tour. Res.*, *26*(4), 792-816.

Wildlife Institute of India (2019). Community Reserves. http://wiienvis.nic.in/Database/cri_8228.aspx. Accessed 21 November, 2019.

World Animal Protection (2019). Checking out of cruelty. https://www.worldanimalprotection.org/sites/default/files/int_files/pdfs/checking_out_of_cruelty.pdf. Accessed 15 November 2019.

World Bank (1991). *India - Structural Adjustment Credit Project (English)*. Washington, DC: World Bank. http://documents.worldbank.org/curated/en/999451468260069468/India-Structural-Adjustment-Credit-Project. Accessed 30 December, 2019.

12

Diatoms from Indian Himalaya (Renuka Lake) Responses to 20th Century Global Warming and Climate Change

Pranaya Diwate[1*], Narendra Kumar Meena[2], Sundeep Pandita[3], Khayingshing Luirei[2], Sumedh Humane[4] and Ravi Bhushan[5]

[1]*Department of Geology, School of Basic and Applied Sciences, MGM University, Chhatrapati Sambhajinagar, Maharashtra, India*
[2]*Wadia Institute of Himalayan Geology, Dehardun, India*
[3]*Department of Geology, University of Jammu, Jammu, India*
[4]*Department of Geology, RTMN University, Nagpur, India*
[5]*Physical Research Laboratory, Navrangpura, Ahmedabad, India*

Abstract

In this study, climatic and ecological changes over the past 170 years (1839–2007) were investigated from a sediment core of Renuka Lake of the NW Indian Himalaya using proxies, such as diatoms, Chysophycean Cyst (C.C), and Chemical Index of Weathering (CIW). Seven benthic and one planktonic diatom species were identified in the lake. The diatoms and C.C first appeared in 1903 AD in low yields, suggesting the initiation of changes in lake ecology. Their populations increased dramatically from 1936 to 1972 AD and then started declining until 2007 AD. The continued increase in diatom and C.C populations suggests a breach of the ecological threshold in the lake due to changes in temperature and productivity. The mean grain size and elemental geochemistry results also support the response of the diatom. Aluminum also ascended and reached its maxima in the core (7.76%), exhibiting high terrigenous input from the catchment. The mean grain size showed an increasing trend, and the average mean grain size during this period was 13.34 µm, which was slightly higher than the core average value. The correlation of these records with global warming indices, Global Temperature Mean Anomaly (GTMA), Northern Hemisphere Mean Temperature Anomaly (NHMTA), global

**Corresponding author*: pranayadiwate7@gmail.com

Shruti Kanga, Gowhar Meraj, Suraj Kumar Singh, Majid Farooq and M. S. Nathawat (eds.) Geospatial Technology for Natural Resource Management, (303–320) © 2024 Scrivener Publishing LLC

methane (CH_4) concentration, and global carbon dioxide (CO_2) concentration suggests that the diatom population in the lake is due to global warming.

Keywords: Himalaya, Renuka Lake, sediment core, diatom, climate change, global warming

12.1 Introduction

Climate change is considered the most severe threat to ecosystems worldwide. The mean global temperatures are widely documented to be increasing, particularly since the mid-19th century (Singh *et al.*, 2020). Numerous studies have shown that mean global temperatures are increasing (Farooq *et al.*, 2022). One of the most severe effects of climate change is its impact on water resources worldwide. Water directly or indirectly beneficiates us on social issues such as health, food supply, transportation, industry, and ecosystem integrity. Global warming and other human-induced climatic changes are raising severe concern about their immediate impact on the earth's environment and alarming the ability of the earth in the future to offer a favorable environment for human civilization. Especially in the last 60-year human rapidly changed the world ecosystem more than in entire human history (Shruti *et al.*, 2022). The 1940s and the 1950s marked the start of these changes that occurred in the earth environment due to enhanced human activities. These are also known as "The Great Acceleration" which is distinct by the large-scale increase in population and changes in natural processes (Shruti *et al.*, 2022).

Studies from different parts of the world have recommended that climatic warming has many impacts on ecosystems. Nath *et al.* (2023) showed that abrupt changes in the late 1930 AD and late 1940 AD caused an ecological impact in the Swedish sub-Arctic. In recent times, the temperature in the Northern Hemisphere has been highest in the last 2 Ka (Kanga *et al.*, 2023). In the recent past, glacial retreat has also been considered an effect of warming (Gulati *et al.*, 2023). In particular, ecologically sensitive mountainous regions such as the Himalayas are more susceptible to these climatic changes than land (Tomar *et al.*, 2023) due to physiographic conditions, their unique landscape, and fragile ecosystems. Diatoms and Chrysophytes are highly sensitive to climatic changes and are essential components of phytoplankton in mountain lakes (Singh *et al.*, 2022). The diatoms and pollen study in the peat profile from Kumaon Higher Himalayas, Northwest India shows climatic changes from this region. The significant impact of warming has led to the complete turnover of diatom assemblages over the last 200 years. Based on the record above, diatoms have been proven to be an excellent indicator of climatic shift, especially in mountainous regions such as the Himalayas, to assess the effect of global warming on the ecosystem. Paleolimnological studies of lakes from such remote areas are a suitable tool to gain an understanding of the effects of global warming-related climate change on aquatic systems. Lakes are warming at an average of 0.3 °C over each decade (Lavane *et al.*, 2023). This warming already shows a significant shift in the ecosystem, such as algal blooms and a decrease in the fish population. This high-resolution lake record provides an excellent opportunity to extend the monitoring record of the temporal range of recent warming.

Himalayan lakes are sensitive as they receive rainfall from both the Indian summer monsoon and westerlies and provide an excellent archive to reconstruct high-resolution paleoclimate data (Upadhyay *et al.*, 2023). The 800-year tree-ring record from Himalayas shows the 1944–1953 AD record warmest 10 years mean in the above period (Sud *et al.*, 2023). However, high-resolution climatic records from the high altitude of the Himalayas are poorly documented because of logistical barriers such as the setup, maintenance, and observation of the station (Tripathi *et al.*, 2023).

In lakes with changes in physicochemical conditions, their assemblages show dramatic effects. Therefore, they play a vital role in deciphering past limnological conditions, such as nutrient availability (P and N), pH, and conductivity, and are widely used as paleoecological indicators (Singh *et al.*, 2023). A characteristic feature of diatoms is siliceous frustules that do not break easily and thus contain a record of changes, which is coupled with their ability to react rapidly to changes in the surrounding environment, and their abundance, distribution, and variation at the species level (Rathore *et al.*, 2023; Singh *et al.*, 2023).

Increases in temperature and climate change severely impact agriculture and water resources (rivers, lakes, and glaciers) and cause an ecological imbalance worldwide (Singh *et al.*, 2023). The monsoon-dominated Indian subcontinent also experienced these changes, as the glaciers of the Himalayas showed signs of negative mass balance (Kumar *et al.*, 2023), an increase in extreme events (Debnath *et al.*, 2023), changes in river discharge, and changes in lake ecologies (Rai *et al.*, 2023). However, most studies on global warming and its impact on ecology are in their initial stages. Therefore, there is a need for an ecological indicator that can provide a relationship between global warming and related climate change in the Himalayas (Choudhury *et al.*, 2023). In this study, diatoms and chrysophytes were used as ecological indicators of a climatically sensitive Himalayan Lake (Renuka Lake).

12.2 Study Area

Renuka Lake (latitude 30°36′36″ and longitude 77°27′45″) is located at an altitude of 620 msl in the Sirmour District of Himachal Pradesh, India (Figure 12.1a). The oblong-shaped lake is flanked by steep hills on the northern and southern sides, covering an area of 254 ha [36]. The lake receives most of its water from rain-runoff and underground seepage. Geologically, the catchment area contains Proterozoic rocks belonging to the Blaini, Infra Krol, Krol, and Jaunsar groups that consist mainly of carbonaceous shale, slate, siltstone, purple dolomite, and limestone (Figure 12.1b) [31].

Climatically, the study area falls in the subtropical zone where winters are mild and dry, and summers are hot (average annual temperature 26 °C). The lake receives an average annual rainfall of 1,500 mm; 80% is contributed by the Indian Summer Monsoon (Tropical Rainfall Measurement Mission data 1998–2015). Phytoplankton are the primary producers in this lake and constitute the primary food source for aquatic animals. Aquatic and marshy vegetation near the lake controls the nutrient supply to the lake (Saqib *et al.*, 2023; Sharma *et al.*, 2023). The primary nutrients, phosphorus and nitrogen are added to the lake from agriculture, animal dung, and sewage discharge. As the lake is rich in nutrients, it supports a large biological diversity (Ahmad *et al.*, 2023). The growth and decline of algae in Renuka Lake are concurrent with favorable temperatures (Rani *et al.*, 2023).

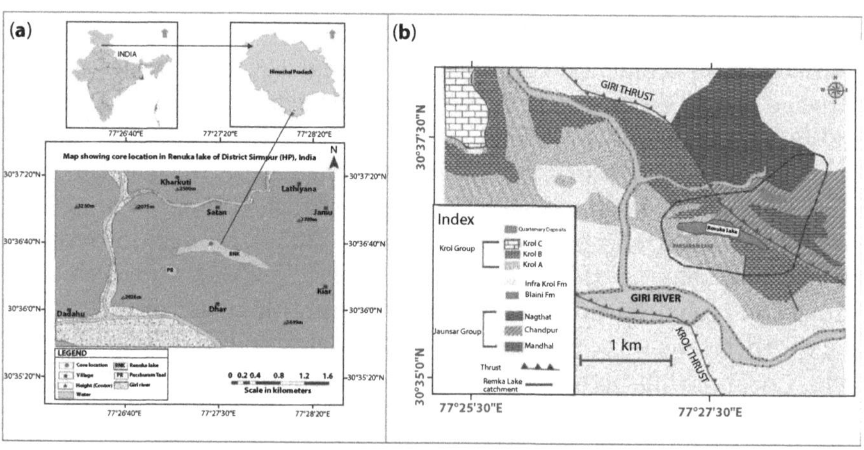

Figure 12.1 Location and geological map of the Renuka Lake (study area).

12.3 Material and Methods

12.3.1 Sampling and Diatom Analysis

A 13-meter sediment core was recovered from the central part of Renuka Lake at 11-meter depth using a piston corer with a jacket that could be split vertically into two equal half-U channels. The jacket was removed, and the core was sliced perpendicular to the length at a continuous 2-cm interval.

In the present study, a top two-meter core was used, from which a total of 22 sediment samples were analyzed from various depths (06 cm resolution). All the samples were processed and studied at the Wadia Institute of Himalayan Geology, Dehradun. For the diatom slide preparation, sediment samples were prepared using standard digestion procedures. The method involves the following steps. Approximately 2 g of the sample was taken and washed thoroughly with distilled water to make it free from atmospheric contamination. The samples were treated with 30% H_2O_2 and 10% HCl to remove the organic carbon and carbonate, respectively. The samples were left overnight until all organic material had been completely removed. The samples were then centrifuged at 3,000 rpm to remove the acid from the samples. Washing and centrifugation were repeated at least three to four times to remove all chemicals. After digestion, the samples were air-dried on coverslips and mounted onto microscope slides using the mounting medium Distyrene Plasticizer Xylene (DPX). The diatoms were identified using an Olympus bx-61 microscope with an attached digital camera having a ×100 objective.

12.3.2 Chronology

The chronology of the Renuka Lake sediment core was determined from the sedimentation rate determined by the ^{210}Pb and ^{137}Cs radionuclide methods at the Physical Research Laboratory (PRL), Ahmedabad, India. The activities of ^{137}Cs and ^{210}Pb were measured according to the described in. The chronology of this study was adopted from, and a detailed dating methodology was described in. Based on the mean sedimentation rate, the entire 126 cm core was dated 170 years old from ca 1839 AD to 2010 AD.

The Constant Rate of Supply (CRS) model was used to determine the age with respect to depth [36]. The CRS model is used for most of the ^{210}Pb dated sedimentary sequences. This is especially true in archives, where sediment supply may differ in response to anthropogenic and climatic changes. Since ^{210}Pb limits age determination from 150 to 200 years, most anthropogenic activity has dominated in recent decades. This model assumes a constant supply of lead-210 flux but a variation in the sediment supply. The fact that lead-210 flux has been constant over the last 100 years validates this model.

12.3.3 Diatom

The samples were treated with 30% H_2O_2 and 10% HCl to remove the organic carbon and carbonate, respectively. A total 25 diatom slides were prepared according to the standard protocol. Diatoms were identified using an Olympus bx-61 microscope with an attached digital camera and a ×100 objective. Dominant diatom taxa were classified according to taxonomic references.

12.3.4 Grain Size

The grain size analysis of the samples was performed using a Laser Particle Size Analyzer (Malvern Mastersizer 2000) at the Sedimentology Lab, Wadia Institute Himalayan Geology, Dehradun. A sample of approximately 5 g was used for particle size analysis. The samples were sieved through a 1 mm sieve to remove particles that were above the upper limit of the Mastersizer sensitivity. The samples were treated with hydrogen peroxide (30%) and HCL (1N) to remove the organic material and carbonate, respectively. The result of the analysis was the volume percentage of the size distribution. The volume % of particle sizes was classified mainly in three, i.e., sand, silt, and clay on the basis of the logarithmic Udden–Wentworth grade scale [36–37] clay (<0.0039 mm), silt (0.0039–0.0625 mm), and sand particles (>0.0625 mm).

12.3.5 Global Linkage

Renuka Lake records were compared with global forcing factors that indicate global warming and associated climate change. As the sun is the primary source of energy, variations in solar activity are directly associated with the Earth's surface temperature, which ultimately influences the climate system. We used the variation in Total Solar Irradiance (TSI) values reconstructed from different TSI observation models to evaluate the contribution of direct short-term solar warming [38]. We also used the global Temperature Mean Anomaly (GTMA) based on analyses of global surface temperature changes that are routinely carried out by several groups, including the NASA Goddard Institute for Space Studies (GISS), the NOAA National Climatic Data Center (NCDC), and a joint effort of the

UK Met Office Hadley Centre and the University of East Anglia Climatic Research Unit (HadCRUT). The Northern Hemisphere Temperature Mean Anomaly (NHTMA) data are based on land-based meteorological station data and fixed-position weather ship data. The methane (CH_4) concentration was taken as average data from ice cores of the Antarctic (three cores), Greenland (two cores), Antarctic fern layer, and archived air from Tasmania, Australia and CO_2 from the Law Dome DE08, DE08-2, and DSS ice cores. In addition, the All-India Annual Rainfall (AIAR) also considers establishing regional correlations.

12.4 Results

12.4.1 ^{137}Cs Isotope, ^{210}Pb Isotope

^{137}Cs vertical profiles were used to validate the chronologies established using the ^{210}Pb dating method. The ^{137}Cs vertical profile in Renuka Lake showed a clear maximum value in the 37.5 cm (7 dpm/g) layer (Figure 12.1). The ^{137}Cs inventory exhibited minimum values of 0 dpm/g to a maximum of 7 dpm/g with an average of 4.55 dpm/g. The minimum and maximum ^{210}Pb excess values ranged from 0.5 to 9 dpm/g. The ^{210}Pb ex (on a logarithmic scale) in the sediment cores from Renuka Lake is shown in (Figure 12.1). Excess ^{210}Pb (^{210}Pb ex) activity showed an exponential decrease with depth in the Renuka core profile. The top sample at a depth of 2 cm showed a value of 9 dpm/g followed by 4 dpm/g at 19 cm. The sample at 37.5 cm possesses 0.5 dpm/g and exhibits exponential decay. The sample below 37.5 cm did not show any ^{210}Pb ex activity.

The sedimentation rate for Renuka Lake was estimated using ^{210}Pb and ^{137}Cs radioactive isotope techniques. According to ^{210}Pb the lake indicated two different sedimentation rates, i.e., 0.51 cm/year upper part (0–19 cm depth) for the period of ca. 2011 to 1974 AD and 0.78 cm/year lower part (19–40 cm depth) from 1974 to 1950 AD, whereas the average sedimentation of the lake was measured to be 0.64 cm/year.^{137}Cs profile shows a distinctive peak at 37.5 cm (Figure 12.2) and was considered as ^{137}Cs fallout in the atmosphere, which is referred to as maximum ^{137}Cs fallout due to 1963 nuclear weapon test. The age-depth model was calculated based on the above data. The CRS model requires excess lead-210 measurements for age determination. Based on the model described above, the age of Renuka Lake ranges from 1839 to 2010 AD corresponding to 126-0 cm depth, which is used for the top two (2) m cores.

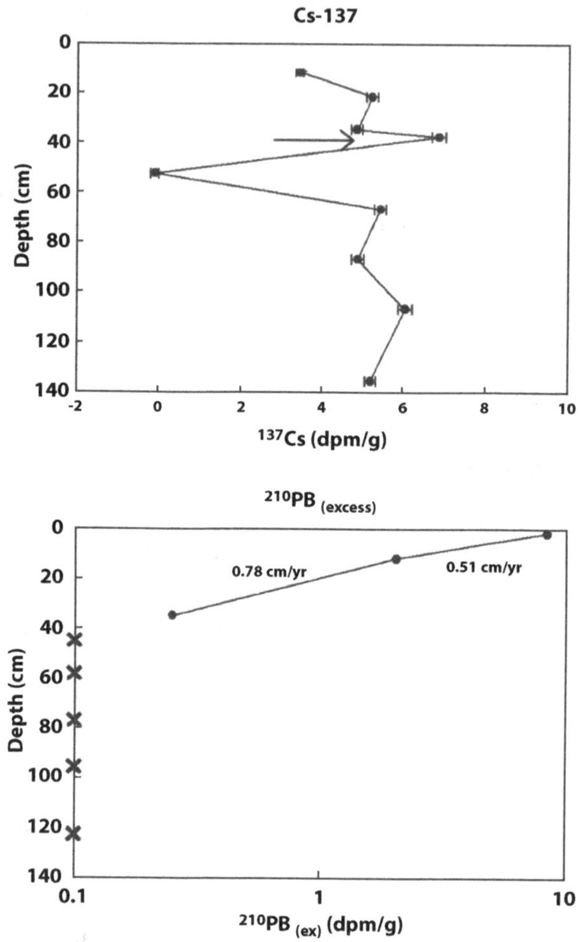

Figure 12.2 ^{210}Pbex (dpm/g) and ^{137}Cs activities with a 1963 peak (dpm/g) in the sediment core of Renuka Lake.

12.4.2 Diatoms and Grain Size

The present study contains diatom data with other proxies ranging from AD 1840 to 2010. These data represent the changes occurring in the lake and its catchment and productivity during this period. In Renuka Lake, the most pronounced change over the past ca.170 years occurred in 1940 AD, with a marked shift in the diatom population characterized by a higher taxon. Morphological study of different diatoms at the generic and species levels was performed according to the classification proposed by Singh *et al.* (2023). Eight taxa were identified in the samples. Based on their shape, the recovered diatoms were subdivided into centric and seven-pennate forms. A total of eight diatom genera were identified, mainly represented by the genera *Aulacoseira*, *Navicula*, *Gomphonema*, *Cocconeis*, *Rhopalodia*, *Cymbella*, *Eunotia*, and *Mastogloia*. The Al% was taken into consideration when studying terrigenous influx. The magnetic susceptibility (*xlf*) and grain size data were plotted against diatoms to determine the catchment input to the lake.

Present climatic records span a time interval of 170 years. In the starting period from AD 1840 to 1902 (126–74 cm), there was a complete absence of diatoms and C.C. The aluminum and xlf values in this time interval show fluctuations with an increasing trend (Figure 12.4). In AD 1903 the first appearance of diatoms and (C.C) in the Renuka Lake had marked with *Navicula leptostriata Jørgensen* and *Mastogloia grevillei W. Smith* species with very low abundance (Figure 12.3). The former is considered to live in oligo-mesotrophic waters with low organic carbon [43], whereas the latter occurs in temperate conditions [35]. The mean grain size also increased with another parameter, i.e., *xlf* and Al after AD 1903 and AD 1905, respectively.

After AD 1915, the C.C increased, which is referred to primarily freshwater algae. They are primarily found in temperate and high-altitude lakes. Interestingly, from AD 1936 to AD 1954, the population of *Aulacoseira ambigua* (Grunow) Simonsen and C.C. The abundance of both planktonic forms increased drastically. *A. ambigua* reflects climate change. The percentage of alumina also increased during this time interval. *A. ambigua* and C.C contributed to 65% of all phytoplankton, with averages of 140 and 408, respectively. The diatoms species found followed by decreasing order, *Rhopalodia gibba* (Ehrenberg) Otto Müller (27%), *A. ambigua*, *Cymbella cistula* (Ehrenberg) (24%) O. Kirchner (21%), *Eunotia pectinalis* (Kützing) Rabenhorst (7%), *Cocconeis placentula* Ehrenberg (6%), *Gomphonema brasiliense* Grunow (6%), *Navicula leptostriata Jørgensen* (5%), and *Mastogloia grevillei* W. Smith (4%) (Figure 12.3).

The mean grain size also shows an increasing trend and the average mean grain size in this period is 13.34 um slightly higher than the average throughout the core, i.e., 13.11um. After 1959–1972, it increased for a small-time interval with *Navicula leptostriata Jørgensen*, contributing >50% of all diatoms and representing a major part of all diatom taxa.

From AD 1972 to 2007, the abundance of diatoms and C.C has decreased in recent years. The average C.C decreased from 850 to 253 during this period. Whereas the average diatom population also decreased from AD

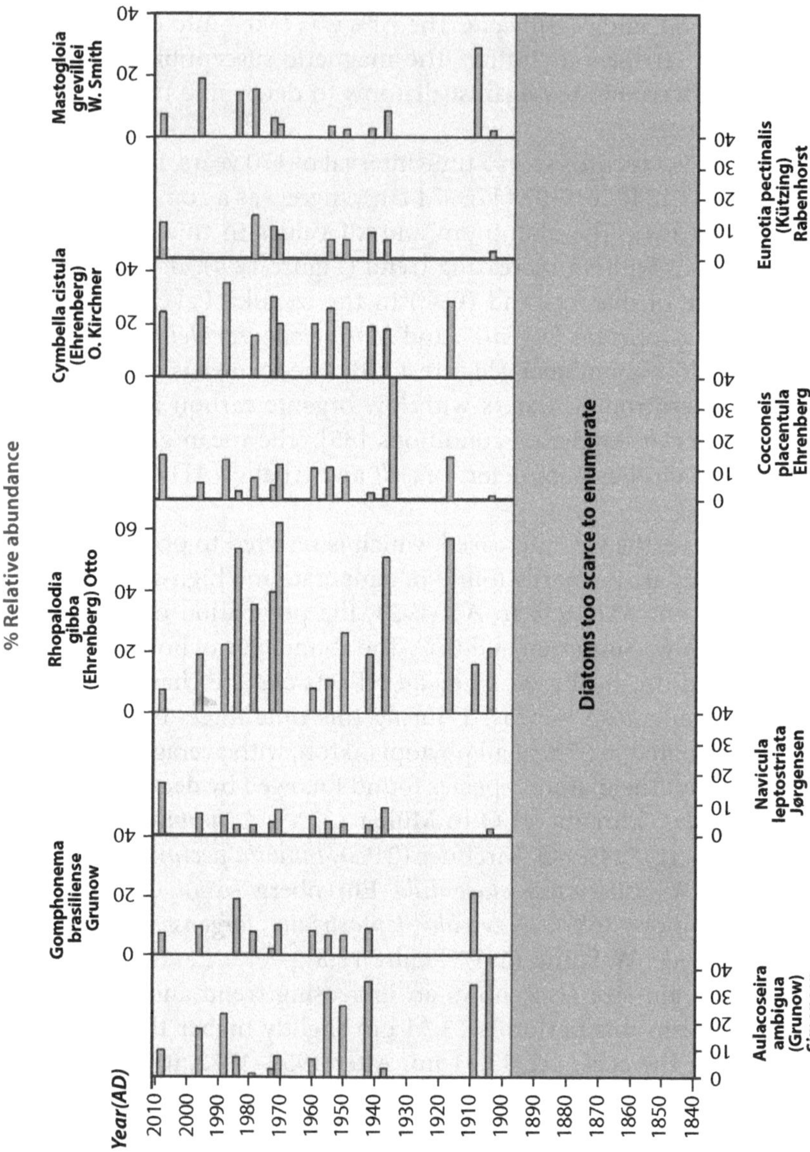

Figure 12.3 Relative abundance of the diatoms in the Renuka Lake.

1972 to 2007 from 454 to 246. *Rhopalodia gibba* (Ehrenberg) Otto Müller (29%) and *C. cistula* (Ehrenberg) O. Kirchner (21%) is the highest populated diatoms in this period. The other proxy means that grain size increases up to AD 1970–1999 and then increases up to AD 2007, whereas xlf increases at the same time that the percentage of Al is also slightly decreased.

Overall, the diatoms and C.C were absent up to AD 1903, and after the first appearance in AD 1903, the population of this phytoplankton increased up to AD 1972, and after AD 1972, it decreased again in recent times.

12.4.3 Classification and Quantification of Diatoms and Chrysophycean Cysts

Diatoms and Chrysophycean cysts can provide indirect evidence of climatic change, as they are known to respond to the physical and chemical changes associated with various components. Aquatic systems in high-altitude regions are very sensitive to climatic changes, even in the absence of catchment modifications. In this study, we used different regional and global datasets to determine the correlation and effects of the forcing mechanism. The dominant diatom taxa were classified according to the taxonomic references. A total of eight diatom species were identified and subdivided into centric (one species) and pennate (seven species) types based on their shape. Taxonomically, the diatoms were represented by the genera *Aulacoseira*, *Navicula*, *Gomphonema*, *Cocconeis*, *Rhopalodia*, *Cymbella*, *Eunotia*, and *Mastogloia* (Figure 12.3). The quantity of the various diatom species is in 126 cm core are as follows, *Rhopalodia gibba* (Ehrenberg) Otto Müller (27%), *A. ambigua*, *C. cistula* (Ehrenberg) (24%), O. Kirchner (21%), *Eunotia pectinalis* (Kützing) Rabenhorst (7%), *Cocconei splacentula* Ehrenberg (6%), *Gomphonema brasiliense* Grunow (6%), *Navicula leptostriata* Jørgensen (5%), and *Mastogloia grevillei* W. Smith (4%) (Figure 12.3). Chrysophycean algae produce a siliceous resting stage (cyst), which is known as a chrysophycean cyst. The chrysophyte classes (Chrysophyceae and Synurophyceae) are a group of freshwater algae known as golden algae [35]. The C.C is a hollow siliceous structure that is globose in shape and has a single pore. The diameter of the cysts ranges from 2 μm to 35 μm, and their shape ranges from spherical to oval and pyramidal. *A. ambigua* and C.C species account for approximately 65% of all phytoplankton.

12.5 Discussion

12.5.1 Temporal Variation in Diatoms and its Ecological Implications

The diatom and C.C records of the Renuka Lake sediment core indicate the ecological and productivity changes that occurred in the lake during the past 170 years (AD 1839 to D 2010). However, the population of diatoms and C.C was negligible or absent from AD 1839 to AD 1902 (Figure 12.3). The total disappearance of diatoms and C.C presumably resulted from low productivity and the absence of a favorable environment for preservation, possibly because of the colder surface water of the lake [42]. It is evident in the higher Himalayan region the low diatom concentration is associated with a cold and dry climate under weak Indian Summer Monsoon (ISM) conditions, whereas warm and wet conditions under strong ISM yield more diatoms in the Himalayas as well as elsewhere [2006]. C.Cs are better indicators of environmental change than diatoms because they respond more quickly to acidification and act as early warning indicators of limnological changes [43].

The first appearance of diatoms and C.C in Renuka Lake was marked by *N. leptostriata* Jørgensen and *Mastogloia grevillei* W. Smith in AD 1903, although their abundance was very low (Figure 12.3). The former lives in oligo-mesotrophic waters with low organic carbon, whereas the latter occurs in temperate conditions. The C.C also increased after AD 1915; they generally indicate a contribution of high planktonic algae blooms during spring. They are mostly found in temperate and high-altitude lakes. The appearance of diatoms and C.C indicates slight changes in the lake ecology since AD 1903. Similar trends have also been recorded in an approximately 100-year-long East Svalbard, high Arctic Lake sediment record, and from a small coastal lake where multiple records of warm years have been observed since AD 1912. This seems to be the beginning of the warming trend in Renuka Lake. Interestingly, from AD 1936 to AD 1954, the population of both planktonic forms of *A. ambigua* (Grunow) Simonsen and C.C increased dramatically, suggesting that an ecological threshold may have been crossed. The genus *Aulacoseira* is a widely distributed diatom commonly found in meso-eutrophic and stagnant waters with high photic conditions. This species favors relatively warm waters, so an increase suggests an increase in water temperature. Therefore, the increase in *A. ambigua* suggests an increase in the average temperature in the Renuka Lake area during this time period. The volume of diatoms increased from AD 1954 to AD 1972 for short intervals, with *N. leptostriata* Jørgensen contributing

>50% of all diatoms and representing a significant part of all diatom taxa. The *N. leptostriata* Jørgensen is considered an acidophilic diatom which is a mobile biraphid that migrates close to sediment nutrient sources. The abundance of this species was probably influenced by reduced turbulence (compared to shallower areas), adequate light, and higher nutrient availability from organic sediments. Hence, turbidity was reduced, and water became more stagnant during this period. Water turbidity is mostly

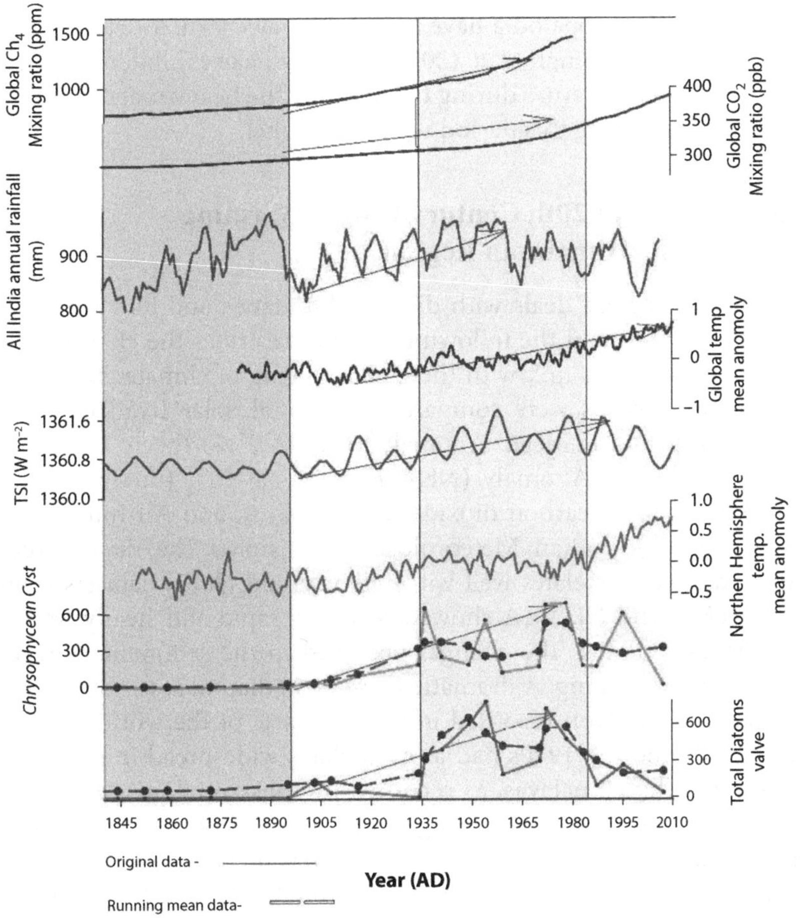

Figure 12.4 Diatom and Chrysophycean cyst records from Renuka Lake in comparison with Total Solar Irradiance (TSI; [38]), Global Temperature Mean Anomaly (GTMA; [39]), Northern Hemisphere Mean Temperature Anomaly (NHMTA; https://www.ncdc.noaa.gov/monitoring-references/faq/anomalies.php#anomalies), Global CH_4 (Methane mixing ratio), Global CO_2 (Carbon dioxide mixing ratio) and All India Annual Rainfall (AIAR; Indian Meteorological Division).

controlled by the presence of macrophytes in lakes. Lower turbidity, mostly in shallow lake conditions, refers to the reduction of macrophytes. From AD 1952 to AD 1972, decreased turbidity suggests a reduction in phytoplankton, and the C.C follows the diatom trend in this zone. From 1972 to 2007, the average C.C decreased from 850 to 253 valves per slide, and the average diatom population decreased from 454 to 246. *Rhopalodia gibba* (Ehrenberg) Otto Müller (29%) and *C. cistula* (Ehrenberg) O. Kirchner (21%) were the most abundant of the diatoms during this period. These two species were found in eutrophic, warm waters. The taxa belonging to the Müller genus rhopalodia have a complex valve structure and symmetry, as described by Singh *et al.* (2023). Renuka Lake exhibited the highest surface water temperature during this period. The heavy metal concentrations recorded during this period were also higher.

12.5.2 Traces of 20th Century Global Warming in the Himalayan Region

The previous section deals with diatom abundance and its vis-à-vis link with lake ecology, and the following section describes the change in lake ecology and vis-à-vis link with the global forcing of climate. The Renuka Lake diatom records were compared with Total Solar Irradiance (TSI), Global Temperature Mean Anomaly (GTMA), Northern Hemisphere Mean Temperature Anomaly (NHMTA), Global CH_4 (methane mixing ratio), Global CO_2 (carbon dioxide mixing ratio), and All India Annual Rainfall (AIAR) (Indian Meteorological Division). The diatom record of Renuka Lake correlates well with regional and global datasets (Figure 12.4). GTMA and NHMTA showed relatively rapid and steady warming from the early 1940s. The changes observed in the sedimentary records reflect climate warming. A dramatic increase in diatom assemblages in the 19th century has been recorded in various parts of the world. Warming in the 1930s and the 1940s had a particularly widespread impact, which also affected the Himalayas, as recorded by substantial glacier retreat in this region. Tree ring-based spring temperature reconstruction from the western Himalayas indicates that warming has increased in recent times, especially since the 1920s. Besides, the period AD 1945–1974 was considered the warmest 30-year mean, while 1944–1953 recorded the warmest 10-year mean. The IPCC states that the reduction in glacial mass since the mid-19th century is evident and persistent in many parts of the world IPCC-AR5 (2014). The Gangotri Glacier has been continuously receding since 1935. The total area decreased by 0.58 km^2 from 1935 to 1996.

Most of the aforementioned changes in Himalayan ecology and glacial mass balance agree with the Renuka Lake diatom and C.C population yields.

12.6 Conclusions

- A total of eight species of diatoms were found in Renuka Lake, among which one species is centric and seven species belong to the pennate form.
- The biological changes in Renuka Lake occurred after AD 1903 until the presence of diatom species was very low, with the first appearance of diatoms and C.C, because major changes in lake ecology were initiated here. The further intensification of diatoms and C.C growth from AD 1936 to AD 1954 and from AD 1954 to AD 1972 resulted from the increased productivity and temperature in the lake. The lake reached its highest ecological threshold from AD 1936 to AD 1954. From 1972 to 2007, the temperature of the surface water was higher in the lake.
- The diatom abundance of Renuka Lake shows strong coherence with global warming indices such as GTMA, NHMTA, and Global CH_4 and CO2. Hence, the populations of the seven benthic diatoms (*Acicula*, *Gomphonema*, *Cocconeis*, *Rhopalodia*, *Cymbella*, *Eunotiaand*, and *Mastogloia*), the planktonic diatom (*Aulacoseira*) and C.C flourished in the lake due to global warming.
- Major climatic (glacier retreat) and ecological (tree ring) changes in other parts of the Himalayas also occurred simultaneously.
- Based on the present study, we conclude that Renuka Lake has responded to climatic changes, especially warming. The widely distributed warming of the 1930s and the 1940s also affected Himalayan aquatic ecosystems. Diatoms reflect climate variability, such as direct effects of temperature on aquatic systems.
- There are more possibilities if the rate of warming continues to accelerate in the future, as suggested by several atmospheric general circulation models. We should expect ecological and biological changes in Himalayan lakes to continue and monitor this record by various observations,

which is equally essential for the establishment of a better balance between ecology and society.

Acknowledgments

The authors are grateful to the Director, WIHG, Dehradun for providing all the necessary facilities. We also sincerely thank Prof. G. T. Thong, Department of Geology, Nagaland University Kohima, India, for improving the manuscript. This is the WIHG contribution No. WIHG/0066.

References

Singh, S., Singh, S. K., Kumar, R., Shrama, A., & Kanga, S. (2022). *Finding Alternative to River Sand in Building Construction.*

Singh, S., Meraj, G., Kumar, P., Singh, S. K., Kanga, S., Johnson, (2023). Decoding Chambal River Shoreline Transformations: A Comprehensive Analysis Using Remote Sensing, GIS, and DSAS. *Water*, 15(9), 1793.

Singh SK, Kanga S, Gulati B, Raič M, Sajan B, Đurin B, Singh S (2023) Spatial and Temporal Analysis of Hydrological Modelling in the Beas Basin Using SWAT+ Model. *Water* 15, 3338. https://doi.org/10.3390/w1519333

Rani A, Gupta SK, Singh SK, Meraj G, Kumar P, Kanga S, Đurin B, Dogančić D (2023) Predicting Future Land Use Utilizing Economic and Land Surface Parameters with ANN and Markov Chain Models. *Earth* 4(3):728-751.

Ahmad T, Gupta SK, Singh SK, Meraj G, Kumar P, Kanga S (2023) Unveiling Nature's Resilience: Exploring Vegetation Dynamics during the COVID-19 Era in Jharkhand, India, with the Google Earth Engine. *Climate* 11(9):187.

Sharma M, Upadhyay RK, Tripathi G, Kishore N, Shakya A, Meraj G, Kanga S, Singh SK, Kumar P, Johnson BA, Thakur SN (2023) Assessing Landslide Susceptibility along India's National Highway 58: A Comprehensive Approach Integrating Remote Sensing, GIS, and Logistic Regression Analysis. *Conservation* 3(3):444-459.

Saqib N, Rai PK, Kanga S, Kumar D, Đurin B, Singh SK (2023) Assessment of Ground Water Quality of Lucknow City under GIS Framework Using Water Quality Index (WQI). *Water* 15(17):3048.

Choudhury U, Singh SK, Kumar A, Meraj G, Kumar P, Kanga S (2023) Assessing land use/land cover changes and urban heat island intensification: A case study of Kamrup Metropolitan District, Northeast India (2000–2032). *Earth* 4(3):503-521.

Rai NK, Mathur S, Singh SK, Tiwari M, Singh VK, Haque R, Tiwari S, Sharma LK (2023) Differential regulation of mitochondrial complex I and oxidative stress based on metastatic potential of colorectal cancer cells. *Oncol. Lett.* 26(1):1-1.

Debnath J, Sahariah D, Lahon D, Nath N, Chand K, Meraj G, Kumar P, Singh SK, Kanga S, Farooq M (2023) Assessing the impacts of current and future changes of the planforms of river Brahmaputra on its land use-land cover. *Geosci. Front.* 14(4):101557. Elsevier.

Kumar P, Singh SK, Gupta SK, Kanga S, Meraj G (2023) Uncovering the hydro-meteorological drivers responsible for forest fires utilizing geospatial techniques. *Theor. Appl. Climatol.*. https://doi.org/10.1007/s00704-023-04497. Springer Nature.

Singh SK, Kotecha MJ, Kanga S, Pankhaniya SK, Agrawal S, Meraj G (2023) Urban Heat Island (UHI) Resilience Plan in Varying Climatic Conditions Using Geospatial Approach: A Case Study of Rajkot City. In: *Climate Change, Agriculture and Society*, vol 1. pp 307-338.

Singh S, Singh SK, Prajapat DK, Pandey V, Kanga S, Kumar P, Meraj G (2023) Assessing the Impact of the 2004 Indian Ocean Tsunami on South Andaman's Coastal Shoreline: A Geospatial Analysis of Erosion and Accretion Patterns. *J. Mar. Sci. Eng.* 11(6):1134.

Rathore SS, Mathur SC, Mathur S, Sharma S, Singh SK (2023) Late Cretaceous–Early Paleocene Geodiversity of Fatehgarh Formation of Petroliferous Barmer Basin, Western Rajasthan, India: A Potential Geopark Site. *Geoheritage* 15:54.

Singh S, Meraj G, Kumar P, Singh SK, Kanga S, Johnson BA, Prajapat DK, Debnath J, Sahariah D (2023) Decoding Chambal River Shoreline Transformations: A Comprehensive Analysis Using Remote Sensing, GIS, and DSAS. *Water* 15(9):1793.

Tripathi G, Shakya A, Upadhyay RK, Singh SK, Kanga S, Pandey SK (2023) Landslide Susceptibility Mapping of Tehri Reservoir Region Using Geospatial Approach. In: *Climate Change Adaptation, Risk Management and Sustainable Practices in the Himalaya*, pp 135-156.

Sud A, Kanga R, Singh SK, Meraj G, Kanga S, Kumar P, Ramanathan AL, Bhardwaj V (2023) Simulating Groundwater Potential Zones in Mountainous Indian Himalayas—A Case Study of Himachal Pradesh. *Hydrology* 10(3):65.

Upadhyay RK, Tripathi G, Đurin B, Šamanović S, Cetl V, Kishore N, Sharma M, Singh SK, Kanga S, Wasim Md, Rai PK, Bhardwaj V (2023) Groundwater Potential Zone Mapping in the Ghaggar River Basin, North-West India, Using Integrated Remote Sensing and GIS Techniques. *Water* 15(5):961.

Lavane K, Kumar P, Meraj G, Han TG, Ngan LHB, Lien BTB, Ty TV, Thanh NT, Downes NK, Nam NGD, Minh HVT, Singh SK, Kanga S (2023) Assessing the effects of drought on rice yields in the Mekong Delta. *Climate* 11(1):13.

Nath N, Sahariah D, Meraj G, Debnath J, Kumar P, Lahon D, Chand K, Farooq M, Chandan P, Singh SK, Kanga S (2023) Land use and land cover change monitoring and prediction of a UNESCO world heritage site: Kaziranga eco-sensitive zone using cellular automata-Markov model. *Land* 12(1):151.

Kanga S, Upadhyay RK, Sharma M, Sharma S, Tripathi G, Singh SK (2023) Geotechnical Stability Assessment of Road-cut Slopes: A Case Study of Srinagar, Garhwal Himalaya, India. *Biol. Forum* 15(5):1071-1080.

Gulati B, Sharma R, Kanga S, Singh SK, Sajan B, Meraj G, Kumar P, Ramanathan AL (2023) Unraveling the Relationship Between Stubble Burning and Air Quality Degradation in Punjab: A Temporal and Spatial Analysis (2019-2022). *J. Clim. Change* 9(2):43-53, IOS Press.

Tomar P, Singh SK, Kanga S, Pattanaik A (2023) Water Bodies mapping and monitoring using high-resolution satellite images. *Sustainability, Agri, Food and Environmental Research (SAFER)* 11.

Singh H, Meraj G, Singh S, Shrivastava V, Sharma V, Farooq M, Kanga S, Singh SK, Kumar P (2022) Status of Air Pollution during COVID-19-Induced Lockdown in Delhi, India. *Atmosphere* 13(12):2090.

Farooq M, Mushtaq F, Meraj G, Singh SK, Kanga S, Gupta A, Kumar P, Singh D, Avtar R (2022) Strategic Slum Upgrading and Redevelopment Action Plan for Jammu City. *Resources* 11(12):120.[11] Singh, S., Singh, S. K., Prajapat, D. K., Pandey, V., Kanga, S., Kumar, P., & Meraj, G. (2023). Assessing the Impact of the 2004 Indian Ocean Tsunami on South Andaman's Coastal Shoreline: A Geospatial Analysis of Erosion and Accretion Patterns. *J. Mar. Sci. Eng.*, 11(6), 1134.

Shruti, S., Singh, P. K., & Ohri, A. (2020). Evaluating the environmental sustainability of smart cities in India: The design and application of the Indian smart city environmental sustainability index. *Sustainability*, 13(1), 327.

Shruti, S., Singh, P. K., Ohri, A., & Singh, R. S. (2022). Development of Environmental Decision Support System for Sustainable Smart Cities in India. *Environ. Prog. Sustain. Energy*, e13817.

Singh, R. P., Kumar, S., & Rana, P. S. (2020). UN SDGs and context of Holy-Heritage cities in India: a study of Ayodhya and Varanasi. In *Geoecology of Landscape Dynamics* (pp. 187-206). Springer, Singapore.

13

Beas Basin Snow Area Health Monitoring Utilizing Remote Sensing with Elevation Zones and Aspect

Roy K.*, Kant C. and Meena R.S.

Department of Civil Engineering, National Institute of Technology, Hamirpur, India

Abstract

Civilizations have historically centered around rivers, springs, and watersheds, with snowmelt from the Himalayas significantly influencing seasonal water availability. Snow Coverage Area (SCA) serves as a vital indicator of the health of the Beas Basin, impacting summer snowmelt runoff across the basin's spring sheds and mountain watersheds. This study employs temporal and spatial satellite data over 22 years (February 2000 to July 2022) using ASTER GDEM for elevation zoning and MODIS 8 Day L3 Global 500 m Grid (MOD09A1) for snow signal determination to analyze snow cover and its contributions to streams. The Beas River basin, covering approximately 11,280 km^2, shows an average SCA of 17%, with monthly variations peaking at 37% in February 2019 and a minimum of 2.2% in August 2016. The highest rate of SCA change during the accumulation period was 5.42% per 100 m elevation (3,500-4,000 m), while during the depletion period, it was 4.39% per 100 m (4,500-5,000 m). Aspect analysis reveals that the north to west-north-west and south to south-west aspects (380-3,000 m) accumulated ~0.69% SCA, while higher elevations (3,000-5,000 m and 5,000-6,585 m) show greater variability. Specifically, the south-south-east to south-south-west aspect had higher SCA (~2.01%) compared to the east-north-east aspect (~0.88%) in the 3,000-5,000 m range. In the 5,000-6,585 m range, the south-east to south-south-west aspect had higher SCA (~0.39%) compared to the east-north-east aspect (~0.19%), with both ranges showing maximum SCA in February and minimum in August (August 2021 to July 2022). These findings shows the importance of continuous monitoring and analysis of snow cover health in the Beas Basin for

**Corresponding author*: 21mce312@nith.ac.in

Shruti Kanga, Gowhar Meraj, Suraj Kumar Singh, Majid Farooq and M. S. Nathawat (eds.) Geospatial Technology for Natural Resource Management, (321–342) © 2024 Scrivener Publishing LLC

assessing basin health, managing water resources, and supporting sustainable practices crucial for maintaining the overall health of the basin.

***Keywords*:** Snow cover health, ablation and depletion, snow cover area, water resources, beas basin

13.1 Introduction

Snow Cover Area (SCA) in mountainous or hilly regions provide a source of water to natural spring sheds during summer season while it is being accumulated during winter season (Tripathi *et al.*, 2023; Sud *et al.*, 2023). It contributes to the summer runoff from the basin, which, when used in contour farming, becomes a resource for various fruit growers such as apples and civilizations that have learned to sustain in higher elevation regions by using this summer runoff for domestic purposes on elevation zones and aspects, combining community and government initiatives to provide the region's water supply (Singh *et al.*, 2022; Lavane *et al.*, 2023).

Abstract Forest loss may exacerbate droughts, fires, and floods in Western Europe. The decrease in snow cover may have been caused by the collapse of alpine forests (which boosted mountain precipitation) through a biotic moisture pump (Upadhyay *et al.*, 2023; Nath *et al.*, 2023). Natural and cultural influences affect watershed land use and cover. Population increase affects the Himalayan LULC. Long-term ecological stability and sustainability are issues in the Himalayan watersheds. Land use/cover is affected by height, slope, geology, and cultural factors including technology, socioeconomics, and institutions (Rani *et al.*, 2023). The Swiss Alps features three climatic zones: north, inner, and south. The southern slopes are sunnier and rainier (Ahmad *et al.*, 2023). Snow covers 80% of the Himalayas. The Western Himalayas receive more snow (Sharma *et al.*, 2023). The Himalayas include the Outer Himalaya, Lesser Himalaya, and Greater Himalaya (Rai *et al.*, 2023). The Himalayas affect air and water circulation. Himalayan weather in the Indian subcontinent of northern Asia (Sqib *et al.*, 2023). April warms the Himalayas. Solar warmth and reduced cloud cover make May the warmest month on the northern subcontinent. This raises rivers and streams (Singh *et al.*, 2023). Rivers are fed by snow/glacier melts (Debnath *et al.*, 2023; Kanga *et al.*, 2023). They provide water security in North India (Gaddam *et al.*, 2020). Snow cover affects land surface radiation, water, and heat transport (Kumar *et al.*, 2023). SCA research in the Himalayas forecasts monsoons, balances glaciers, and models melt-runoff. Himalayan seasonal and monsoonal rainfall affects river flow and water distribution (Choudhury *et al.*, 2023). Each SCA system requires long-term planning (Singh *et al.*, 2023). Snow data collection is expensive, time-consuming, and dangerous. The Beas Basin has few data-collection sites. Satellites provide superior spatiotemporal data. Field observations confirm the satellite data (Rathore *et al.*, 2023). Satellites measure the snowfall. Satellites monitor the snow distribution daily. High visible-band reflectivity helps snow stand out in the photos. Height affects snowfall. Elevation affects winter snow accumulation, whereas topography affects summer snow accumulation. The MODIS 8 Day L3 Global 500 m Grid (MOD09A1) was used to explore the Beas River Basin. This article estimated the snow-extent of the Beas River basin over 22 years (February 2000 to July 2022). The DEM and snow cover data were geographically compared. The snow direction impacts coverage.

13.2 Materials and Methods

13.2.1 Study Area and Data Source

In our study, the Beas River basin (area ≈ 11,280 km^2) which originates from High Hills of Manali to Maharana Pratap Sagar and bounded between

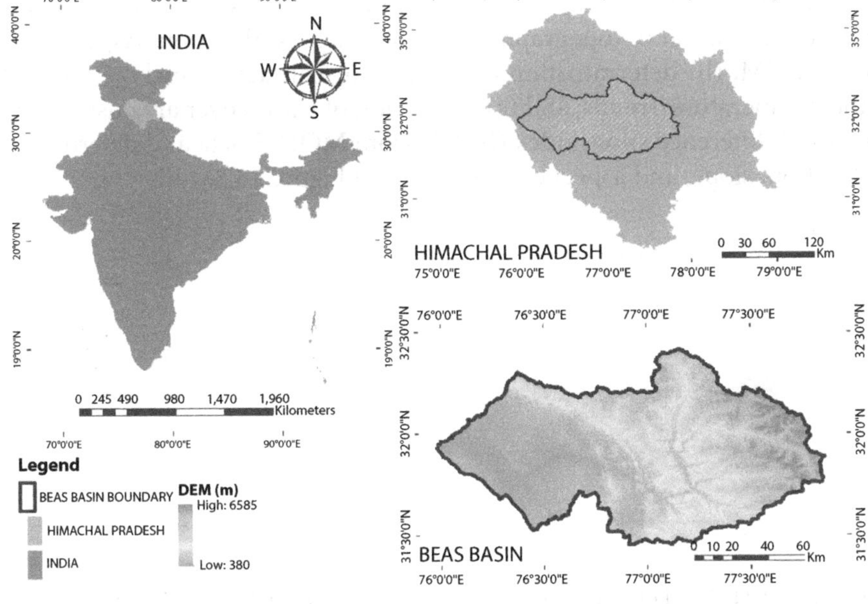

Figure 13.1 Map of Beas Basin.

31.25 N–32.25 N and 75.50 E–77.50 E. The location map of the Beas River basin is shown in Figure 13.1.

13.2.2 Data Collection

Spatial DEM data are saved in GeoTiff format, which contains a grid. Depending on the resolution of the DEM, each pixel represents a point or an area, and their placements disclose the feature heights relative to the MSL shown in Figure 13.2. ArcGIS 10.2 is used to calculate the catchment boundary and split the elevation range into zones using ASTER-GDEM DEM 30 m data (Space Systems, 2019).

Enhanced MODIS Terra MOD09A1 Version 6.1 product images (Singh *et al.*, 2023) from 2000 to 2022 for the estimation and analysis of snow cover area by generating Normalized Difference Snow Index (NDSI) maps from Bands 1 through 7 (seven) shown in Figure 13.2 of 500-meter (m) spatial resolution bands that were pre-corrected for atmospheric conditions such as Cloudless, Rayleigh scattering, gases, and zenith angle corrections. This allowed for estimation and analysis and made it possible to make estimates and conduct analyses, as shown in Figure 13.2.

13.2.3 Selection of Input Parameters and Method

The methodology of the proposed work comprises three basic steps: (a) determination of topographical factors such as Elevation, Aspect class from DEM; (b) determination of meteorological factors such as Rainfall and Temperature Trends; and (c) estimation of snow cover area using normalized difference snow index (NDSI) from MODIS data and determined elevation, slope, and aspect classes shown in Figure 13.3.

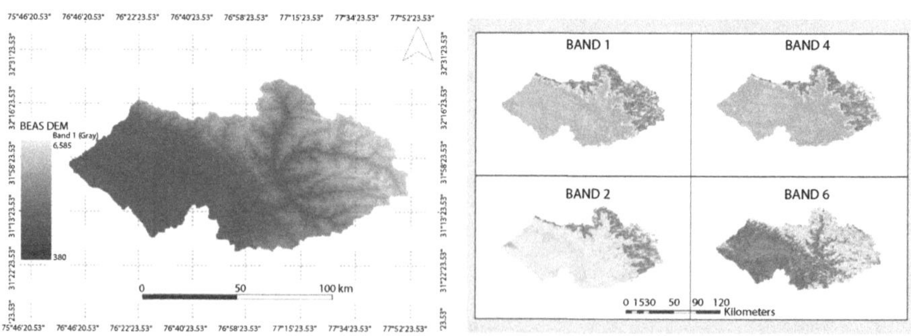

Figure 13.2 (1) DEM of Beas Basin, (2) Map of Beas Basin visible as MODIS band (a) band 1, (b) band 4, (c) band 2, and (d) band 6.

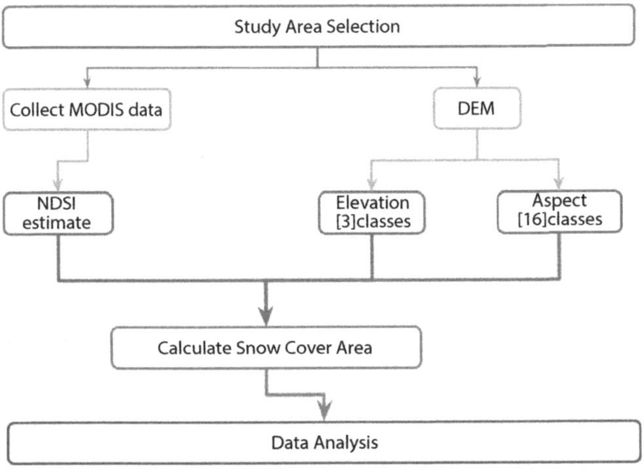

Figure 13.3 Methodology of proposed work.

13.2.4 Topographical Factors

The elevation of the terrain plays a decisive role in snow accumulation. High elevation, in general, occurs at lower temperatures, which in turn leads to higher snow accumulation. The DEM was classified into 14 elevation zones from (380 to 6,585 m) shown in Figure 13.4(a), with a regular interval of 500 m. The areas with elevations between (3,000 and 5,000 m) and those between (5,000 and 6,858 m) were classified at 2,000 m intervals, from the three zones as lower elevation zones, which show higher sensitivity to snow cover with time, as shown in Figure 13.4(d) zone (380 m to 3,000 m), where fresh snow and fresh depletion are to be calculated. The average percent snow cover with respect to the three elevation classes was plotted against elevation classes in graphical form, as shown in Table 13.1, to establish a relationship between elevation and snow cover.

The Aspect, defined as the orientation of the slope face, is an important component of topography. The detailed Aspect map display the general orientation of each area of land, typically according to 16 zones: north (N), north-west-north (NNW), north-west (NW), west-north-west (WNW), west (W), west-south-west (WSW), south-west (SW), south-south-west (SSW), south (S), south-south-east (SSE), south-east (SE), east-south-east (ESE), east (E), east-north-east (ENE), north-east (NE), and north-north-east (NNE). The aspect map of the study area was prepared from ASTER-GDEM using ArcGIS 10.2 in Raster Calculator. The aspect map was crossed with the NDSI map to determine the distribution of aspect classes with respect to different Aspect zones, as shown in

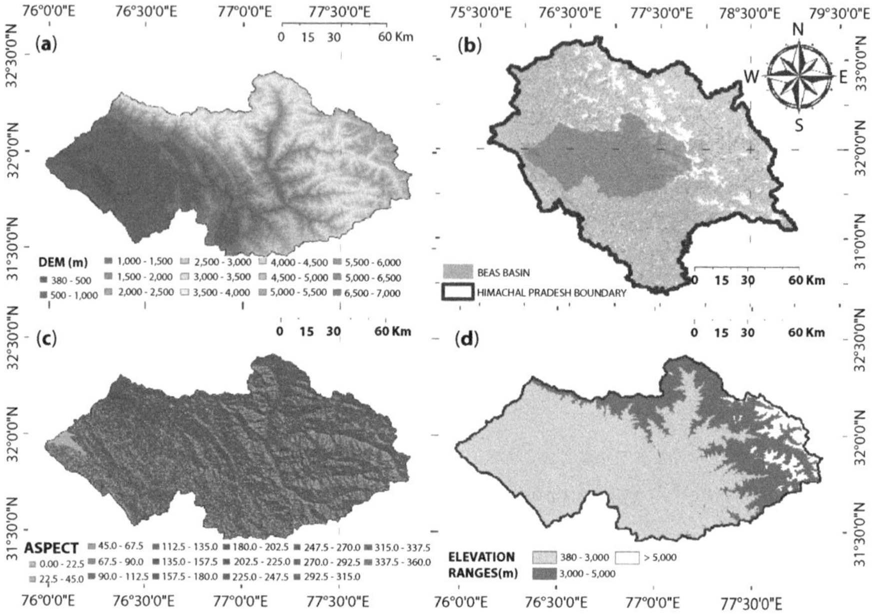

Figure 13.4 (a) 14 Elevation zones Map of Beas Basin; (b) Map of Beas Basin location in Himachal Pradesh; (c) 16 Aspect Zones Map of Beas Basin for detailed study; and (d) Grouped of 3 Elevation Zones Map of Beas Basin for detailed study.

Figure 13.4(d). The percentage of snow-covered areas in terms of different aspect classes and elevation zones was calculated for a period of 22 years (February 2000 to July 2022) with 8-day satellite product intervals.

Table 13.1 Sources and description of the input variables used for the study.

S. no.	Data type	Source	Resolution (Spatial/ Temporal)	Description
1.	Topography	LP DAAC (https://appeears.earthdatacloud.nasa.gov/)	1 arc second (~30 m)	ASTER Global Digital Elevation Model (GDEM) Version 3 in GeoTiff format
2.	MODIS Data	NASA Earth Observing System Data and Information System (EOSDIS) data provider LP DAAC. Land Processes Distributed Active Archive Center within the EODIS (https://appeears.earthdatacloud.nasa.gov/)	500 m/8-Day	MOD09A1 v061 for Snow cover area Estimation and analysis from Bands 1 to 7 at 500 m spatial resolution with applied corrections in GeoTiff format

13.2.5 Estimation of Snow Cover Area Using NDSI

To generate snow cover maps, NDSI was implemented on MODIS reflectance imagery with a spatial resolution of 500 m using GDAL and ArcMap 10.2. NDSI utilizes two spectral bands, i.e., visible (green) as MODIS bands four (0.545–0.565 μm) and near-infrared (NIR) or short-wave IR (SWIR) as MODIS bands six (1.628–1.652 μm) range, as shown in Equation (13.1).

$$NDSI = \frac{MODIS_{(green)} - MODIS_{(SWIR)}}{MODIS_{(green)} + MODIS_{(SWIR)}} \tag{13.1}$$

the code to process the required NDSI in GDAL.

```
#creating file NDSI
!gdal_calc.py \
--overwrite \
--type=Float32 \
```

```
--NoDataValue=0 \
-A {pathb4} \
--A_band 1 \
-B {pathb6} \
--B_band 1 \
-C {pathb2} \
--C_band 1 \
--outfile={temp_dir}"BothCheck_result_final.tif" \
--calc="(((A.astype(float) - B)/(A.astype(float) + B))>=0.4)*(C.astype(float)/10000>0.11)"
```

To avoid water bodies, a pixel may be allocated to the snow category if (a) its NDSI value is more than 0.4 and (b) its near-infrared reflectance (band 2) is greater than 11%. If neither criterion is satisfied, the same pixel is assigned to non-snow. Figure 13.5 shows Beas basin's monthly snow cover.

$$\text{SC MAP} = (\text{NDSI} \geq 0.4) \cap \left(\frac{\text{MODIS}_{(\text{Band2})}}{10{,}000} > 0.11 \right) \quad (10.2)$$

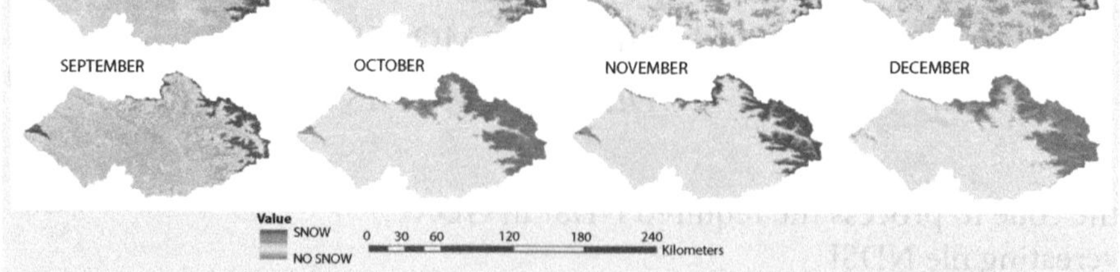

Figure 13.5 NDSI Map of the Beas basin from August 2021 to December 2021 and January 2022 to July 2022.

13.2.6 Rate of Change of SCA with Elevation

- R_n = nth standard zone for the change of SCA % per meter rise in elevation in the snow

$$R_n = \frac{S_n - S_{n-1}}{ME_n - ME_{n-1}} \tag{13.3}$$

- S_n = SCA (%) for *n*th standard elevation zone
- S_{n-1} = SCA (%) for (n – 1)th standard elevation zone
- ME_n = middle elevation value of the *n*th elevation zone
- ME_{n-1} = middle elevation value of the previous elevation zone.

13.3 Results and Discussions

13.3.1 SCA Variation with Elevation Bands

The SCA was intersected with the classified DEM of 15 elevation zones between (380 m–6,585 m) with a 500 m difference for the Beas basin. The first three elevation zones had almost negligible SCA in the Beas basin, as shown in Table 13.2.

The whole elevation zone (2,000 m–3,000 m) of the Beas basin is just (approximately 2,000 km^2), despite the average monthly SCA rate peaking in February 2019 (approximately 24.66%) In this region, Table 13.3 shows that SCA filled most months each year. This zone also catches snow that melts into the streams. This observation supports the concept that snow cover at higher altitudes may decline proportionately with seasonal changes, as is the case in this zone. This zone shows that the planetary boundary layer, more than 1,500 m above the ground, controls the deposition of healthy snow cover. This extensive investigation classified elevation zones as follows: (380 m–3,000 m, 3,000 m–5,000 m, and 5,000 m–6,585 m) zones, respectively shown in Figure 13.6 as zone (380 m–3,000 m) mostly will contain fresh snow accumulation and depletion trend, (3,000 m–5,000 m) range shows snows accumulation changes from previous years, and (5,000 m–6,585 m) range shows stable snow layers which on change of area may result from Figure 13.6 and Table 13.4 shows the accumulation and depletion phases throughout the basin from February 2000 to June

Table 13.2 Detailed information about the area of Beas basin in different elevation zones, Grouped elevation zones and Aspects along with the percentage with respect to the total snow cover area and geographical area of February 2022.

Elevation ranges (m)	Beas basin Snow Cover Area (SCA) in km² (% snow in each zone of total snow)	Aspect classes	(380 m–3,000 m)	(3,000 m–5,000 m)	(5,000 m–6,585 m)
380–500	0 (0 %)	**N (0°–22.5°)**	71	175	26
500–1,000	0 (0 %)	**NNE (22.5°–45°)**	61	158	23
1,000–1,500	1 (0.02 %)	**NE (45°–67.5°)**	55	118	21
1,500–2,000	85 (1.92 %)	**ENE (67.5°–90°)**	53	100	21
2,000–2,500	396 (8.93 %)	**E(90°–112.5°)**	45	105	23
2,500–3,000	672 (15.15 %)	**ESE (112.5°–135°)**	59	147	28
3,000–3,500	631 (14.23 %)	**SE (135°–157.5°)**	70	163	38
3,500–4,000	635 (14.32 %)	**SSE (157.5°–180°)**	72	195	42
4,000–4,500	737 (16.62 %)	**S (180°–202.5°)**	76	227	44
4,500–5,000	703 (15.85 %)	**SSW (202.5°–225°)**	78	210	42
5,000–5,500	453 (10.22 %)	**SW (225°–247.5°)**	73	162	29
5,500–6,000	63 (1.42 %)	**WSW (247.5°–270°)**	62	148	28

(Continued)

Table 13.2 Detailed information about the area of Beas basin in different elevation zones, Grouped elevation zones and Aspects along with the percentage with respect to the total snow cover area and geographical area of February 2022. (*Continued*)

Elevation ranges (m)	Beas basin Snow Cover Area (SCA) in km^2 (% snow in each zone of total snow)	Aspect classes	(380 m–3,000 m)	(3,000 m–5,000 m)	(5,000 m–6,585 m)
6,000–6,500	5 (0.11 %)	**W (270°–292.5°)**	59	144	29
6,500–6,585	0 (0 %)	**WNW (292.5°–315°)**	68	164	31
		NW (315°–337.5°)	66	167	31
		NNW (337.5°–360°)	73	178	36
Grouped Elevation ranges (m)	**SCA in km^2 (% in each zone of total area of the zone)**				
380–3,000	1,130 (14.57%)				
3,000–5,000	2,677 (94.76%)				
5,000–6,585	519 (100%)				

2022 for grouped zones that retained the month as August until July of the subsequent year.

Assigning April–September (ablation/depletion period) and October–March (accumulation period) with snowfall events at higher elevation zones, as the temperature decreases with altitude (Ahluwalia *et al.*, 2016). The melting period or the ablation/depletion period is highly related to the sun–earth distances.

The SCA rate increases within the area of elevation zones, although a negative rate of change is found (>6,000 m) elevation during the accumulation

Table 13.3 Distribution of SCA (%) on monthly basis in elevation zone (2,000 m–3,000 m), "X" represents no satellite coverage period and "Y" represents satellite still yet to be launched.

.	JAN	FEB	MAR	APR	MAY	JUN	JUL	AUG	SEP	OCT	NOV	DEC
2000	Y	4.01	6.42	0.09	0.01	0.02	0.39	0.12	0.03	–	1.32	0.12
2001	2.24	1.04	0.84	0.11	–	2.80	0.19	0.05	–	–	–	2.95
2002	0.35	X	0.93	0.34	–	0.03	-	0.11	0.01	–	–	0.14
2003	2.75	11.79	2.67	0.15	–	0.01	0.18	0.23	0.04	–	0.61	1.69
2004	17.99	9.14	0.18	0.04	–	0.03	0.21	0.19	–	0.63	0.21	1.79
2005	16.93	18.68	5.58	0.29	–	–	0.51	0.17	0.08	–	–	–
2006	5.95	1.48	3.19	0.04	0.01	0.06	0.09	0.15	0.18	0.02	0.13	2.67
2007	0.27	12.88	9.47	0.10	–	0.08	–	0.06	–	–	–	3.54
2008	14.13	12.83	0.83	0.28	–	0.03	0.19	0.08	–	–	–	0.69
2009	3.41	2.26	1.47	0.27	0.01	0.02	0.04	0.39	–	–	0.92	0.26
2010	0.94	8.59	0.79	–	–	0.01	0.71	1.40	0.02	0.08	–	0.60
2011	11.72	15.33	5.39	1.16	0.01	0.03	0.01	0.44	0.01	–	–	0.23
2012	20.58	15.24	2.50	0.40	–	–	0.07	0.33	–	–	0.24	3.30
2013	10.67	19.65	3.81	0.16	–	0.01	0.07	0.56	0.02	0.01	0.11	1.36
2014	10.54	23.73	7.21	1.16	0.06	–	0.34	0.09	0.08	–	–	5.06
2015	13.04	16.82	11.46	0.84	–	0.02	0.08	0.78	–	0.02	0.19	2.76
2016	1.23	2.63	3.07	0.06	–	0.01	0.05	0.02	0.04	–	–	0.01
2017	16.24	4.59	4.39	0.02	0.01	0.09	0.39	0.24	–	–	0.42	1.79
2018	0.23	4.66	0.63	0.06	–	0.01	0.25	0.05	–	0.18	1.58	1.43
2019	19.19	24.66	9.73	0.59	–	–	0.02	0.13	0.03	0.06	1.32	6.99
2020	19.76	4.34	5.37	0.73	0.01	0.03	0.08	0.04	–	–	4.88	2.16
2021	2.26	1.76	0.46	0.77	0.03	0.01	0.01	0.01	2.08	0.06	–	1.49
2022	12.84	15.11	0.94	–	0.06	0.03	X	X	X	X	X	X

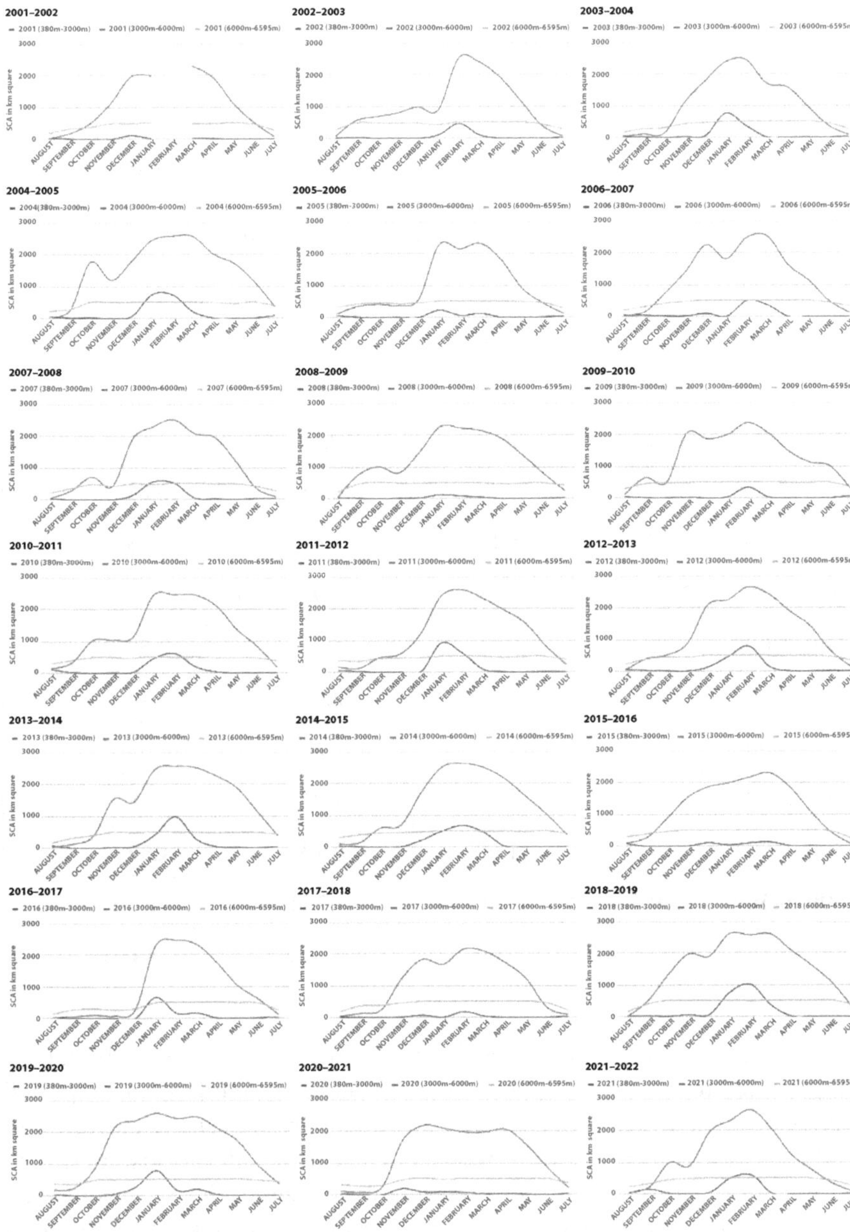

Figure 13.6 Monthly mean SCA in km^2 variation for 22 years from August 2000 to July 2022 of three elevation zones for the Beas basin analysis.

Table 13.4 Change of SCA % (per 100 m height) during April–September (ablation/depletion period), and October–March (accumulation period) for Beas basin.

	Corrected values		**Calculated values**	
Elevation ranges (m)	**Accumulation period**	**Ablation/ depletion period**	**Accumulation period**	**Ablation/ depletion period**
380–500	0.00	0.00	0.54	0.93
500–1,000	0.00	0.00	−0.54	−0.92
1,000–1,500	0.00	0.00	0.00	0.00
1,500–2,000	0.13	0.01	0.13	0.01
2,000–2,500	0.71	0.01	0.71	0.01
2,500–3,000	1.67	0.05	1.67	0.05
3,000–3,500	3.93	0.69	3.93	0.69
3,500–4,000	5.42	2.57	5.42	2.57
4,000–4,500	2.85	3.43	2.85	3.43
4,500–5,000	2.61	4.39	2.61	4.39
5,000–5,500	1.25	4.28	1.25	4.28
5,500–6,000	0.13	1.62	0.13	1.62
6,000–6,500	0.00	0.00	−2.58	−2.36
6,500–6,585	0.00	0.00	−16.12	−14.69

period in the Beas basin, which is due to the spatial accuracy of the sensor being limited to (500 m) spatial resolution; hence, corrections were applied to consider that snow as permanent snow also areas over the elevation of (6,000 m) is less in magnitude than zones at lower elevations, as shown in Table 13.4. However, in the higher elevation zone, the SCA showed a higher rate of change on annual average basis. Corrected values in zones (380 m–1,500 m) were required because lower elevation water bodies sometimes signal snow to the satellite sensor. Calculation of the monthly average of 22 years SCA exhibited a maximum in February 2019, February 2014, February 2013, and January 2020 and a minimum in August of years (2016, 2018, 2001, and 2017), as shown in Table 13.5.

Table 13.5 Maximum and minimum monthly computed average SCA % for Beas basin in 6 sorted years in decreasing order.

Year-month	Maximum SCA % of the Basin	Year-month	Minimum SCA % of the Basin
2019-February	58.80	**2016-August**	2.25
2014-February	58.61	**2018-August**	2.52
2013-February	56.70	**2001-August**	2.71
2020-January	54.81	**2017-August**	2.71
2019-January	54.41	**2003-August**	2.72
2015-February	53.69	**2013-August**	3.08

SCA with Respect to Aspect

Table 13.6 Pearson correlation coefficient values for direction in two consecutive years from Monthly Computed Average SCA for Beas basin in 22 years. "X" represents no satellite coverage period and "Y" represents satellite still yet to be launched.

YEAR	JAN	FEB	MAR	APR	MAY	JUN	JUL	AUG	SEP	OCT	NOV	DEC
00-01	y	0.99	1.00	0.99	0.99	0.99	0.98	0.97	0.98	0.95	0.76	0.97
01-02	0.99	x	1.00	0.99	0.98	0.99	0.97	0.97	0.94	0.97	0.97	0.94
02-03	0.94	x	0.99	1.00	1.00	0.99	0.98	0.97	0.96	0.93	0.95	0.95
03-04	0.85	1.00	0.90	1.00	0.99	1.00	0.99	0.99	0.95	0.83	0.96	1.00
04-05	1.00	0.99	0.89	1.00	0.99	0.98	0.98	0.99	0.98	0.94	0.94	0.97
05-06	1.00	0.98	1.00	1.00	0.99	0.99	0.96	0.95	0.96	0.97	0.96	0.95
06-07	0.94	0.99	1.00	0.99	1.00	0.98	0.99	0.92	0.98	0.99	0.78	1.00
07-08	0.93	1.00	0.98	0.99	0.99	0.99	0.99	0.97	0.98	0.95	0.95	0.99
08-09	0.99	1.00	0.99	1.00	0.98	0.98	0.99	0.93	1.00	0.99	0.81	1.00
09-10	0.99	1.00	0.99	0.99	1.00	0.99	0.99	0.95	0.99	0.99	0.73	0.91
10-11	0.99	1.00	0.98	0.98	0.99	0.99	0.98	0.96	0.98	0.98	0.96	0.99
11-12	1.00	1.00	1.00	1.00	0.99	1.00	1.00	0.96	0.97	0.98	0.96	0.93

(Continued)

Table 13.6 Pearson correlation coefficient values for direction in two consecutive years from Monthly Computed Average SCA for Beas basin in 22 years. "X" represents no satellite coverage period and "Y" represents satellite still yet to be launched. (*Continued*)

YEAR	JAN	FEB	MAR	APR	MAY	JUN	JUL	AUG	SEP	OCT	NOV	DEC
12-13	1.00	1.00	1.00	0.99	1.00	1.00	0.99	0.96	0.93	0.97	0.99	0.96
13-14	1.00	1.00	1.00	0.99	0.99	0.99	0.98	0.95	0.96	0.98	0.98	0.96
14-15	1.00	1.00	1.00	1.00	1.00	1.00	0.99	0.98	0.99	0.91	0.98	0.98
15-16	0.98	0.99	1.00	1.00	0.99	0.99	0.97	0.98	0.97	0.90	0.93	0.97
16-17	0.99	0.99	1.00	1.00	1.00	1.00	0.96	0.96	0.98	0.98	0.94	0.96
17-18	0.93	0.99	1.00	0.99	0.99	0.99	0.98	0.96	0.94	0.86	0.99	1.00
18-19	0.88	0.99	0.99	1.00	0.99	0.99	0.97	0.92	0.98	0.97	0.99	0.97
19-20	1.00	0.99	1.00	1.00	1.00	1.00	0.99	0.96	0.96	0.98	1.00	0.99
20-21	0.99	0.97	0.98	1.00	1.00	0.99	0.99	0.99	0.88	0.98	0.70	1.00
21-22	0.99	0.95	1.00	0.92	0.96	0.98	0.96	x	x	x	x	x

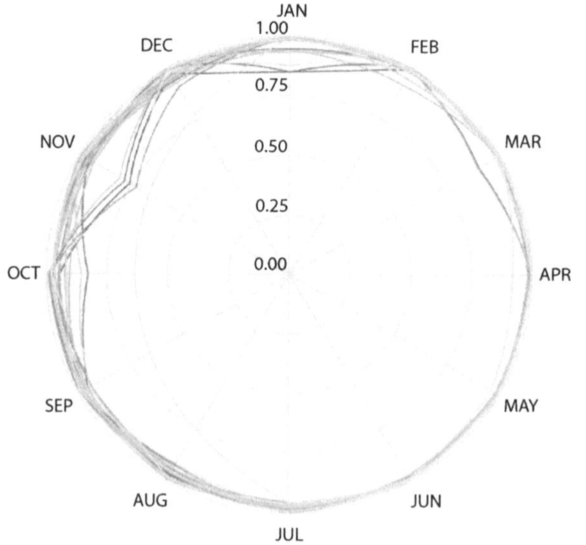

Figure 13.7 Comparison of Pearson correlation coefficient values plotted for all the years combined to one shows all aspect receive the snow every year are highly correlated in every direction with each other in the basin.

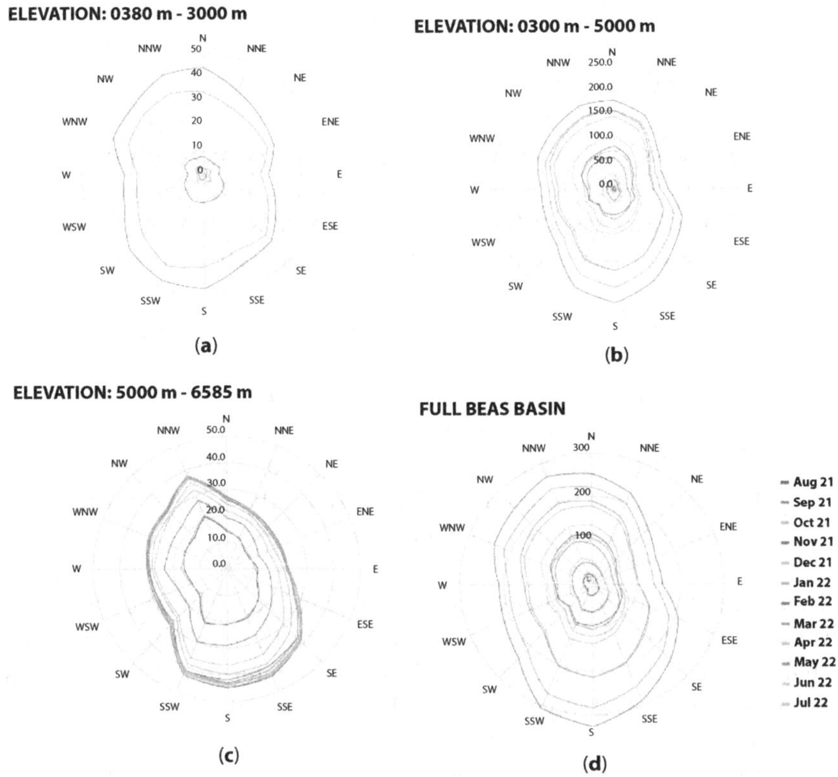

Figure 13.8 Comparison of monthly mean SCA in km^2 variation in three elevation zones and complete basin with respect to aspect for the period of (August 2021–July 2022) in the Beas basin. (a) aspect classes of elevation range (380 m–3,000 m), (b) aspect classes of elevation range (3,000 m–5,000 m), (c) aspect classes of elevation range (5,000 m–6,585 m), and (d) aspect classes for complete Beas basin.

13.4 Conclusions

This research compares the basin's aspect and elevation zones with 22 years of 8-day satellite data to demonstrate that Aspect affects shows the Pearson correlation coefficient values in Figure 13.7 and snow accumulation in the lower altitude zones higher than higher altitude zone in Figure 13.8. Lower elevation zones receive snow from various directions, but in fewer months than the middle elevation zone, which receives snow from the south and north. The final elevation zone receives snow for most of the month, and the direction remains the same for all months of the year. The elevation zone (3,000 m–5,000 m) has less convection energy than the surface below (3,000 m) which insulates the upper atmosphere by providing a broad surface of previous and continuous accumulation of snow, which favors snow accumulation in this elevation zone rather than the zones below it.

The following conclusions were drawn from this study:

1. The elevation zone (2,000 m–3,000 m) is more prone to receive SCA and is likely a suitable location for a more detailed analysis of the presence of biomass changes below the new snow accumulation trends as well as depletion trends.
2. The prolonged snow presence in the year 2020–2021 SCA showed a flatter peak until April.
3. The maximum change of SCA % (per 100 m height) was 5.42 for the accumulation period.
4. The evaluation for every 22-year showed Pearson correlation coefficient values more than 0.70 on the Aspects for the complete basin shown in Table 13.6.

Therefore, this study shows the ability to quickly evaluate the aspect and elevation zones for snow cover area changes in the basin by correlating annual data, which can be considered as a parameter to check the health of the basin as SCA changes generate both temperature energy changes effecting the weathering of the below biomass and mass changes affecting the sediment transportation from streams generated over the surface of the basin. Hence, the health of the basin can be regularly checked to determine if there is a deviation from natural changes in snow.

Acknowledgments

The authors acknowledge the support received from the Department of Civil Engineering (NIT HAMIRPUR), Government of India, for conducting the present work. The authors are also thankful to the Land Processes Distributed Active Archive Center (LPDAAC) and the National Aeronautics and Space Administration (NASA) for providing the necessary data to conduct the present study.

References

Ahluwalia, R. S., Rai, S. P., Gupta, A. K., Dobhal, D. P., Tiwari, R. K., Garg, P. K., & Kesharwani, K. (2016). Towards the understanding of the flash flood through isotope approach in Kedarnath valley in June 2013, Central Himalaya, India. *Nat. Hazard.*, 82(1), 321-332.

Cohen, J., & Rind, D. (1991). The effect of snow cover on the climate. *J. Clim.*, 4(7), 689-706.

Dimri, A. P., & Dash, S. K. (2012). Wintertime climatic trends in the western Himalayas. *Clim. Change*, 111(3), 775-800.

DiPietro, J. A., & Pogue, K. R. (2004). Tectonostratigraphic subdivisions of the Himalaya: A view from the west. *Tectonics*, 23(5).

Gaddam, V. K., Kulkarni, A. V., & Gupta, A. K. (2020). Assessment of the Baspa basin glaciers mass budget using different remote sensing methods and modeling techniques. *Geocarto Int.*, 35(3), 296-316.

Kumar, S., Singh, V., & Saroha, J. (2022). Interpretation of land use/land cover dynamics with the application of geospatial techniques in sarbari khad watershed of Himachal Pradesh, India. *GeoJournal*, 1-11.

Lang, H., & Rohrer, M. (1987). Temporal and spatial variations of the snow cover in the Swiss Alps. *Large Scale Effects of Seasonal Snow Cover*. International Association of Hydrological Sciences Press, Institute of Hydrology, Wallingford, Oxfordshire UK. IAHS Publication, (166).

Makarieva, A. M., & Gorshkov, V. G. (2007). Biotic pump of atmospheric moisture as driver of the hydrological cycle on land. *Hydrol. Earth Syst. Sci.*, 11(2), 1013-1033.

Meetei, P. N., Ahluwalia, R. S., Rai, S. P., Khobragade, S., Sarangi, S., Goel, M., & Kumar, S. (2022). Spatio-temporal analysis of snow cover and effect of terrain attributes in the Upper Ganga River Basin, central Himalaya. *Geocarto Int.*, 37(4), 1139-1159.

NASA/METI/AIST/Japan Spacesystems and U.S./Japan ASTER Science Team (2019). <i>*ASTER Global Digital Elevation Model V003*</i> [Data set]. NASA EOSDIS Land Processes DAAC. Accessed 2022-08-11 from https://doi.org/10.5067/ASTER/ASTGTM.003

Vermote, E. (2021). MODIS/Terra Surface Reflectance 8-Day L3 Global 500m SIN Grid V061 [Data set]. *NASA EOSDIS Land Processes DAAC.* Accessed 2022-08-11 from https://doi.org/10.5067/MODIS/MOD09A1.061

Singh, S., Singh, S. K., Kumar, R., Shrama, A., & Kanga, S. (2022). *Finding Alternative to River Sand in Building Construction.*

Singh, S., Meraj, G., Kumar, P., Singh, S. K., Kanga, S., Johnson, (2023). Decoding Chambal River Shoreline Transformations: A Comprehensive Analysis Using Remote Sensing, GIS, and DSAS. *Water*, 15(9), 1793.

Singh SK, Kanga S, Gulati B, Raiˇc M, Sajan B, Đurin B, Singh S (2023) Spatial and Temporal Analysis of Hydrological Modelling in the Beas Basin Using SWAT+ Model. *Water* 15, 3338. https://doi.org/10.3390/w1519333

Rani A, Gupta SK, Singh SK, Meraj G, Kumar P, Kanga S, Đurin B, Dogančić D (2023) Predicting Future Land Use Utilizing Economic and Land Surface Parameters with ANN and Markov Chain Models. *Earth* 4(3):728-751.

Ahmad T, Gupta SK, Singh SK, Meraj G, Kumar P, Kanga S (2023) Unveiling Nature's Resilience: Exploring Vegetation Dynamics during the COVID-19 Era in Jharkhand, India, with the Google Earth Engine. *Climate* 11(9):187.

Sharma M, Upadhyay RK, Tripathi G, Kishore N, Shakya A, Meraj G, Kanga S, Singh SK, Kumar P, Johnson BA, Thakur SN (2023) Assessing Landslide Susceptibility along India's National Highway 58: A Comprehensive Approach Integrating Remote Sensing, GIS, and Logistic Regression Analysis. *Conservation* 3(3):444-459.

Saqib N, Rai PK, Kanga S, Kumar D, Đurin B, Singh SK (2023) Assessment of Ground Water Quality of Lucknow City under GIS Framework Using Water Quality Index (WQI). *Water* 15(17):3048.

Choudhury U, Singh SK, Kumar A, Meraj G, Kumar P, Kanga S (2023) Assessing land use/land cover changes and urban heat island intensification: A case study of Kamrup Metropolitan District, Northeast India (2000–2032). *Earth* 4(3):503-521.

Rai NK, Mathur S, Singh SK, Tiwari M, Singh VK, Haque R, Tiwari S, Sharma LK (2023) Differential regulation of mitochondrial complex I and oxidative stress based on metastatic potential of colorectal cancer cells. *Oncol. Lett.* 26(1):1-1.

Debnath J, Sahariah D, Lahon D, Nath N, Chand K, Meraj G, Kumar P, Singh SK, Kanga S, Farooq M (2023) Assessing the impacts of current and future changes of the planforms of river Brahmaputra on its land use-land cover. *Geosci. Front.* 14(4):101557. Elsevier.

Kumar P, Singh SK, Gupta SK, Kanga S, Meraj G (2023) Uncovering the hydro-meteorological drivers responsible for forest fires utilizing geospatial techniques. *Theor. Appl. Climatol.*. https://doi.org/10.1007/s00704-023-04497. Springer Nature.

Singh SK, Kotecha MJ, Kanga S, Pankhaniya SK, Agrawal S, Meraj G (2023) Urban Heat Island (UHI) Resilience Plan in Varying Climatic Conditions Using Geospatial Approach: A Case Study of Rajkot City. In: *Climate Change, Agriculture and Society*, vol 1. pp 307-338.

Singh S, Singh SK, Prajapat DK, Pandey V, Kanga S, Kumar P, Meraj G (2023) Assessing the Impact of the 2004 Indian Ocean Tsunami on South Andaman's Coastal Shoreline: A Geospatial Analysis of Erosion and Accretion Patterns. *J. Mar. Sci. Eng.* 11(6):1134.

Rathore SS, Mathur SC, Mathur S, Sharma S, Singh SK (2023) Late Cretaceous–Early Paleocene Geodiversity of Fatehgarh Formation of Petroliferous Barmer Basin, Western Rajasthan, India: A Potential Geopark Site. *Geoheritage* 15:54.

Singh S, Meraj G, Kumar P, Singh SK, Kanga S, Johnson BA, Prajapat DK, Debnath J, Sahariah D (2023) Decoding Chambal River Shoreline Transformations: A Comprehensive Analysis Using Remote Sensing, GIS, and DSAS. *Water* 15(9):1793.

Tripathi G, Shakya A, Upadhyay RK, Singh SK, Kanga S, Pandey SK (2023) Landslide Susceptibility Mapping of Tehri Reservoir Region Using Geospatial

Approach. In: *Climate Change Adaptation, Risk Management and Sustainable Practices in the Himalaya*, pp 135-156.

Sud A, Kanga R, Singh SK, Meraj G, Kanga S, Kumar P, Ramanathan AL, Bhardwaj V (2023) Simulating Groundwater Potential Zones in Mountainous Indian Himalayas—A Case Study of Himachal Pradesh. *Hydrology* 10(3):65.

Upadhyay RK, Tripathi G, Đurin B, Šamanović S, Cetl V, Kishore N, Sharma M, Singh SK, Kanga S, Wasim Md, Rai PK, Bhardwaj V (2023) Groundwater Potential Zone Mapping in the Ghaggar River Basin, North-West India, Using Integrated Remote Sensing and GIS Techniques. *Water* 15(5):961.

Lavane K, Kumar P, Meraj G, Han TG, Ngan LHB, Lien BTB, Ty TV, Thanh NT, Downes NK, Nam NGD, Minh HVT, Singh SK, Kanga S (2023) Assessing the effects of drought on rice yields in the Mekong Delta. *Climate* 11(1):13.

Nath N, Sahariah D, Meraj G, Debnath J, Kumar P, Lahon D, Chand K, Farooq M, Chandan P, Singh SK, Kanga S (2023) Land use and land cover change monitoring and prediction of a UNESCO world heritage site: Kaziranga eco-sensitive zone using cellular automata-Markov model. *Land* 12(1):151.

Kanga S, Upadhyay RK, Sharma M, Sharma S, Tripathi G, Singh SK (2023) Geotechnical Stability Assessment of Road-cut Slopes: A Case Study of Srinagar, Garhwal Himalaya, India. *Biol. Forum* 15(5):1071-1080.

14

Evaluation of Groundwater Potential Zones in Godavari Sub-Basin Using Analytical Hierarchy Process (AHP) and GIS

Manoj Pawar[1], Pavankumar Giri[2], Pranali Kathe[3] and Pranaya Diwate[4]*

[1]*Chhatrapati Sambhajinagar, Maharashtra, India*
[2]*Shri Shivaji College of Arts, Commerce and Science, Akola, India*
[3]*Centre for Climate Change and Water Research, Suresh Gyan Vihar University, Jaipur, India*
[4]*Department of Geology, School of Basic and Applied Sciences, MGM University, Chhatrapati Sambhajinagar, Maharashtra, India*

Abstract
Groundwater is an important natural resource that is a basic need for human survival. However, overexploitation has considerably depleted groundwater availability. Assessing the potential zones of groundwater is extremely important for the appropriate management of groundwater systems. Remote sensing and Geographic Information System (GIS) techniques were utilized to assess the groundwater potential zones in this study, and a composite groundwater potential map of the study area was generated. The use of this methodology in the present study was demonstrated in a selected study area in Maharashtra. Various influencing parameters, such as slope, drainage density, geomorphology, and lineament density were hierarchically used and integrated with a weighted overlay in ArcGIS by assigning suitable ranks for each category of these parameters. The weightage factor for each category was decided based on its expected capability towards groundwater potential, based on a knowledge-driven approach. The resultant groundwater potential zones were classified into five categories: very poor, poor, moderate, good, and very good, which were validated to some extent using the field/ground database of CGWB, and the results were found to be encouraging. This groundwater potential

**Corresponding author*: pranayadiwate7@gmail.com

Shruti Kanga, Gowhar Meraj, Suraj Kumar Singh, Majid Farooq and M. S. Nathawat (eds.) Geospatial Technology for Natural Resource Management, (343–362) © 2024 Scrivener Publishing LLC

information will be useful for the effective identification of suitable locations for the future exploitation of water resources, and the methodology adopted can be substantiated with suitable ground/field inputs to enhance the credibility of the results obtained from the above applied laboratory-based study.

***Keywords*:** Godavari, groundwater potential zone, AHP, MCDM, watershed

14.1 Introduction

Remote Sensing and GIS techniques have been to be proven efficient tools for the delineation, updating, and morphometric analysis of drainage basins (Singh *et al.*, 2022). Any hydrological investigation, including an assessment of groundwater potential and management, requires drainage basin analysis (Kanga *et al.*, 2020). The physiographic characteristics of drainage basins, such as their size, shape, drainage density, tributary size, length, and slope, can be correlated with several important hydrological phenomena (Singh *et al.*, 2023; Lavane *et al.*, 2023). Remote sensing data can be used in conjunction with conventional data for the delineation of ridgelines, characterization, priority evaluation, problem identification, assessment of potential and management needs, identification of erosion-prone areas, evolving water conservation strategies, and selection of sites for check dams and reservoirs (Rani *et al.*, 2023; Singh *et al.*, 2023). This study describes the drainage characteristics of the Aurangabad district obtained through Remote Sensing and GIS-based morphometric analysis. It is believed that this study will be useful for understanding the hydrological behavior of the basin (Ahmad *et al.*, 2023; Kanga *et al.*, 2022).

Groundwater is a globally important and valuable renewable resource for human lives and economic development. It constitutes a major portion of the Earth's water circulatory system, known as the hydrological cycle, and occurs in permeable geological formations known as aquifers (Nath *et al.*, 2023; Kanga *et al.*, 2023). In recent decades, the exploitation of groundwater has increased greatly, particularly for agricultural purposes, because large parts of the country have little access to rainfall due to the frequent failures of monsoons. Currently, 85% of the water requirements for domestic use in rural areas, 55% for irrigation, and over 50% for industrial and urban requirements are met by groundwater (Sharma *et al.*, 2023; Gulati *et al.*, 2023). Groundwater is the primary source of water for domestic, agricultural, and industrial use in many countries, and its contamination has been recognized. Industrial wastewater, sewage sludge, and solid waste materials are currently indiscriminately discharged into the environment.

Water quality is a vital concern for humans because it is directly linked to human health (Sqib *et al.*, 2023; Tomar *et al.*, 2023).

Groundwater is the most abundant natural resource and an essential ingredient in animals and plants. More than 98% of the freshwater on Earth lies below its surface. Groundwater acts as a reservoir by the large pore space in the material constituting the Earth, as a conduit that can transport water over long distances and acts as a mechanical filter that improves water quality by removing suspended solids and bacterial contamination (Choudhury *et al.*, 2023; Singh *et al.*, 2022). Generally, groundwater quality depends on the quality of the recharged water, atmospheric precipitation, inland surface water, and surface geochemical processes. These materials enter subsurface aquifers, resulting in pollution of irrigation and drinking water (Rai *et al.*, 2023; Farooq *et al.*, 2022).

Temporal changes in the origin and constitution of recharged water, hydrologic, and human factors may cause periodic changes in groundwater quality. These changes may be related to both human and natural phenomena. Generally, shallow groundwater is affected more by contamination than deep aquifers (Debnath *et al.*, 2023; Kumar *et al.*, 2023).

Most industries discharge their effluents without proper treatment into nearly open channels, resulting in groundwater contamination. In general, the main cause of this study was groundwater pollution due to effluent discharge (Singh *et al.*, 2023). Groundwater pollution occurs in urban areas, where large volumes of waste are concentrated and discharged into small hydrological basins. Hydrogeochemical conditions are also responsible for significant variations in groundwater quality (Rathore *et al.*, 2023).

14.2 Study Area and Location

The study area was the Godavari sub-basin of Maharashtra, India. The Godavari River basin spans the latitudes of 160 16' 00" North and 22' 36" North as well as the longitudes of 73' 26" East and 83' 07" East (Figure 14.1). The Godavari River basin is important in the Marathwada region and other parts of the Maharashtra state. This river is important for the growth of Maharashtra crops because it is the largest basin. The digital elevation model indicates that the highest elevation above the mean sea level is 1,067 m, which is approximately 38 km from the river basin source. Numerous major crops, including sugarcane, cotton, horticulture, and soybeans, are available in the study area. Compared to other basin regions, this area has a lot of land that is cultivated, whereas, in the semi-arid region, many crops use only groundwater and rainwater (Singh *et al.*, 2023).

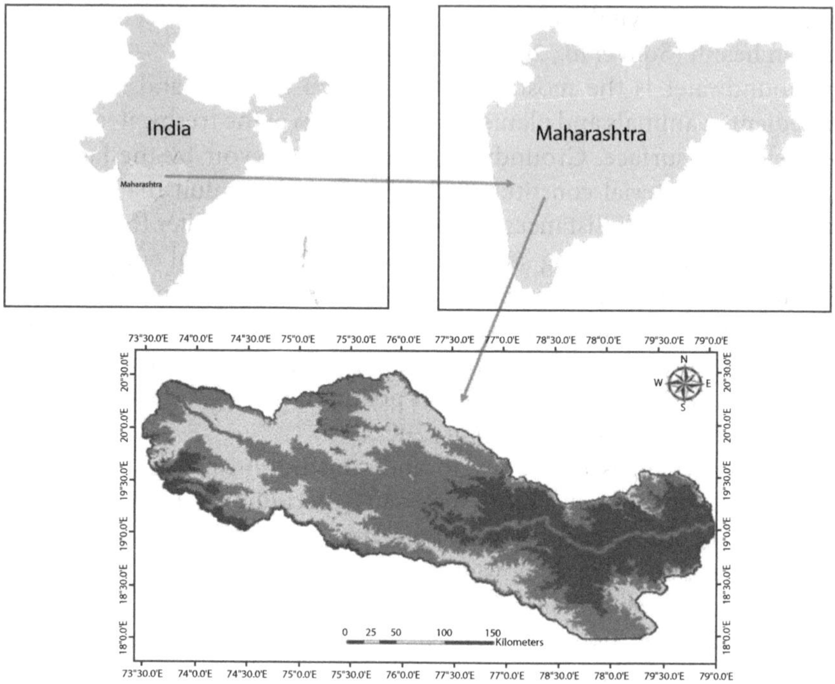

Figure 14.1 Location map of the study area.

14.3 Material and Methodology

The present work is based on a quantitative analysis of geomorphic indices extracted from the SRTM DEM (30 M resolution). All topographical maps and satellite images were mosaicked and georeferenced using the ArcGIS software. After georeferencing the images, they were rectified and projected onto the Universal Transverse Mercator (UTM) projection WGS 1984, Zone 43 North.

Various thematic parameters influencing the groundwater potential of an area must be considered when evaluating the groundwater potential of an area. In this study, four thematic parameters were used: Slope, Geomorphology, Drainage Density and Lineament Density, which consider arriving at the groundwater potential of the study area based on its expected influence on terrain characteristics (Tripathi *et al.*, 2023).

14.4 Results and Discussion

14.4.1 Slope

The freely available 30m resolution SRTM DEM of the area was downloaded and processed to yield a derivative-like slope map of the area (Sud *et al.*, 2023; Upadhyay *et al.*, 2023). The resultant map was classified into five discrete classes based on slope characteristics: <20 nearly flat slopes, 20–60 very gentle slopes, 60–130 gentle slopes, 130–230 moderate slopes, and >230 steep slopes (Table 14.1).

The slope map of the area generally depicts flat to very gentle slopes. A steeper slope was encountered in the western part of the area, whereas a moderate slope was represented in the southern fringe of the area. Some sporadic moderate gentle slopes were noted in the northern and eastern extremities of the study area. A graph depicting the percentage distribution of the area versus the slope is plotted (Figure 14.2). The graph shows that approximately 55% of the area falls under the nearly flat slope category, 34.5% of the area falls under the very gentle slope category, 6.8% of the area falls under the gentle slope category, 2.5% of the area falls under the moderate slope, and approximately 6% falls under the steep slope category

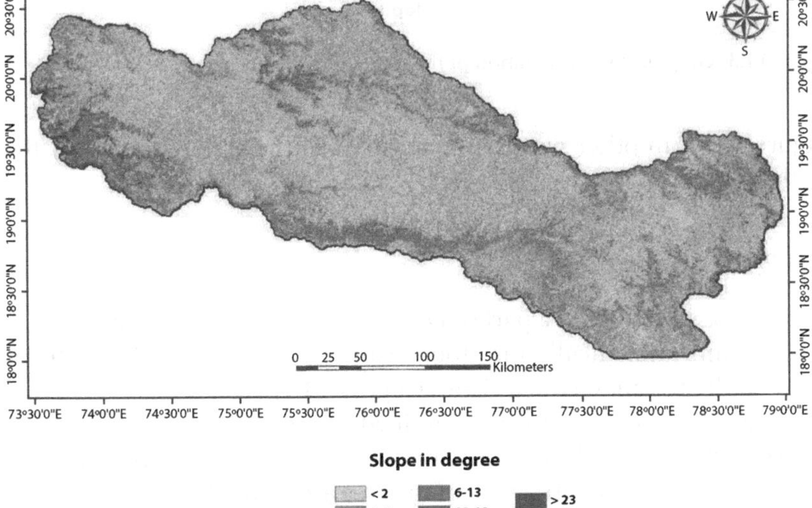

Figure 14.2 Slope map of study area.

Table 14.1 Slope gradient and category wise area distribution.

Class	Degree	Area %	Slope category	Ground water potential
1	<2	55.80	Nearly flat	Very good
2	2–6	34.46	Very gentle	Good
3	6–13	6.48	Gentle	Moderate
4	13–23	2.62	Moderate	Poor
5	>23	0.64	Steep	Very Poor

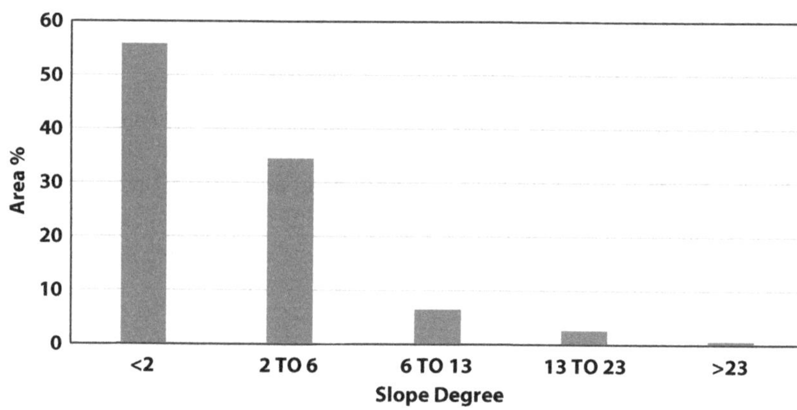

Figure 14.3 Graphical representation of the slope of the study area.

(Figure 14.3). In other words, a major part of the area has a nearly flat to gentle slope.

14.4.2 Geomorphology

The present study area is a part of Deccan Basalt Provenance has been classified geomorphologically into four distinct geomorphological categories based on the terrain characteristics (Figure 14.4) [10].

The geomorphological unit identified in the area is Pedi plain in the eastern part of the study area or either side of the stream bank, a weathered plateau covering the central part of the area, bordered by low dissected plateau on the southern, western and northern fringes. The moderately dissected plateau unit is restricted to the western fringe of the area. A graph

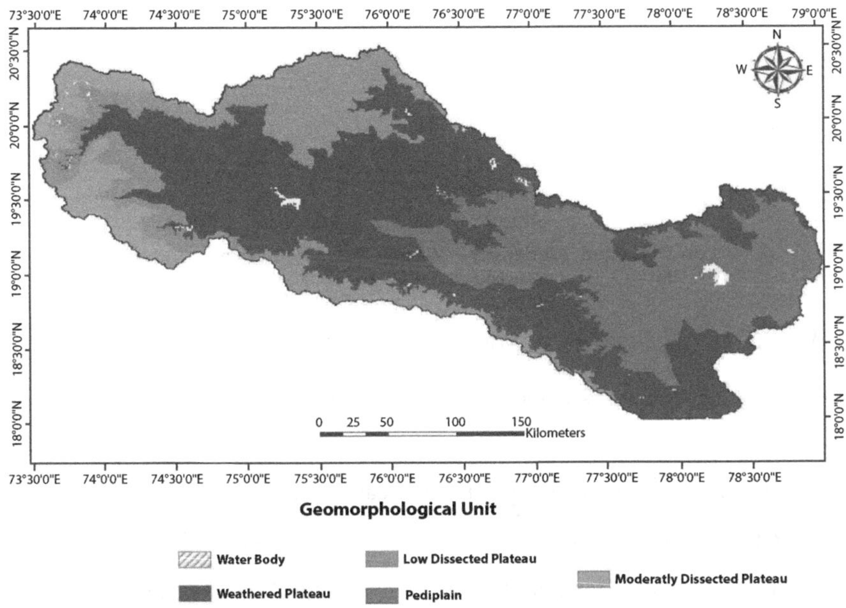

Figure 14.4 Geomorphology map.

Table 14.2 Geomorphology category.

Class	Class name	Category
1	Low Dissected Plateau	Very Good
2	Moderately Dissected Plateau	Good
3	Weathered Plateau	Moderate
4	Water Body	Poor
5	Pedi plain	Very Poor

depicting the percentage distribution of the area versus the geomorphological unit is presented (Table 14.2).

A graph depicts that about 18.9% of the area falls under the low dissected geomorphological category, 6.6% of the area falls under the moderately dissected plateau geomorphological category, 48.2% area falls under the weathered plateau geomorphological category and 26.1% of the area falls under Pedi plain geomorphological category (Figure 14.5). In other words, a major part of the area was represented by a weathered plateau (Table 14.3).

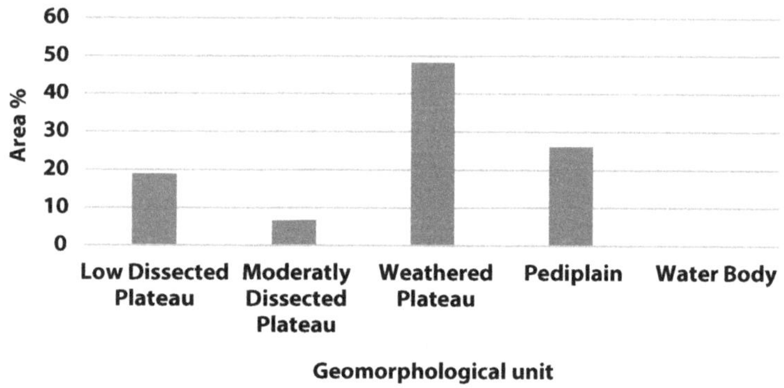

Figure 14.5 Graphical representation of geomorphology of study area.

Table 14.3 Geomorphology area and category.

Class	Class name	Area (Km^2)	Area %
1	Low Dissected Plateau	14,621.97	18.9
2	Moderately Dissected Plateau	5,042.04	6.6
3	Weathered Plateau	37,005.63	48.2
4	Water Body	334.16	0.2
5	Pedi plain	20,254.27	26.1

14.4.3 Drainage Density

The stream order map for the study area was derived from the processed DEM data and converted to a feature class with due classification of the drainage order. Eight stream orders were identified in the area. Most of the area exhibits a fine dendritic pattern, which locally changes to a sub-dendritic or sub-parallel pattern due to the influence of lineaments, as identified in the study area (Figure 14.6).

The drainage map of stream ordering was utilized to generate a drainage density map of the area using (Figure 14.7) various tools such as intersection, spatial join, join, and interpolation techniques. Thiessen polygon method was adopted to know the area of influence of each point. A point is generated on the stream network-to-point by using feature-to-point tools.

A graph depicting the percentage distribution of the area versus drainage density.

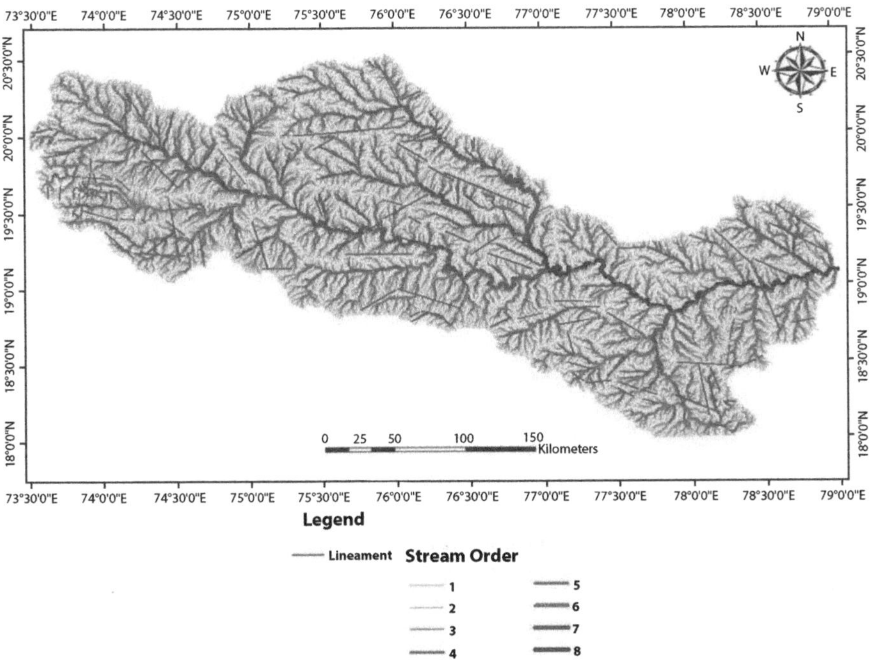

Figure 14.6 Stream order with lineament.

Table 14.4 Drainage density category.

Class	Drainage density/Km²	Area %	Drainage density category
1	<0.6	19.09	Very Good
2	0.6–0.8	31.70	Good
3	0.8–1.0	22.78	Moderate
4	1.0–1.3	19.11	Poor
5	>1.3	7.33	Very Poor

The graph shows that about 19.09% of the area falls under the very low drainage density category, 31.70% of the area falls under the low drainage density category, 22.78% falls under the moderate drainage density category, 19.11% falls under the high drainage density category, and 7.33% falls under the very high drainage density category (Table 14.4). This is to say that a major part of the area is represented by low to moderate drainage density.

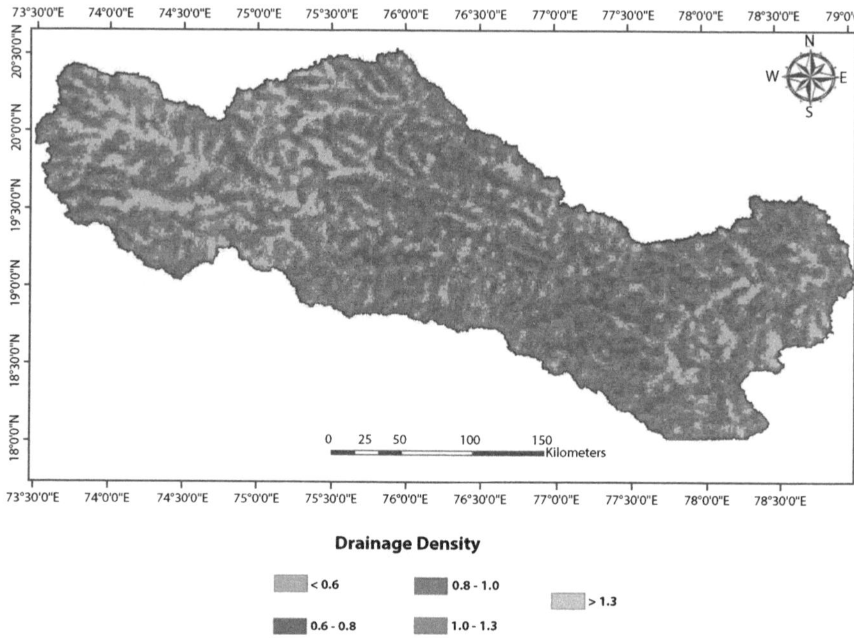

Figure 14.7 Drainage density map.

14.4.4 Lineament Density

A lineament density map can be used to quantitatively evaluate the relationship between the lineaments and the occurrence of groundwater. The lineament density map of the study area was derived from the process of digitalization of stream order, hillshade, and OLI data of the study area.

The feature class of lineaments was utilized to generate the lineament density map of the area using various tools such as intersection, spatial join, join, and interpolation techniques. The graph shows the percentage distribution of the area versus lineament density (Figure 14.8) [11].

The graph shows that approximately 4.33% of the area falls under the very high lineament density category, 42.90% of the area falls under the high lineament density category, 35.77% of the area falls under the moderate lineament density category, 11.58% falls under the low lineament density category, and 5.40% falls under the very low lineament density category (Figure 14.9). In other words, a major part of the area is represented by high to moderate lineament density (Table 14.5).

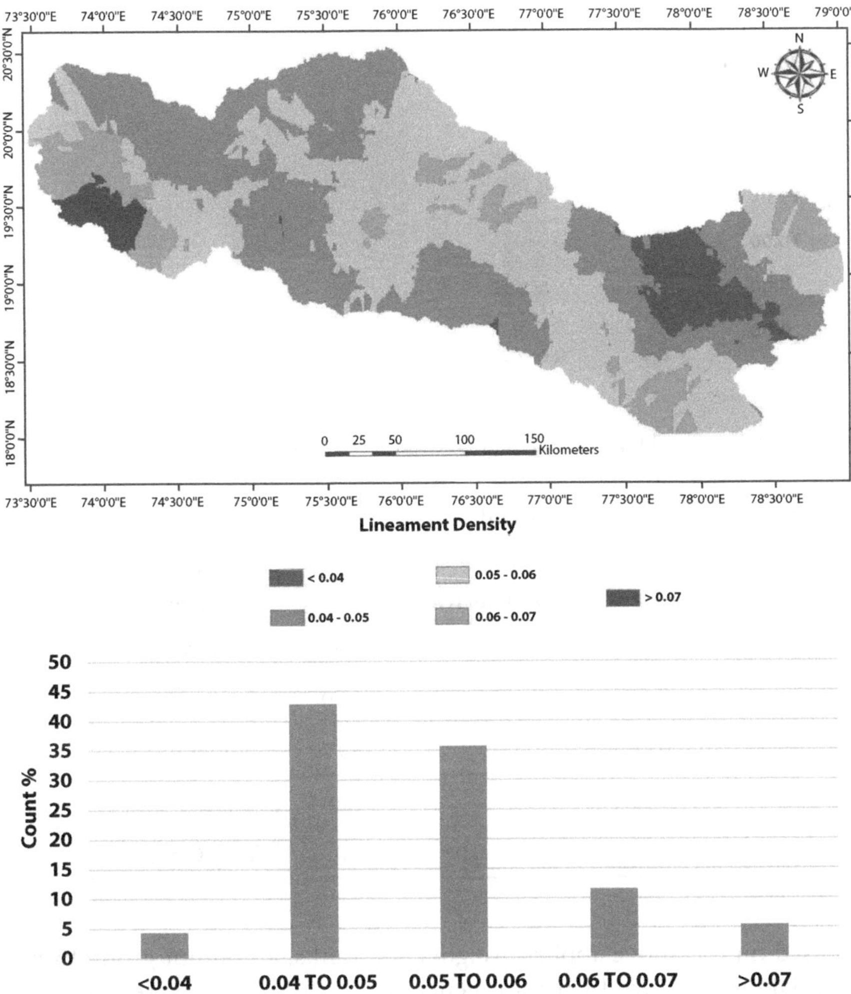

Figure 14.8 Graphical representation of lineament density of study area.

14.4.5 Integration

The various factors influencing the groundwater potential of the area, as discussed above, were considered as per the knowledge-driven weightages and ranks for the study area were utilized in arriving at the groundwater potential of the area using the weighted average overlay analysis raster tool of the ARC GIS. The different parameters considered in this study along with the relevant weightage and ranks applied are listed in Table 14.6 [12–14].

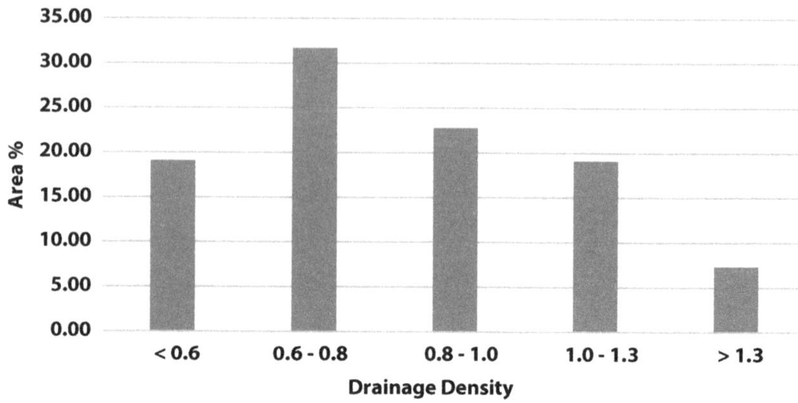

Figure 14.9 Graphical representation of drainage density of study area.

Table 14.5 Lineament density and category.

Class	Lineament density Km/Km²	Area %	Lineament density category
1	<0.04	4.33	Very Poor
2	0.04–0.05	42.90	Poor
3	0.05–0.06	35.77	Moderate
4	0.06–0.07	11.58	Good
5	>0.07	5.40	Very Good

Based on the above applied weights and ranks, an output groundwater potential map (Figure 14.10) was generated for further interpretation.

Figure 14.10 depicts the very high groundwater potential zone restricted to the central and southeastern parts of the area, which is dominated by a peneplain geomorphic unit with high to moderate lineament density, low to very low drainage density, and flat to gentle slope.

High groundwater potential zones have been noted mostly in the central, southeastern, and northeastern parts of the study area, which are generally dominated by pediplain and weathered plateau geomorphic units with very high to moderate lineament density, high to moderate drainage density, and very gentle to nearly flat slopes.

A moderate groundwater potential zone is found in the central, north, southwestern, and south-east, and parts of the study area, which is dominated by weathered plateau/low dissected plateau and Pedi plain

Table 14.6 Rank and AHP rating and weights for the different parameters of the groundwater potential zone.

Parameter	Classes	Groundwater prospect	Weight (%)	Rank
Geomorphology	Pedi plain Water body Weathered Plateau Moderate dissected Plateau Low dissected Plateau	Very Good Good Moderate Poor Very Poor	30	5 4 3 2 1
Lineament density (Km/Km^2)	<0.04 0.04–0.05 0.05–0.06 0.06–0.07 >0.07	Very poor Poor Moderate Good Very good	25	5 4 3 2 1
Drainage density	<0.6 0.6–0.8 0.8–1.0 1.0–1.3 >1.3	Very good Good Moderate Poor Very poor	25	5 4 3 2 1
Slope Classes	Nearly level <2° Very gently sloping 2°-6° Gently sloping 6°–13° Moderately sloping 13°–23° Steeply sloping >23°	Very good Good Moderate Poor Very poor	20	5 4 3 2 1

geomorphic unit with moderate to low lineament density, high to very high drainage density, and very gentle to nearly flat slope.

The poor groundwater potential zone can be traced in the western, northwestern, and southwestern parts of the study area, which is dominated by low dissected plateau geomorphic units with very low to low lineament density, very high to high drainage density, and a very gentle to nearly flat slope [15, 16].

A very poor groundwater potential zone can be identified in a sporadic distribution in the northwestern parts of the study area, which is dominated by a moderately dissected plateau geomorphic unit with a low lineament density, very high drainage density, and steep slope.

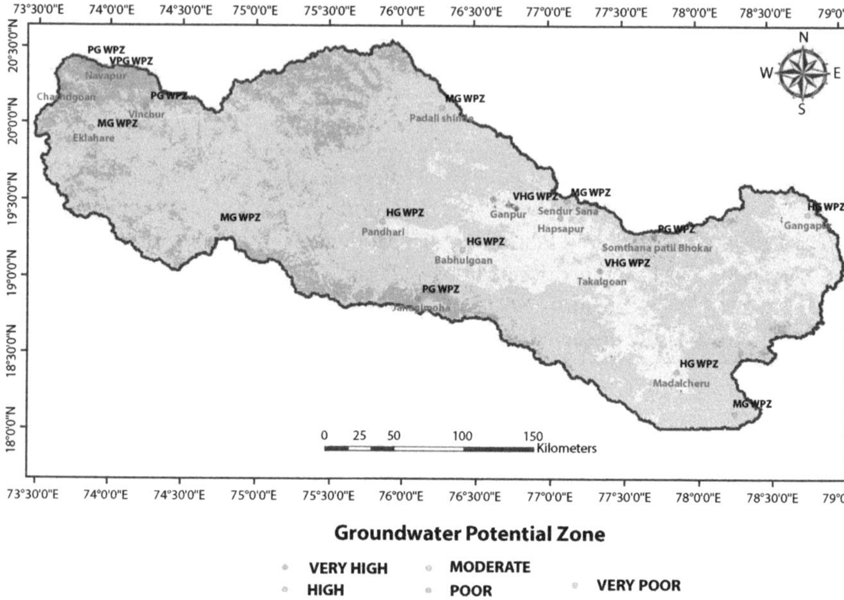

Figure 14.10 Groundwater potential zone of the study area.

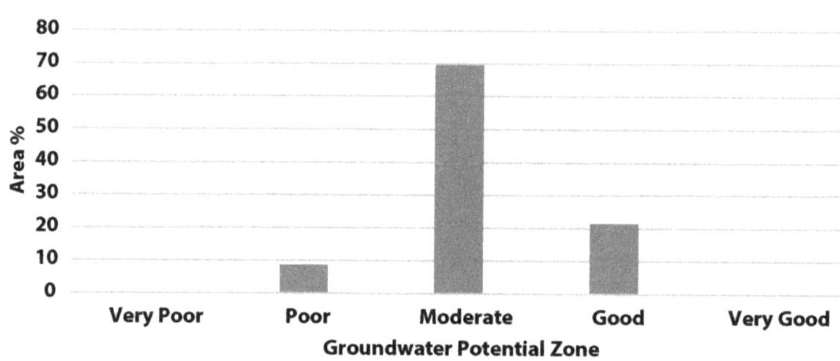

Figure 14.11 Graphical representation of groundwater potential zone.

Based on the above analysis and interpretation the very high groundwater potential has been adjacent to the following villages: 1.22 km north-east of Adgoan garage, 2.37 km south-west of Pimpri, 0.67 km north-east of Gangapur, 0.72 km south-east of Takalgoan and the very high water potential has been adjacent to the following villages: 1.46 km south-east Pandhari, near Godavari river, 0.47 km north-west of Babhulgoan, 0.88 km north of Hapsapur, 1.14 km north north-east of Madalcheru, and 0.84 km south-west of Gangapur.

Table 14.7 Groundwater potential zone category.

Class	Area in Sq. Km	Area %	Ground water potential zone
1	5.09	0.01	Very poor
2	6,698.61	8.7	Poor
3	53,624.98	69.52	Moderate
4	16,782.39	21.5	Good
5	22.13	0.27	Very good

Some of the above findings could be validated by the ground information and field report published by the Central Ground Water Board in their District Wise Report in 2011 (Parbhani, Nanded, beed, Maharashtra, Nizamabad, Adilabad Telangana, and the adjoining districts). The locations where the validation was achieved are tabulated below:

It can be correlated from the above table (Table 14.7) that most of the locations which have been identified as having very high groundwater potential are exhibiting a comparatively shallow water table of generally between 2 and 5 m, whereas in some cases the water table has been recorded to fluctuate between 5 and 10 m. This coincides well with the observations made as part of this study. Another noteworthy aspect is that the locations woth very high groundwater potential are generally aligned adjacent to the identified distinct lineaments in the study area. Hence, these aspects arrived at based on the laboratory study can be duly considered, while arriving at the ground water potential and exploration of an area, which needs to be coupled with sufficient ground information to arrive at the most credible estimate of groundwater resource estimation of a given area.

14.5 Conclusion

Geographical information system and remote sensing are powerful and cost-effective methods for determining groundwater potential of an area. Four influencing thematic maps, slope, geomorphology, drainage density, and lineament density were utilized to determine the groundwater potential of the area. This kind of study can provide firsthand information to local authorities and planners about areas suitable from the groundwater point of view.

Table 14.8 Central ground water board (*Source: CGWB*).

Sl. no.	Longitude	Latitude	State	District	Taluka	Village	Sp_yeild post monsoon (m bgl)	Sp_yeild pre monsoon (m bgl)
1	76.3723	19.3017	Maharashtra	Parbhani	Selu	Adgoan	2–5	5–10
2	76.4643	19.2640	Maharashtra	Parbhani	Sonpeth	Pimpri	2–5	5–10
3	76.4350	19.2754	Maharashtra	Parbhani	Jintur	Ganpur	5–10	5–10
4	77.2025	19.2301	Maharashtra	Nanded	Nanded	Takalgoan	2–5	10–15
5	75.5154	19.2141	Maharashtra	Beed	Georai	Pandhari	2–5	5–10
6	76.2435	19.1022	Maharashtra	Parbhani	Pathri	Babhulgoan	2–5	5–10
7	77.4260	19.2242	Maharashtra	Hingoli	Basmat	Hapsapur	2–5	5–10
8	77.5113	18.2157	Telangana	Nizamabad	Banswada	Madalcheru	2–5	10–20
9	78.4537	19.2236	Telangana	Adilabad	Uthnur	Gangapur	5–10	10–20

The present study on the groundwater potential of the Godavari sub-basin, with the aid of remote sensing data and GIS technology, proved to be an important input for evaluating the role and significance of different influencing parameters attributed to the groundwater potential of the area. The study area is composed of various Deccan basalt flows. Generally in Deccan Basalt, terrain groundwater occurs under phreatic conditions in the exposed basaltic rocks and under semi-confined conditions in the flows at a deeper level. The groundwater in such terrain is generally controlled by primary porosity in the form of vesicles and secondary porosity in the form of fractures and joints.

The four influencing parameters of slope, geomorphology, drainage density, and lineament density were studied in detail and the study area was classified into very good, good, moderate, poor, and very poor groundwater potential zones (Figure 14.11). With regard to slope, it can be seen that the areas with slopes less than 2° over the pediplain geomorphic unit with drainage density less than 0.6 and lineament density greater than 0.07 falls in a very good groundwater potential zone and cover an area of 0.27 km^2. Whereas the areas having slope ranging from 2° to 6° with drainage density of 0.6 to 0.8 and lineament density of 0.06 to 0.07 has been classified as good groundwater potential zone, covering an area of 21.5 km^2. Areas with slopes ranging from 6° to 13° covered by a weathered plateau with a drainage density of 0.8 to 1.0, and a lineament density of 0.05 to 0.06 has been classified as a moderate groundwater potential zone covering an area of 69.52 km^2. Areas with a slope ranging from 13° to 23° covered by a moderate dissected plateau with a drainage density of 1.0 to 1.3 and lineament density of 0.05 to 0.04 has been classified as poor groundwater potential zone covering an area of 8.7 km^2. The areas having slope greater than 23° covered by low dissected plateau with drainage density greater than 1.3 and lineament density less than 0.04, were classified as very poor groundwater potential zones covering an area of 0.1 km^2. The ground truth data from CGWB (Table 14.8) showed a good correlation with the present study, which enhances the reliability of the results.

References

Singh, S., Singh, S. K., Kumar, R., Shrama, A., & Kanga, S. (2022). *Finding Alternative to River Sand in Building Construction.*

Singh, S., Meraj, G., Kumar, P., Singh, S. K., Kanga, S., Johnson, (2023). Decoding Chambal River Shoreline Transformations: A Comprehensive Analysis Using Remote Sensing, GIS, and DSAS. *Water*, 15(9), 1793.

Singh SK, Kanga S, Gulati B, Raič M, Sajan B, Đurin B, Singh S (2023) Spatial and Temporal Analysis of Hydrological Modelling in the Beas Basin Using SWAT+ Model. *Water* 15, 3338. https://doi.org/10.3390/w1519333

Rani A, Gupta SK, Singh SK, Meraj G, Kumar P, Kanga S, Đurin B, Dogančić D (2023) Predicting Future Land Use Utilizing Economic and Land Surface Parameters with ANN and Markov Chain Models. *Earth* 4(3):728-751.

Ahmad T, Gupta SK, Singh SK, Meraj G, Kumar P, Kanga S (2023) Unveiling Nature's Resilience: Exploring Vegetation Dynamics during the COVID-19 Era in Jharkhand, India, with the Google Earth Engine. *Climate* 11(9):187.

Sharma M, Upadhyay RK, Tripathi G, Kishore N, Shakya A, Meraj G, Kanga S, Singh SK, Kumar P, Johnson BA, Thakur SN (2023) Assessing Landslide Susceptibility along India's National Highway 58: A Comprehensive Approach Integrating Remote Sensing, GIS, and Logistic Regression Analysis. *Conservation* 3(3):444-459.

Sqib N, Rai PK, Kanga S, Kumar D, Đurin B, Singh SK (2023) Assessment of Ground Water Quality of Lucknow City under GIS Framework Using Water Quality Index (WQI). *Water* 15(17):3048.

Choudhury U, Singh SK, Kumar A, Meraj G, Kumar P, Kanga S (2023) Assessing land use/land cover changes and urban heat island intensification: A case study of Kamrup Metropolitan District, Northeast India (2000–2032). *Earth* 4(3):503-521.

Rai NK, Mathur S, Singh SK, Tiwari M, Singh VK, Haque R, Tiwari S, Sharma LK (2023) Differential regulation of mitochondrial complex I and oxidative stress based on metastatic potential of colorectal cancer cells. *Oncol. Lett.* 26(1):1-1.

Debnath J, Sahariah D, Lahon D, Nath N, Chand K, Meraj G, Kumar P, Singh SK, Kanga S, Farooq M (2023) Assessing the impacts of current and future changes of the planforms of river Brahmaputra on its land use-land cover. *Geosci. Front.* 14(4):101557. Elsevier.

Kumar P, Singh SK, Gupta SK, Kanga S, Meraj G (2023) Uncovering the hydro-meteorological drivers responsible for forest fires utilizing geospatial techniques. *Theor. Appl. Climatol.* https://doi.org/10.1007/s00704-023-04497. Springer Nature.

Singh SK, Kotecha MJ, Kanga S, Pankhaniya SK, Agrawal S, Meraj G (2023) Urban Heat Island (UHI) Resilience Plan in Varying Climatic Conditions Using Geospatial Approach: A Case Study of Rajkot City. In: *Climate Change, Agriculture and Society*, vol 1. pp 307-338.

Singh S, Singh SK, Prajapat DK, Pandey V, Kanga S, Kumar P, Meraj G (2023) Assessing the Impact of the 2004 Indian Ocean Tsunami on South Andaman's Coastal Shoreline: A Geospatial Analysis of Erosion and Accretion Patterns. *J. Mar. Sci. Eng.* 11(6):1134.

Rathore SS, Mathur SC, Mathur S, Sharma S, Singh SK (2023) Late Cretaceous–Early Paleocene Geodiversity of Fatehgarh Formation of Petroliferous Barmer Basin, Western Rajasthan, India: A Potential Geopark Site. *Geoheritage* 15:54.

Singh S, Meraj G, Kumar P, Singh SK, Kanga S, Johnson BA, Prajapat DK, Debnath J, Sahariah D (2023) Decoding Chambal River Shoreline Transformations: A Comprehensive Analysis Using Remote Sensing, GIS, and DSAS. *Water* 15(9):1793.

Tripathi G, Shakya A, Upadhyay RK, Singh SK, Kanga S, Pandey SK (2023) Landslide Susceptibility Mapping of Tehri Reservoir Region Using Geospatial Approach. In: *Climate Change Adaptation, Risk Management and Sustainable Practices in the Himalaya*, pp 135-156.

Sud A, Kanga R, Singh SK, Meraj G, Kanga S, Kumar P, Ramanathan AL, Bhardwaj V (2023) Simulating Groundwater Potential Zones in Mountainous Indian Himalayas—A Case Study of Himachal Pradesh. *Hydrology* 10(3):65.

Upadhyay RK, Tripathi G, Đurin B, Šamanović S, Cetl V, Kishore N, Sharma M, Singh SK, Kanga S, Wasim Md, Rai PK, Bhardwaj V (2023) Groundwater Potential Zone Mapping in the Ghaggar River Basin, North-West India, Using Integrated Remote Sensing and GIS Techniques. *Water* 15(5):961.

Lavane K, Kumar P, Meraj G, Han TG, Ngan LHB, Lien BTB, Ty TV, Thanh NT, Downes NK, Nam NGD, Minh HVT, Singh SK, Kanga S (2023) Assessing the effects of drought on rice yields in the Mekong Delta. *Climate* 11(1):13.

Nath N, Sahariah D, Meraj G, Debnath J, Kumar P, Lahon D, Chand K, Farooq M, Chandan P, Singh SK, Kanga S (2023) Land use and land cover change monitoring and prediction of a UNESCO world heritage site: Kaziranga eco-sensitive zone using cellular automata-Markov model. *Land* 12(1):151.

Kanga S, Upadhyay RK, Sharma M, Sharma S, Tripathi G, Singh SK (2023) Geotechnical Stability Assessment of Road-cut Slopes: A Case Study of Srinagar, Garhwal Himalaya, India. *Biol. Forum* 15(5):1071-1080.

Gulati B, Sharma R, Kanga S, Singh SK, Sajan B, Meraj G, Kumar P, Ramanathan AL (2023) Unraveling the Relationship Between Stubble Burning and Air Quality Degradation in Punjab: A Temporal and Spatial Analysis (2019-2022). *J. Clim. Change* 9(2):43-53, IOS Press.

Tomar P, Singh SK, Kanga S, Pattanaik A (2023) Water Bodies mapping and monitoring using high-resolution satellite images. *Sustain. Agri Food Environ. Res.* 11.

Singh H, Meraj G, Singh S, Shrivastava V, Sharma V, Farooq M, Kanga S, Singh SK, Kumar P (2022) Status of Air Pollution during COVID-19-Induced Lockdown in Delhi, India. *Atmosphere* 13(12):2090.

Farooq M, Mushtaq F, Meraj G, Singh SK, Kanga S, Gupta A, Kumar P, Singh D, Avtar R (2022) Strategic Slum Upgrading and Redevelopment Action Plan for Jammu City. *Resources* 11(12):120.[11] Singh, S., Singh, S. K., Prajapat, D. K., Pandey, V., Kanga, S., Kumar, P., & Meraj, G. (2023). Assessing the Impact of the 2004 Indian Ocean Tsunami on South Andaman's Coastal Shoreline: A Geospatial Analysis of Erosion and Accretion Patterns. *J. Mar. Sci. Eng.*, 11(6), 1134.

Kanga, S., Singh, P. K., & Ohri, A. (2020). Evaluating the environmental sustainability of smart cities in India: The design and application of the Indian smart city environmental sustainability index. *Sustainability*, 13(1), 327.

Kanga, S., Singh, P. K., Ohri, A., & Singh, R. S. (2022). Development of Environmental Decision Support System for Sustainable Smart Cities in India. *Environ. Prog. Sustain. Energy*, e13817.

Singh, R. P., Kumar, S., & Rana, P. S. (2020). UN SDGs and context of Holy-Heritage cities in India: a study of Ayodhya and Varanasi. In *Geoecology of Landscape Dynamics* (pp. 187-206). Springer, Singapore.

15

Analyzing the Trend of LULC Change Over Five Decades in Dhanbad District, Jharkhand (India) Using Geospatial Techniques

Gaurav Chakrabarty[1], M.S. Nathawat[2] and Suman Sinha[3]*

[1]Department of Remote Sensing, Birla Institute of Technology, Mesra, Ranchi, India
[2]School of Science, Indira Gandhi National Open University, New Delhi, India
[3]Department of Geography, Amity Institute of Social Sciences, Amity University, Kolkata, India

Abstract

Assessment of changes in Land Use/Land Cover (LULC) provides significant indicators for decision makers to understand the dynamics of the changing environment, their associated reasons, and future actions. Dhanbad District, which is known for the famous Jharia Coalfield (JCF) and its mining activities, has experienced vital changes in LULC due to extensive coal mining in the last hundred years. Coal mining has not only changed the LULC of Dhanbad district, but has also had an enormous effect on air and water quality, both surface and sub-surface.

The present study used geospatial techniques to assess the changes in LULC over the past five decades, ranging from 1972 to 2019. LANDSAT satellite images of five different time periods, i.e., 1972, 1989, 1999, 2009, and 2019, were acquired and used to quantify the land use land cover changes. ERDAS Imagine 2014 was used to carry out supervised classification using a maximum likelihood technique along with NDVI classifications to assess a clear change in vegetation cover. Furthermore, the Overall Accuracy and Kappa coefficient were determined for the entire dataset.

The LULC changes, as noticed from the statistics obtained, are mainly due to deforestation for anthropogenic activities, such as agriculture, coal mining, and urbanization. Dense forests were converted to moderately dense forests, which

**Corresponding author*: sumanrumpa.sinha@gmail.com

Shruti Kanga, Gowhar Meraj, Suraj Kumar Singh, Majid Farooq and M. S. Nathawat (eds.) Geospatial Technology for Natural Resource Management, (363–392) © 2024 Scrivener Publishing LLC

were later converted to scrub and barren land. Forests were harvested for agriculture, the creation of various household usages of wooden materials, shelter, and fuel, along with mining activities in the earlier decades. In recent years, with the rapid increase in population, there has been steady industrial growth and urbanization with a massive escalation of settlements around and near mining areas. This resulted in a considerable change in the LULC dynamics in the Dhanbad region.

Keywords: Land use land cover, classification, NDVI, JCF, accuracy assessment, kappa coefficient

15.1 Introduction

Land cover, such as forests and vegetation, offer to human well provides fundamental protection of soil and water resources as well as multiple ecosystem services and cultural or spiritual values (Costanza *et al.*, 2017; Miura *et al.*, 2015). Human pressure over the biosphere and geosphere is significant, manifesting itself in accelerated biodiversity loss, acidification of oceans, climate change, alteration of biogeochemical cycles, and LULC changes (Steffen *et al.*, 2015). LULC changes cause significant alterations in the organization and function of ecosystems (Betts *et al.*, 2017; Lambin *et al.*, 2003), disturbing the quality of life of people, as human societies depend on the appropriate operation of natural systems (Chiesura and de Groot, 2003; Turner *et al.*, 2007).

Changes in LULC with the passage of time are expected, and the reasons may include one or more factors, such as climatic, political, or socio-economic circumstances of a region (Aboelnour *et al.*, 2018; Kafi *et al.*, 2014). Land surfaces have been altered by human beings depending on their need for natural land cover for anthropogenic land use in the form of agriculture, industrialization, embankments, settlements (Sinha *et al.*, 2018; 2020), mining, etc. The transformation of land cover into pastures, croplands, or urban areas is a global problem (Foley *et al.*, 2005; Kareiva and Marvier, 2011; Santra *et al.*, 2020). The outcome of this alteration is the disruption of natural vegetation, land degradation and desertification, habitat loss, disturbing effects on biodiversity, land use (Zvoleff *et al.*, 2014), environmental and climate change, economic change, outside policy intervention, loss of adaptive capacity, and changes in social organization and attitudes (Lambin *et al.*, 2003). Understanding the spatiotemporal patterns and mechanisms driving LULC changes is of great importance for maintaining the balance between the natural environment and human social systems, and for improving the regional ecological environment (Hu *et al.*, 2020; Kong *et al.*, 2020).

Jharia Coalfield (JCF) located at Dhanbad district in the state of Jharkhand is the largest coalfield in India and is the primary source of coking coal. Commercial coal mining in India started in 1774, which was later nationalized in two phases: the first with coking coal mines in 1971–72 and then with non-coking coal mines in 1973 (https://coal.nic.in/content/historybackground). From the information provided by the Ministry of Mines (GOI), it can be stated that 80% of the mining activities are attributed to coal, and the remaining 20% are attributed to all other metallic and non-metallic minerals (Singh *et al.*, 2007; Sinha *et al.*, 2010). Thus, although coal mining in India is one of the sources of immense revenue for the government, its impact on the environment is severe.

Mining activities cause rock extraction and disruption of rock continuity and groundwater regimes, resulting in the crushing and mixing of various rocks, leading to the formation of new loose anthropogenic deposits that enable changes in the features of the natural landscape (Saranathan *et al.*, 2006). The impacts of mining vary in severity depending on whether the mine is working or abandoned, the mining methods used, and the geological conditions (Bell *et al.*, 2001), and cause immense damage to the landscape and biological communities of the earth. Mining of minerals and fossil fuels causes vast tracts of forest land to be cleared (Aryee *et al.*, 2003). The destruction of forests during mining operations is invariably accompanied by extensive damage and loss of the system. The overburden (OB) of coal mines when dumped, known as OB-dump, in non-mining areas produces mine spoils that affect neighboring natural land cover. The land that is directly hit by mining operations such as excavations and dumping of overburden and coal is converted into new habitats, called colliery spoils, which completely lack the soil structure and whose elemental concentration is greatly altered (Shankar, 1993). The unscientific mining of minerals poses a serious threat to the environment, resulting in the reduction of forest cover; erosion of soil on a greater scale; pollution of air, water, and land; and reduction of biodiversity (UNESCO, 1985).

After Coal Nationalization in 1971–73, along with the increase in mining and its allied activities, there was an increase in urbanization near mining areas because of an increase in population density because a large number of workers migrated to Dhanbad district from all parts of India as a source of employment and business to earn their livelihood. This has resulted in the massive exploitation of natural resources in the upcoming decades, with an increase in population and urbanization.

Geospatial Techniques encompassing Remote Sensing and GIS provide timely and accurate change detection techniques that are important for understanding and measuring LULC dynamics. Remotely Sensed Satellite imagery has the unique ability to provide synoptic views of large areas over a given period, which is not possible using conventional survey methods. Together, Remote Sensing and GIS help in the understanding of the relationships and interactions between different spatial classes, as changes in one affect the other. A land use class is a result of decisions based on environmental, socioeconomic, and cultural factors, and as these factors vary over time, land use changes, for example, agriculture to built-up or forest to mining land use class. These changes in LULC can be assessed using several temporally remotely sensed images for a particular area at a specific period. This process considers the following when monitoring natural resources: detecting changes, identifying the nature of the changes, measuring the areal extent of the change, and assessing the spatial pattern of the change (Macleod *et al.*, 1998). As depicted in several studies, there are many popular methods for performing and estimating changes, such as image differencing, principal component analysis, and post-classification comparison, but no single approach can solve an issue; rather, methods are often compared to obtain optimum results (Daniel *et al.*, 2002; Lu *et al.*, 2004). Furthermore, remote sensing along with GIS plays a vital role in the assessment of LULC dynamics (Sinha *et al.*, 2015) and monitoring of human-induced and natural-derived land degradation with applications of geo-statistical analyses (Prăvălie *et al.*, 2021), time series segments, residual trends (Li *et al.*, 2021), and machine learning algorithms (Basu *et al.*, 2021; Bekele *et al.*, 2021; Rukhovich *et al.*, 2021).

The main objective of the present study is to inspect the Land Use Land Cover temporal changes in Dhanbad district using geospatial techniques from the period of the post-coal nationalization era from 1972 to 2019, determine the accuracy assessment through on-field and off-field checks and kappa coefficient, and assess and infer the reasons for land use land cover changes over the study period of five decades. For this study, satellite images of Landsat 1 (MSS), Landsat 5 (TM), and Landsat 8 Operational Land Imager (OLI/TIRS) were taken for 5 years (1972, 1989, 1999, 2009, and 2019). LULC mapping of Dhanbad District was carried using Remote Sensing and GIS techniques for land suitability study (Saranathan *et al.*, 2006) using the IRS series of satellite images from 1988 to 1998, and Landsat images from 1987 to 2017 (Singh *et al.*, 2018), depicting the impacts of coal mining on LULC, whereas an extensive study of built-up areas from 1972 to 2011 in Dhanbad Urban Agglomeration (DUA) was performed (Lal *et al.*, 2017). The present research considers LULC mapping for a period of five decades, from 1972 to2019 for the entire Dhanbad district, with the aim of understanding and assessing the changes in trends of various land cover/land use classes using geospatial techniques. The study used different image processing techniques, such as image differencing, various band combinations, classification techniques, and proper ground truthing to provide a justified result. This research may aid decision makers in making rational decisions in the near future regarding the unplanned changes in land use patterns in the past five decades in the district due to coal mining, urbanization, and industrialization, which have raised serious concerns regarding forest canopy, agricultural land, air and water quality, urban growth, and the overall ecosystem. Changes in land resources, identifying reasons, and proposing remedies can only be carried out if the changes are predicted, understood, and evaluated for a substantial period. The present study answers these questions by applying an assessment of the changing trends of various land use and land cover classes over five decades.

15.2 Materials and Methods

15.2.1 Study Area

Historically, the name of the Dhanbad district was "Dhanbaid" which has been derived from the words "Dhan" i.e., paddy and "Baid," the paddy land. The Dhanbad district located in the mid-eastern part of Jharkhand State of India has a total area of 2,074 km^2 varying between 23° 37'.3" N and 23° 04' N longitudes 86° 06' 30" E to 86° 50' E latitude (Figure 15.1). The area is included in toposheets 73I/1, 73I/2, 73I/5, 73I/6, 73I/8, 73I/9, 73I/10, 73I/13, and 73I/14 of the Survey of India at a 1:50,000 scale. The district consists of one sub-division, Dhanbad, with 11 development blocks: Baliapur, Baghmara, Gobindpur, Jharia, Dhanbad, Keliyasole, Nirsa, Egarkund, Tundi, PurbiTundi, and Topchanchi (District Survey Report, 2017). The coalfield area was situated in the southern and eastern parts. The northern boundary of the Dhanbad district is marked by the Barakar River and the southern boundary by Damodar River, on which two important reservoirs, Maithon and Panchet, are located.

15.2.1.1 Geology

Topographically, the Dhanbad region has a lower plateau with relatively few undulations. The district consists of two physical district units. The northern area is hilly with forest, whereas the southern area provides the appearance

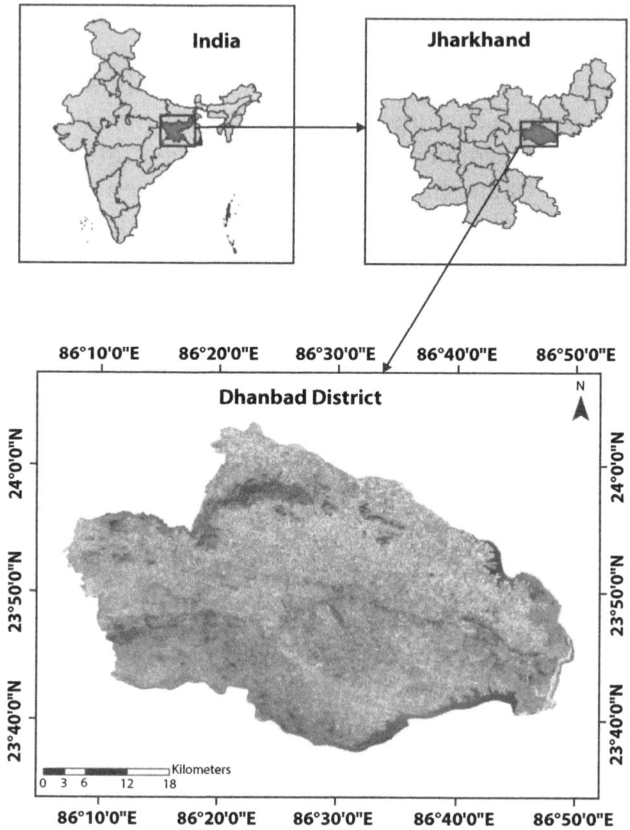

Figure 15.1 Location map of study area.

of a plain. The highest peak in this area was that of Dhangipahar. The general slope ranges from the northwest to the southeast. Geologically, the area comprises Archean granites and gneisses. The majority of the Dhanbad District is covered by metamorphic rocks. These include the rock types of Chotanagpur granite and gneisses of the Achaean age. Sedimentary rocks covering portions of the southern and eastern parts of the district are mainly composed of Gondwana Formations. The metamorphic terrain of the region is underlain by a wide range of geological formations ranging in age from the Archaean to Recent (Saranathan *et al.*, 2006; District Survey Report, 2017).

15.2.1.2 Soil

There are approximately 12 classes of soil subgroups covering the entire Dhanbad district. Most of the soil subgroups are *in situ* in nature and are generally, called sedimentary soils. Based on the origin, these soils are classified into two major soil landscapes: granite-gneiss and Gondwana landscapes. As far as the texture is concerned, the above soil sub-groups have been categorized into four texture types' viz., sandy loam, sandy clayey loam, loam, and clayey loam.

15.2.1.3 Climate

The Dhanbad region can be divided into three broad climatic seasons. The winter season lasts from November to February, in which December and January are the coldest months. During winter, the temperature ranges from 8.4 to 34 °C. Second, the summer season begins in March and continues until May. In April, the wind blows from the west. It remains relatively hot and the temperature ranges from 13.3 to 45.5 °C. Finally, the rainy season normally begins in mid-June when the monsoon winds bring moisture-laden clouds from the Bay of Bengal. During the rainy season, the temperature ranges from 15 to 36 °C.

15.2.2 Data Used

This study was conducted using various primary and secondary data. These include Survey of India (SOI) topographic sheets of 1:50,000 scale, bearing toposheets 73I/1, 73I/2, 73I/5, 73I/6, 73I/8, 73I/9, 73I/10, 73I/13, and 73I/14, and satellite images from Landsat 1 (MSS), Landsat 5 (TM), and Landsat 8 Operational Land Imager (OLI/TIRS). In this study, Landsat satellite images were acquired for 5 years: 1972, 1989, 1999, 2009, and 2019, downloaded freely from the United States Geological Survey (USGS) obtained from NASA's Earth Observing System distribution website (earthexplorer.usgs.gov), whose specifications are provided in Table 15.1.

15.2.3 Methodology

Landsat images comprise many bands. In this study, the required bands were selected (Table 15.1) and Layer Stacking was performed. Mosaicking was performed for Landsat TM data only, and subsetting was carried out to extract the study area using ERDAS Imagine 2014 software. The satellite images were georeferenced and supervised classification was performed using the maximum likelihood algorithm. The study area was classified into 10 major classes viz., Dense Forest (DF), Moderately Dense Forest (MDF), Scrub Land (SL), Agricultural Land (AL), Mining Area (MA),

Table 15.1 Specification of Landsat satellite images.

Satellite	Sensor	Path/row	Acquisition date	Spectral bands used (μm)	Spatial resolution
Landsat 8	OLI_TIRS	140/43	12/11/2019	B2: Blue (0.45–0.51) B3: Green (0.53–0.59) B4: Red (0.64–0.67) B5: NIR (0.85–0.88)	30 m
Landsat 5	TM	140/43 140/44	02/12/2009	B1: Blue (0.45–0.52) B2: Green (0.52–0.60) B3: Red (0.63–0.69) B4: NIR (0.76–0.90)	30 m
Landsat 5	TM	140/43 140/44	07/12/1999	B1: Blue (0.45–0.52) B2: Green (0.52–0.60) B3: Red (0.63–0.69) B4: NIR (0.76–0.90)	30 m
Landsat 5	TM	140/43 140/44	11/12/1989	B1: Blue (0.45–0.52) B2: Green (0.52–0.60) B3: Red (0.63–0.69) B4: NIR (0.76–0.90)	30 m
Landsat 1	MSS	150/43	07/11/1972	B4: Green (0.5–0.6) B5: Red (0.6–0.7) B6: NIR (0.7–0.8)	60 m

Built Up Area (BA), Barren Land (BL), River (R), Sand (SN), and Water Body (WB), based on the knowledge of field survey.

Unsupervised NDVI classification was performed to assess the variation in vegetation cover over time. Vegetation indices, such as NDVI, have been used for a long time in a wide range of fields. NDVI is the most used index in vegetation studies (Breshears *et al.*, 2005; Pan *et al.*, 2003; Pettorelli *et al.*, 2005; Tucker *et al.*, 2005, Abdollahzadeh *et al.*, 2021). These indices are radiometric measures that combine information from the red and near infrared (NIR) portions of the spectrum to enhance the 'vegetation signal' (Kirui, *et al.*, 2021). The values of NDVI show the amount of chlorophyll content present in vegetation, and thus, the amount of greenness can be indicated (Hansen *et al.*, 2003; Sellers, 1985; Zhou *et al.*, 2007). Thus, a greater NDVI value indicates healthy and dense vegetation, whereas a lower NDVI value indicates sparse vegetation. The NDVI was estimated using Eq. (15.1) (Gamon *et al.*, 1995):

$$NDVI = \left(\frac{NIR - RED}{NIR + RED} \right) \tag{15.1}$$

Here, NIR is the Near-infrared band (MSS B6, TM B4, and OLI_TIRS B5), and RED is the Red band (MSS B5, TM B3, and OLI_TIRS B4).

To prepare LULC maps and to assess and compare the changes found in the respective years, the supervised classified raster images were reclassified in ArcGIS 10.3 software. A flowchart of the methodology is shown in Figure 15.2.

The Magnitude Change (MC) for each land-use class was calculated by subtracting the area coverage from the Magnitude of New Year (MNY) and the Previous Year (MPY), as shown in Eq. (15.2).

$$MC = MNY - MPY \tag{15.2}$$

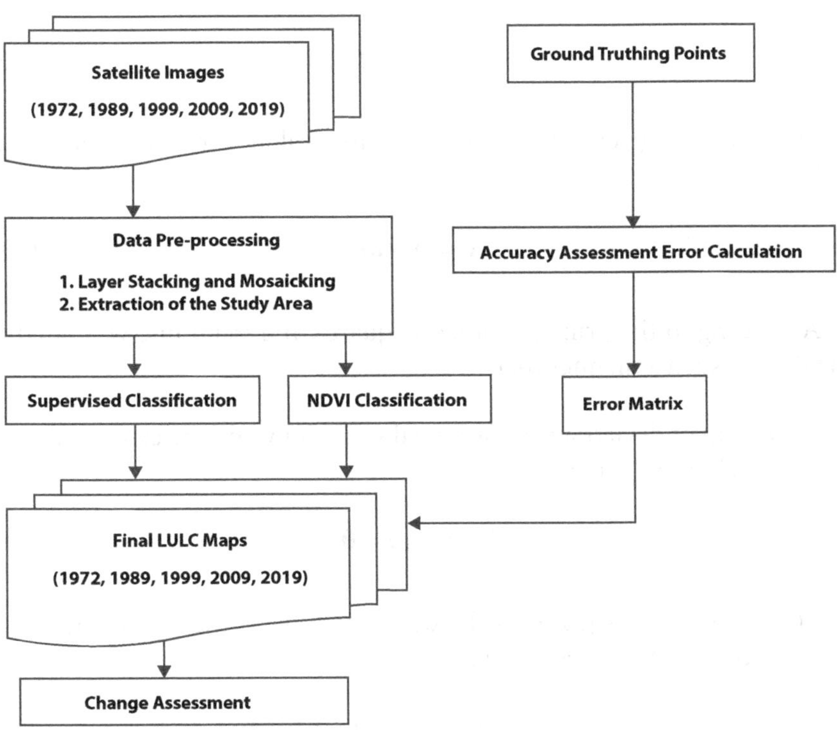

Figure 15.2 Methodology flowchart.

The percentage change (PC) for each land-use type was then calculated by dividing the Magnitude Change (MC) by the Base Year (BY) and multiplied by hundred as shown in Eq. (15.3).

$$PC = \frac{MC}{BY} \times 100 \tag{15.3}$$

15.2.4 Trend Analysis

The time series regression analysis using the sum of least squares statistical method was used to assess the trend and magnitude of the fluctuation for all 10 LULC class categories: Dense Forest (DF), Moderate Dense Forest (MDF), Scrub Land (SL), Agricultural Land (AL), Mining Area (MA), Built Up Area (BA), Barren Land (BL), River (R), Sand (SN), and Water Body (WB) for the period 1972–2019. The least-squares method was applied according to Gupta (2006).

A straight-line trend between the time-series values (y) and time (t) is given by Eq. (15.4).

$$y = a + bt \tag{15.4}$$

Then, for any given time, 't,' the estimated value $\hat{y}$ of y is given by Eq. (15.5).

$$\hat{y} = a + bt \tag{15.5}$$

According to the principle of least squares, the trend line was fitted to the data in such a manner that:

(1) sum of deviations of the actual values of y and the estimated values of $\hat{y}$ is zero, i.e.,

$$\Sigma(y - \hat{y}) = 0 \tag{15.6}$$

(2) sum of the squares of the deviations (E) of the actual and-computed values is least, i.e.,

$$E = \Sigma(y - \hat{y})^2 = \Sigma(y - a - bt)^2 \text{ is minimum} \tag{15.7}$$

For making the square of errors E minimum,

$$\frac{\partial E}{\partial a} = 0 = -2\sum(y - a - bt) \tag{15.8}$$

$$\frac{\partial E}{\partial b} = 0 = -2\sum(y - a - bt) \tag{15.9}$$

Simplifying Eqs. (15.8) and (15.9) provides normal equations for estimating 'a' and 'b'

$$\sum y = na + b\sum t \tag{15.10}$$

$$\sum ty = a\sum t + b\sum t^2 \tag{15.11}$$

Because the time variable 't' is measured as a deviation from its mean, $\sum t = 0$, Eqns. (15.10) and (15.11) are reduced to Eqs. (15.12) and (15.13), respectively.

$$\sum y = na \tag{15.12}$$

$$\sum ty = b\sum t^2 \tag{15.13}$$

From Eqs. (15.12) and (15.13), the values of 'a' and 'b' may be calculated

$$a = \frac{\sum y}{n} = \bar{y} \tag{15.14}$$

$$b = \frac{\sum ty}{\sum t^2} \tag{15.15}$$

In contrast, in case of a second-degree parabolic trend, three values 'a,' 'b,' and 'c' are obtained using the principal of least squares. The second-degree polynomial equation is given by Eq. (15.16).

$$y = a + bt + ct^2 \tag{15.16}$$

In the present study, trend analysis was carried out using Linear and Parabolic trends in Ms Office Excel for all LULC classes.

15.2.5 Accuracy Assessment

Accuracy assessment is considered a significant measure in the assessment of various image-processing procedures in image classification (Lin *et al.*. 2015; Ibharim *et al.*, 2015). Using the ERDAS Imagine 2014 version, an accuracy assessment of the supervised classified LULC raster images was carried out. Using random sampling, 121, 116, 135, 106, and 113 points were randomly generated from the classifier for the 1972, 1989, 1999, 2009, and 2019 supervised images, respectively. The randomly generated points were identified and assigned to different LULC classes. The process was repeated for all supervised classification images. In addition to software-based accuracy assessment, ground truthing was also carried out by randomly generating 60 points using simple random sampling, and field checking was carried out on those points using GPS to evaluate the correctness of the classification process. Additionally, Google Earth images were used to correctly identify the features. The images used in this study have different spatial resolutions specifically Landsat1 MSS image having 60 m compared with other images from Landsat5 TM and Landsat8 OLI_TIRS having 30 m (Table 15.1). To achieve better classification accuracy, the MSS image of 60 m was resampled to 30 m resolution using the nearest neighbor technique so that all datasets had similar resolutions. Next, after post-classification of all the images, the latest classified image of 2019 was vectorized and displayed over the rest of the classified images of 2009, 1999, 1989, and 1972, where rechecking of every class and editing was performed to obtain the final maps and to achieve higher accuracy. From the generated error matrix percentage of the producer's accuracy, user's accuracy, and overall accuracy were calculated. The overall accuracy (*OA*) (Jensen, 1996) was calculated using Eq. (15.17).

$$OA = \frac{\text{number of sampling classes classified correctly}}{\text{number of reference sampling classes}} \tag{15.17}$$

Because of the rough measure of Overall Accuracy (Foody, 2002), to obtain a higher level of confidence in the measure of accuracy, the kappa coefficient of agreement was also calculated using Eqs. (15.18), (15.19), and (15.20) (Afify, 2011).

$$\hat{K} = \frac{P_o - P_c}{1 - P_c} \tag{15.18}$$

$$P_o = \sum_{i=1}^{r} Pii \tag{15.19}$$

$$P_c = \sum_{i=1}^{r} (P_{i+} * P_{+i}) \tag{15.20}$$

Where

- r = The number of rows in the error matrix.
- Pii = The proportion of pixels in row 'i' and column 'i.'
- P_{i+} = The proportion of the marginal total of row 'i.'
- P_{+i} = The proportion of the marginal total of column 'i.'

15.3 Results and Discussion

15.3.1 NDVI Classification

In the case of the district of Dhanbad, NDVI values in 1972 ranged from −0.51 to +0.93; in 1989, the NDVI value indicated a minimum value of −0.55 and a maximum of +0.56. In 1999, it ranged from −0.52 to +0.58, whereas in 2009, the NDVI value indicated a minimum value of −0.40 and a maximum value of +0.53. In 2019, NDVI showed the minimum value which is −0.17 and maximum +0.45 (Table 15.2). Average NDVI values were observed at 0.21, 0.005, 0.03, 0.065, and 0.14 for 1972, 1989, 1999, 2009, and 2019, respectively. In addition, the NDVI for the district of Dhanbad was highest in 1972 (+0.93) and lowest in 1989 (−0.55).

The NDVI classification was performed into five classes: Water, No Vegetation, Low Vegetation, Moderate Vegetation and High Vegetation, using threshold values of 0.0, 0.0 and 0.03, 0.03 and 0.20, 0.20 and 0.35, and greater than 0.35 respectively (Table 15.3). The NDVI maps of the study area clearly depict a decrease in vegetation areas over time (Figures 15.3a–e) from 1972 to 2019 and an increase in non-vegetation areas. NDVI classes represent a spatial pattern of vegetation with three varieties of green

Table 15.2 Maximum and minimum NDVI values.

Year	NDVI min	NDVI max	NDVI average
1972	−0.51	0.93	0.21
1989	−0.55	0.56	0.005
1999	−0.52	0.58	0.03
2009	−0.4	0.53	0.065
2019	−0.17	0.45	0.14

Table 15.3 NDVI threshold values adapted.

Class	Threshold values
Water	**≤0.0**
No Vegetation	**>0.0 and ≤0.03**
Low Vegetation	**>0.03 and ≤0.20**
Moderate Vegetation	**>0.20 and ≤0.35**
High Vegetation	**> 0.35**

shades: high, moderate, and low vegetation areas in the map, and the brown portion shows areas such as mining, built-up, and bare soil.

15.3.2 Land Use/Land Cover Status

The analysis using supervised classification between 1972 and 2019 showed that the study area was covered with various features: Dense Forest (DF), Moderately Dense Forest (MDF), Scrub Land (SL), Agricultural Land (AL), Mining Area (MA), Built-Up Area (BA), Barren Land (BL), River (R), Sand (SN), and Water Body (WB).

In 1972 (Figure 15.4(a)), DF, MDF, SL, Al, MA, BA, BL, R, SN, and WB were 10%, 7.37%, 31.61%, 35.69%, 3.43%, 2.06%, 0.88%, 2.06%, 0.88%, 3.81%, 1.80%, and 3.34% of the total study area respectively, whereas in 1989 (Figure 15.4(b)) they were found to be 7.93%, 8.16%, 24.39%, 41.02%, 5.16%, 4.29%, 1.21%, 3.82%, 1.84%, and 2.18%, respectively.

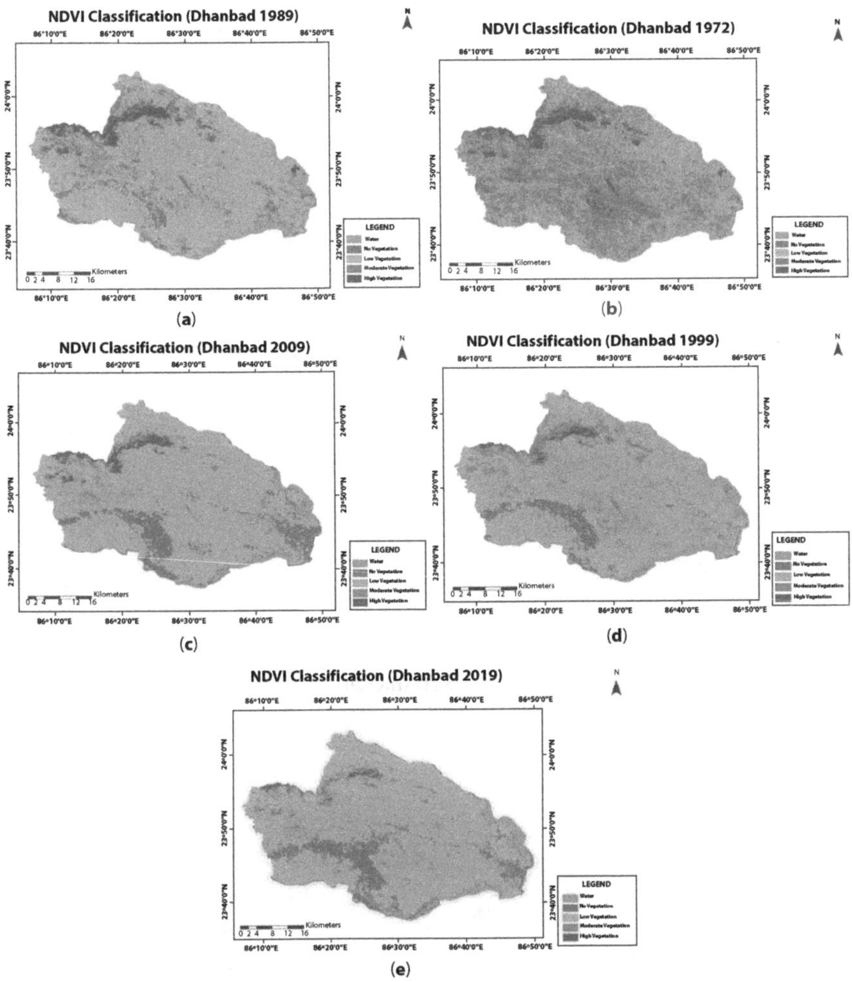

Figures 15.3a–e NDVI maps of the study area.

In 1999 (Figure 15.4(c)), the land features DF, MDF, SL, Al, MA, BA, BL, R, SN, and WB were 5.54%, 7.58%, 22.04%, 43.58%, 4.87%, 6.14%, 2.85%, 3.72% , 2.13%, and 1.55% of the study area whereas in 2009 (Figure 15.4(d)), they were obtained to be 4.99%, 5.91%, 20.88%, 45.89%, 4.58%, 7.34%, 3.02%, 3.76%, 2.32%, and 1.31%, respectively.

In 2019 (Figure 15.4(e)), the DF, MDF, and SL were 4.35%, 3.81%, and 20.80% of the total area, whereas AL, MA, and BA were 46.12%, 4.34%, and 9.20%, respectively. The remaining features, including BL, R, SN, and WB were calculated 4.21%, 3.74%, 2.37%, and 1.37%, respectively (Table 15.4).

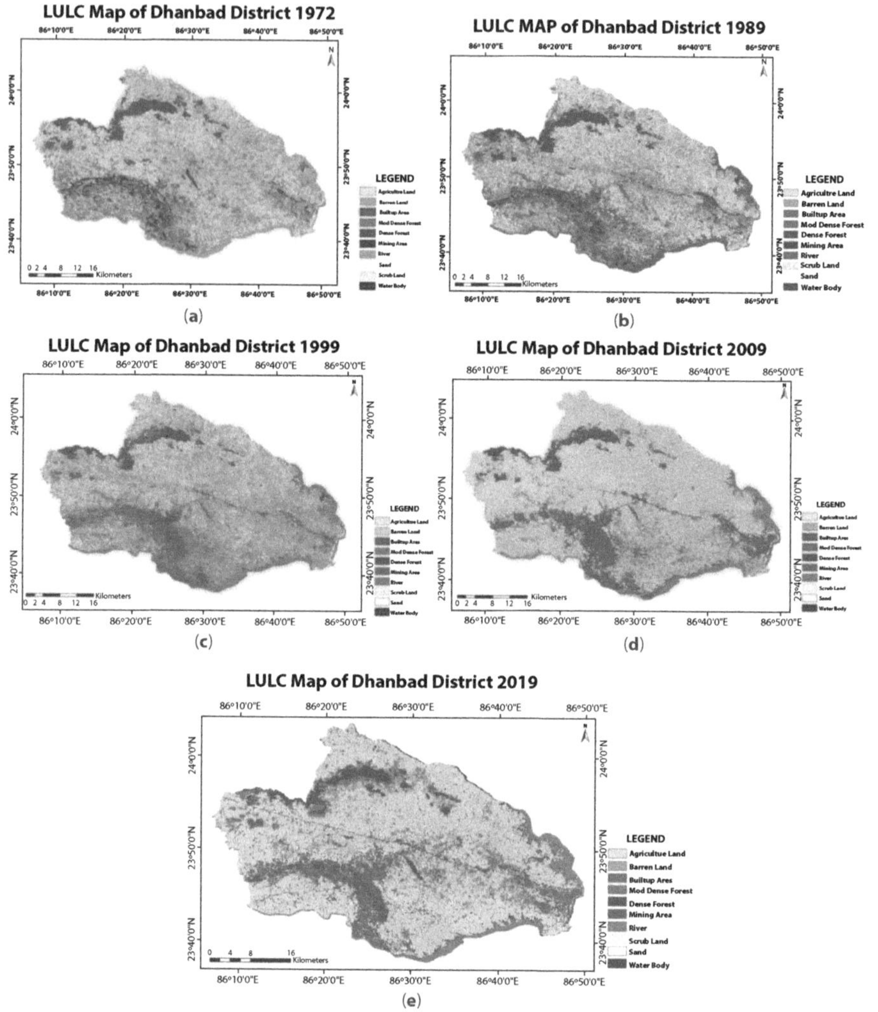

Figures 15.4(a–e) LULC maps of Dhanbad District 1972–2019.

15.3.3 Change Assessment

The areas under different LULC categories in percentages for different periods (1972, 1989, 1999, 2009, and 2019) are shown in Figure 15.5. A comparative chart to analyze the changes in each LULC class over time is shown in Figure 15.6.

From the analysis of the tabular data in Tables 15.4, 15.5, and the graphs in Figures 15.5, 15.6, it was revealed that the DF class steadily decreased during the study period, from 1972 to 2019, with an overall percentage

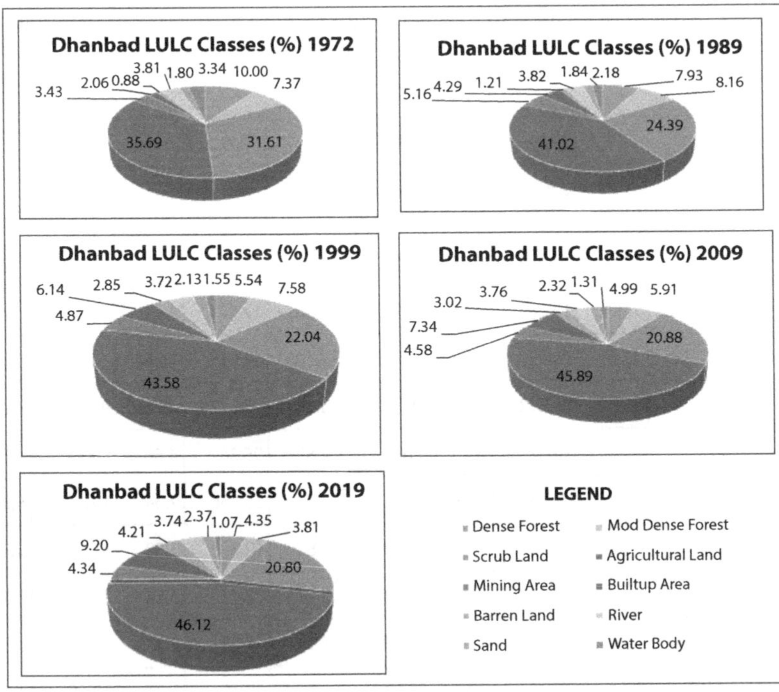

Figure 15.5 LULC classes in percentages at different periods.

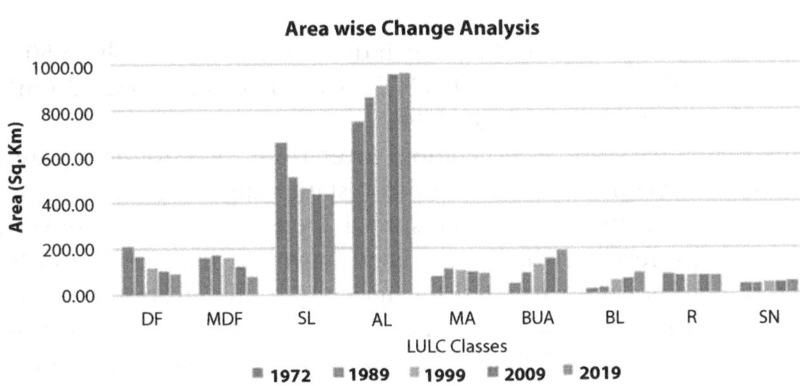

Figure 15.6 Comparative chart.

of 56.49 while the MDF class showed a mixed trend. The MDF initially increased during 1972–1989 from 153.25 km^2 to 169.62 km^2, with an increase of 16.37 km^2 (10.68%), but during later period, it kept on decreasing from 1989 onwards to 79.26 km^2 in 2019. The area of the SL class was

Table 15.4 Area under different LULC categories with respective percentages of total area.

Class names	Year									
	1972		1989		1999		2009		2019	
	Area (Km^2)	%	Area (Km^2)	%	Area (Km^2)	%	Area (Km^2)	%	Area (Km^2)	%
DF	207.89	10.00	164.94	7.93	115.22	5.54	103.76	4.99	90.45	4.35
MDF	153.25	7.37	169.62	8.16	157.54	7.58	122.93	5.91	79.26	3.81
SL	657.28	31.61	507.19	24.39	458.26	22.04	434.25	20.88	432.52	20.80
AL	742.14	35.69	853.22	41.02	906.21	43.58	954.22	45.89	959.12	46.12
MA	71.25	3.43	107.22	5.16	101.25	4.87	95.28	4.58	90.22	4.34
BA	42.81	2.06	89.25	4.29	127.68	6.14	152.56	7.34	191.23	9.20
BL	18.32	0.88	25.22	1.21	59.26	2.85	62.89	3.02	87.52	4.21
R	79.25	3.81	79.47	3.82	77.25	3.72	78.22	3.76	77.84	3.74
SN	37.52	1.80	38.37	1.84	44.33	2.13	48.22	2.32	49.23	2.37
WB	69.52	3.34	45.34	2.18	32.26	1.55	27.15	1.31	22.18	1.07
Total	2079.23	100.00	2079.84	100.00	2079.26	100.00	2079.48	100.00	2079.57	100.00

657.28 km^2 during the year 1972 which decreased continuously to 507.19 km^2 in 1989, 458.26 km^2 in 1999, 434.25 km^2 in 2009 and 432.52 km^2 in 2019, with an overall percentage of 34.2.

This steady decrease in vegetation cover from 1972 to 2019 may be attributed to several factors such as deforestation due to agriculture, mining, population growth, and urban sprawl. Dense forests were converted into low-dense forests, which were later converted into scrub and barren land. Forests were harvested for agriculture, the creation of various household usages of wooden materials, shelter, and fuel, along with mining activities in the decades 70's and the early 80's. In recent years, with the rapid increase in population since the 1980s, there has been a steady and upward rise in industrial growth and urbanization, with a massive escalation of settlements around and near mining areas.

The AL class was progressively increased during the entire study period 1972–2019 from 742.14 km^2 in 1972 to 959.12 km^2 in 2019, with an overall percentage of 29.24. This could be attributed to the conversion of vegetation-covered areas to agricultural land.

Table 15.5 Magnitude Change (MC) and Percentage Change (PC) of each LULC class.

Class names	Year ranges									
	1989–1972		1999–1989		2009–1999		2019–2009		2019–1972	
	Area (Km²)	%	Area (Km²)	%	Area (Km²)	%	Area (Km²)	%	Area (Km²)	%
DF	−42.95	−20.66	−49.72	−30.14	−11.46	−9.95	−13.31	−12.83	−117.44	-56.49
MDF	16.37	10.68	−12.08	−7.12	−34.61	−21.97	−43.67	−35.52	−73.99	−48.28
SL	−150.09	−22.84	−48.93	−9.65	−24.01	−5.24	−1.73	−0.40	−224.76	−34.20
AL	111.08	14.97	52.99	6.21	48.01	5.30	4.90	0.51	216.98	29.24
MA	35.97	50.48	-5.97	−5.57	−5.97	−5.90	-5.06	−5.31	18.97	26.62
BA	46.44	108.48	38.43	43.06	24.88	19.49	38.67	25.35	148.42	346.69
BL	6.90	37.66	34.04	134.97	3.63	6.13	24.63	39.16	69.20	377.73
R	0.22	0.28	−2.22	−2.79	0.97	1.26	−0.38	−0.49	−1.41	−1.78
SN	0.85	2.27	5.96	15.53	3.89	8.78	1.01	2.09	11.71	31.21
WB	−42.95	-20.66	−49.72	−30.14	−11.46	−9.95	−13.31	−12.83	−117.44	−56.49

The MA class was estimated to be 71.25 km² in 1972, increased to 107.22 km² in 1989. However, in later periods, a slight decrement was obtained in this class which was found to be 101.25 km² in 1999, 95.28 km² in 2009, and 90.22 km² in 2019. This decrease might be attributed to the filling of open cast pits with rainwater storage, abandoned mines, and conversion of mined-out area to barren land. However, considering the entire study period, the MA class increased by 26.62%.

The BA class increased rapidly during the study period from 42.81 km² in 1972 to 191.23 km² in 2019, an overall increase of 346.69%. This substantial increase in built-up areas was due to massive population growth, industrialization, and urban sprawl, which resulted in the destruction of forest land, conversion of agricultural land, and soil filling of small water bodies such as ponds.

The BL class increased with a steady rate from 18.32 km² in 1972 to 87.52 km² in 2019, an overall increase of 377.73%. The increase in barren land might be attributed to the increase in mined out wasteland areas, overburden dump areas, and the spread of mine fires and subsidence-affected area.

As seen from the tabulated values and graphs, the classes R and SN had no significant change during the study period; however, the class WB had a substantial decrement from 69.52 km^2 in 1972 to 22.18 km^2 in 2019, an overall decrement of 56.49%. The reason for this reduction in water bodies might be attributed to the filling of water bodies and their conversion to built-up areas. To provide deeper insight into the change detection of various LULC classes, a trend analysis was carried out, as discussed in the next section.

15.3.4 Trend Analysis

The trends, magnitude of fluctuation, and spatial variation of the increase or decrease of all 10 LULC classes from 1972 to 2019 in the Dhanbad district in the state of Jharkhand are shown in Figures 15.7(a–j).

Changes in land use and land cover, whether natural or impacted by anthropogenic activities, do not typically follow a particular trend, such as linear, parabolic, exponential, or logarithmic. However, this analysis was performed to determine whether the changes followed a similar trend. The LULC classes DF, MDF, SL, and AL followed a linear trend, with R^2 values of 0.95, 0.61, 0.86, and 0.95, respectively (Figures 15.7(a–d)). The MA class (Figure 15.7(e)) followed a second-degree parabolic trend, with an R^2 value of 0.87. The classes BA, BL, R, SN, and WB (Figures 15.7(f–j)) followed a linear trend with R^2 values of 0.99, 0.91, 0.53, 0.9, and 0.93, respectively.

15.3.5 Accuracy Assessment

Tables 15.6 and 15.7 present the values of producer's accuracy and user's accuracy for different LULC classes in different periods, respectively whereas Table 15.8 provides the Overall Accuracy and Kappa values for the years 1972, 1989, 1999, 2009, and 2019.

The average producer accuracies for classes DF, LDF, and SL were 91.7%, 85.5%, and 82.6%, respectively, whereas the average user accuracies for the same classes were 82.9%, 84.4%, and 76.6%, respectively. Similarly, the average producer and user accuracies for the Al class were 83% and 83.5%, MA class was 77.6% and 86.3%, BA class was 80.3% and 82.5%, BL class was 84% and 79.5%, R class was 96.4% and 86.4%, SN class was 82.3% and 82.4%, and WB class was 95.6% and 81.5%, respectively.

Average accuracies for producer as well as user are 84.7% and 84.9% for 1972, 85.6% and 79.7% for 1989, 83.2% and 81.7% for 1999, 88.5% and 83% for 2009, and 87.4% and 83.6% for 2019, respectively (Tables 15.6, 15.7).

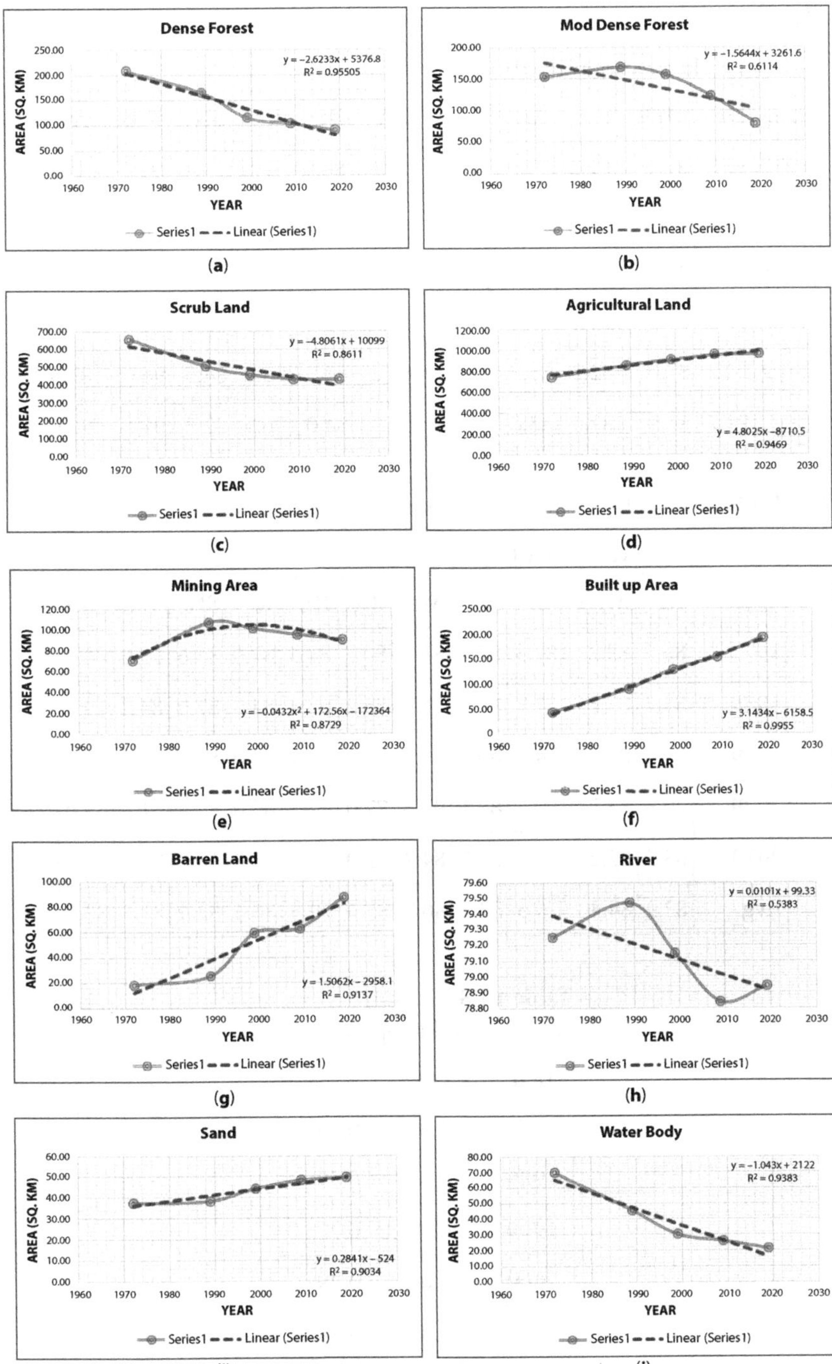

Figure 15.7 Trend line graphs for LULC classes.

Table 15.6 Summary of producer's accuracy.

Year	Producer's accuracy (%)										
	DF	MDF	SL	AL	MA	BA	BL	R	SN	WB	Avg
1972	91.3	87.3	81.2	79.9	81.3	76.3	81.9	97.0	76.4	95.0	84.7
1989	91.3	85.2	84.1	81.3	82.0	77.9	82.7	96.0	80.5	95.0	85.6
1999	89.5	82.3	79.4	83.5	54.0	81.2	83.6	98.0	83.2	97.5	83.2
2009	92.3	85.7	86.2	87.0	84.5	85.0	88.2	95.0	87.0	94.3	88.5
2019	94.2	87.0	82.2	83.2	86.0	81.2	83.7	96.0	84.5	96.0	87.4
Avg	91.7	85.5	82.6	83.0	77.6	80.3	84.0	96.4	82.3	95.6	85.9

Table 15.7 Summary of user's accuracy.

Year	User's accuracy (%)										
	DF	MDF	SL	AL	MA	BA	BL	R	SN	WB	Avg
1972	86.5	83.2	87.3	87.2	88.2	80.7	84.5	87.0	85.6	79.0	84.9
1989	83.2	87.2	57.7	84.1	77.6	85.2	67.3	84.0	87.0	84.2	79.7
1999	81.4	80.7	74.2	80.2	90.0	82.7	82.9	87.0	80.5	77.2	81.7
2009	79.8	83.7	80.5	85.1	91.5	77.0	79.2	89.0	82.5	81.5	83.0
2019	83.5	87.2	83.3	80.7	84.2	87.0	83.5	85.0	76.2	85.6	83.6
Avg	82.9	84.4	76.6	83.5	86.3	82.5	79.5	86.4	82.4	81.5	82.6

Table 15.8 Overall accuracy and kappa values.

Year	OA	K
1972	0.83	0.74
1989	0.81	0.72
1999	0.84	0.77
2009	0.83	0.79
2019	0.85	0.84
Avg	0.83	0.77

The overall classification accuracy was 83% in 1972, 81% in 1989, 84% in 1999, 83% in 2009, and 83% in 2019. Kappa coefficients for 1972, 1989, 1999, 2009, and 2019 were 0.77, 0.74, 0.77, 0.79, and 0.84, respectively.

15.3.6 Findings and Comparison of Trends with Published Literature

The effect of coal mining on LULC changes in the Godavari coalfield area from 1990 to 2014 revealed a decrease in forest area and an increase in agricultural, built-up, and barren land (Garai *et al.*, 2018). This trend in LULC dynamics in the coal belt area was observed in the Dhanbad district in this study. The effect of coal mining on natural resources and its impact on the land resources of Dhanbad District, as discussed, one-fourth of land was found unsuitable for agriculture (Saranathan *et al.*, 2006). These unsuitable portions of agricultural land were later converted into barren land or built-up areas, as observed in the current study, which showed a high rise in these classes. Similarly, in a coal mining area, an increase in built-up area from 1972 to 2011 in the Dhanbad Urban Agglomeration (DUA) was found to be 352.08% with a population growth of 160.07%, along with a significant decrease in vegetation cover of 41.33% (Lal *et al.*, 2017), which matches the trend in the current study. One-third of the world's land surface is currently under agricultural use, and millions of hectares of natural ecosystems are converted to croplands every year (Foley *et al.*, 2005). This matches the present research with the fact that the trend line of the agricultural land class increased steadily throughout the study period of 1972–2019, with conversions from other classes, *viz.*, DF, MDF, and SL. The LULC changes in Dhanbad District from 1987 to 2017 show an increase in agricultural land, barren land, sand, and built-up areas by 119.06%, 211.63%, 234.96%, and 79.85%, respectively (Singh *et al.*, 2018), which matches the trend shown in the present study.

The key changes recorded in the present study are listed as follows (Table 15.5):

1. A big rise in the built-up area from 42.81 km^2 in 1972 to 191.23 km^2 in 2019, an overall increase of 346.69%.
2. A decrease in overall vegetation cover was observed in all three classes of dense forest, moderately dense forest, and scrubland with 56.49%, 48.28%, and 34.2%, respectively.
3. A big increase in the barren land from 18.32 km^2 in 1972 to 87.52 km^2 in 2019, an overall increase of 377.73%.

4. A progressive rise in the agricultural land from 742.14 km^2 in 1972 to 959.12 km^2 in 2019, with an overall increase of 29.24%.

The rise in the built-up area, as depicted at point 1, is a direct consequence of population growth, industrial escalation (Lal *et al.*, 2017), and the huge migration of people post coal nationalization from various parts of India to the Dhanbad district for employment and business purposes. The growth of industrial and urban areas resulted in a constant decrease in vegetation cover, as discussed at point 2. The increase in barren land, as depicted in point 3, illustrates the effects of coal mining, mine fires, subsidence, and an increase in overburden dump that not only affected vegetation cover but also agricultural land (Stracher *et al.*, 2015). The steady rise in agricultural land, as depicted at point 4, is balanced by land consumption through continuous deforestation of dense and moderately dense forests, which were found suitable for agricultural purposes, as a substantial portion of land was found unsuitable for agriculture (Saranathan *et al.*, 2006). Therefore, a continuous transition between different LULC classes occurred over the five decades (1972–2019), as observed in this study.

15.4 Limitations and Future Scope

The present study utilizes Geospatial Science to understand the reasons for Land Cover changes in the Dhanbad district over the last five decades and measures various Land Cover Classes, viz., Dense Forest (DF), Moderately Dense Forest (MDF), Scrub Land (SL), Agricultural Land (AL), Mining Area (MA), Built Up Area (BA), Barren Land (BL), River (R), Sand (SN), and Water Body (WB). It further provides a trend line for each Land Cover class over the study period. However, a further study for the above research could predict future estimates of land use classes using various techniques such as Artificial Neural Networks.

15.5 Conclusion

The present study was conducted in the district of Dhanbad, Jharkhand, India, from 1972 to 2019 using geospatial techniques to assess land use and land cover dynamics. This study revealed significant changes in the study area over the last five decades.

Land is being used by the increasing population to meet its needs, and land resources are continuously being used for developmental purposes such as building industries, commercial shops, and sky-scraper buildings. In this process of development and change, the areas under the classes Dense Forest (DF), Moderately Dense Forest (MDF), Scrub Land (SL), and Water Bodies (WB) decreased significantly, whereas areas under the classes Agriculture Land (AL), Mining Area (MA), Built-up Area (BA), Barren Land (BL), and Sand (SN) increased significantly. The DF class steadily decreased during the study period by 56.49%, whereas the MDF class showed a mixed trend with an increase from 1972 to 89 (10.68%), and then decreased afterwards, with an overall decrease of 48.28% by 2019. This steady decrease in vegetation cover from 1972 to 2019 may be attributed to several factors such as deforestation due to agriculture, mining, population growth, and urban sprawl. The AL class progressively increased during the entire study period (1972–2019), with an overall increase of 29.24%. The MA class increased by 26.62%. The BA and BL classes increased at alarming rates of 346.69% and 377.73%, respectively. This substantial increase in BA class was due to massive population growth, industrialization, and urban sprawl, whereas that of the BL class was the increase in mined out wasteland areas, overburden dump areas, and spread of mine fires and subsidence affected areas. The rapid enhancement in mining and urban built-up areas has resulted in the destruction of vegetation cover, conversion of agricultural land, growth of barren land, and reduction of small water bodies through landfills.

The main focus of this research is to understand and depict the major reasons for the Land Use and Land Cover changes of Dhanbad district over the last five decades using geospatial techniques, to obtain a clear picture, changes in trends of various Land Use Land Cover classes, namely Dense Forest (DF), Moderate Dense Forest (MDF), Scrub Land (SL), Agricultural Land (AL), Mining Area (MA), Built-Up Area (BA), Barren Land (BL), River (R), Sand (SN) and Water Body (WB) for the period 1972–2019. The graphical trend shows linear and parabolic incremental and decremental changes in various classes over the entire period of five decades. This study considers a lengthier study period of five decades to observe the process of Land Cover changes using geospatial techniques in detail and highlights the major causes thereof. The authors believe that this study can be helpful for decision makers in formulating future land use development and management strategies, without affecting the environment.

Acknowledgments

The authors are indebted to NASA for making Landsat datasets freely available under the umbrella of the USGS web server. The authors are grateful to the Birla Institute of Technology, Mesra, Ranchi, India for the facility to conduct this study. This study is part of the M.Sc. thesis of Mr. Gaurav Chakrabarty under the supervision of Prof. M. S. Nathawat, then head, Department of Remote Sensing, BIT Mesra, Ranchi, which was later extended with the help of Dr. Suman Sinha.

References

Aboelnour, M, Engel, B., 2018. Application of Remote Sensing Techniques and Geographic Information Systems to Analyze Land Surface Temperature in Response to Land Use/Land Cover Change in Greater Cairo Region, Egypt. *J. Geogr. Inf. Syst.* 10:57-88. doi: 10.4236/jgis.2018.101003.

Afify, H. A., 2011. Evaluation of change detection techniques for monitoring land cover changes: a case study in new Burg El-Arab area. *Alex. Eng. J.* 50:187–195. http://dx.doi.org/10.1016/j.aej.2011.06.001.

Aryee, N. A. B, Ntibery, B. K, Atorkui, E., 2003. Trends in the small-scale mining of precious minerals in Ghana: a perspective on its environmental impact. *J. Cleaner Prot.* 11: 131–140. http://citeseerx.ist.psu.edu/viewdoc/summary?doi=10.1.1.459.7972.

Basu T, Das A, Pham QB, Al-Ansari N, Linh NTT, Lagerwall G, 2021. Development of an integrated peri-urban wetland degradation assessment approach for the Chatra Wetland in eastern India. *Sci. Rep.* 11(1):4470. https://doi.org/10.1038/s41598-021-83512-6.

Bekele AE, Drabik D, Dries L, Heijman W., 2021. Large-scale land investments, household displacement, and the effect on land degradation in semiarid agro-pastoral areas of Ethiopia. *Land Degrad. Dev.* 32(2):777–791. https://onlinelibrary.wiley.com/doi/abs/10.1002/ldr.3756.

Bell, F.G., Bullock, S.E.T., Halbich, T.F.J, Lindsey, P., 2001. Environmental impacts associated with an abandoned mine in the Witbank Coalfield, South Africa. *Int. J. Coal Geol.* 45:195-216. https://www.sciencedirect.com/science/article/abs/pii/S0166516200000331.

Betts, M. G., Wolf, C., Ripple, W. J., Phalan, B., Millers, K. A., Duarte, A., Butchart, S. H. M., Levi, T., 2017. Global forest loss disproportionately erodes biodiversity in intact landscapes. *Nature.* 547: 441–444. https://www.nature.com/articles/nature23285.

Breshears, D. D., Cobb N. S, Rich P. M., Price K. P., Allen C. D., Balice R. G., Romme W. H., Kastens J. H., Floyd M. L., Belnap J., 2005. Regional vegetation die-off in response to global-change-type drought. *PNAS. USA*. 102: 15144–15148. https://www.pnas.org/content/102/42/15144.

Chiesura, A., de Groot, R., 2003. Critical natural capital: a socio-cultural perspective. *Ecol. Econ*. 44:219–231. https://doi.org/10.1016/S0921-8009(02)00275-6.

Costanza, R., de Groot, R., Braat, L., Kubiszewski, I., Fioramonti, L., Sutton, P., Farber, S., Grasso, M., 2017. Twenty years of ecosystem services: how far have we come and how far do we still need to go? *Ecosyst. Serv.* 28:1–16. https://doi.org/10.1016/j.ecoser.2017.09.008.

Daniel, L.C., J.D. Hurd, E.H. Wilson, M. Song and Z. Zhang, 2002. A comparison of land use and land cover change detection methods. *Proceedings of the ASPRS-ACSM Annual Conference and FIG XXII Congress*, Apr. 22-26.

District Survey Report, District Dhanbad, Jharkhand. 2017. http://jharkhand-minerals.gov.in/Application/uploadDocument/Notification/1506083111_DISTRICT%20SURVEY%20REPORT%20FOR%20SAND%20DHANBAD.pdf.

Foley, J.A., DeFries, R., Asner, G.P., Barford, C., Bonan, G., Carpenter, S.R., Chapin, F.S., Coe, M.T., Daily, G.C., Gibbs, H.K., Helkowski, J.H., Holloway, T., Howard, E.A., Kucharik, C. J., Monfreda, C., Patz, J. A., Prentice, I.C., Ramankutty, N., Snyder, P.K. , 2005. *Global consequences of Land Use*. 309: 570-574. DOI: 10.1126/science.1111772.

Foody G. M., 2002. Status of land cover classification accuracy assessment. *Remote Sens. Environ.* 80:185–201. https://doi.org/10.1016/j.rse.2019.111630

Gamon, J. A., Field, C. B., Goulden, M. L., Griffin, K. L. 1995. Relationship between NDVI, canopy structure and photosynthesis in three Californian vegetation types. *Ecol. Appl.* 5:28–41. doi: 10.2307/1942049.

Hu, C., L., Zhang, Q., Wu, S. Soomro, and S. Q. Jian., 2020. Response of LUCC on runoff generation process in middle Yellow River Basin. *Water* 12 (5):1237–50. doi: 10.3390/w12051237.

Garai., D., Narayana, A.C., 2018. Land use/land cover changes in the mining area of Godavari coal fields of southern India, *The Egypt. J. Remote Sens. Space. Sci.*, Volume 21, Issue 3, Pages 375-381, ISSN 1110-9823, https://doi.org/10.1016/j.ejrs.2018.01.002.

Gupta, S.P., 2006. *Fundamental of Statistics*. Sultan Chand and Sons Educational publishers, New Delhi.

Hansen M., C., DeFries R., S., Townshend J., Carroll M., Dimiceli C., Sohlberg R. A., 2003. Global percent tree cover at a spatial resolution of 500 meters: first results of the MODIS vegetation continuous fields algorithm. *Earth Interact.* 7: 1–15. https://doi.org/10.1175/1087-3562(2003)007<0001:GPTCAA>2.0.CO;2.

Ibharim N. A., Mustapha M. A., Lihan T., Mazlan A. G., 2015. Mapping mangrove changes in the Matang Mangrove Forest using multi-temporal satellite imageries. *Ocean Coastal Manage.* 114:64–76. https://doi.org/10.1016/j.ocecoaman.2015.06.005.

Jensen J. R., 1996. *Introductory Digital Image Processing: A Remote Sensing Perspective.* 2nd ed. New Jersey: Prentice-Hall.

Kafi, K., M., Shafri, H., Z., M., Shariff, A., B., M., 2014. An Analysis of LULC Change Detection Using Remotely Sensed Data: A Case Study of Bauchi City. *IOP Conf. Ser.: Earth Environ. Sci.* 20: 12056. https://doi.org/10.1088/1755-1315/20/1/012056.

Kareiva, P., Marvier, M., 2011. *Conservation Science: Balancing the Nature of People and Nature.* Robers and company, Greenwood Village, Colorado.

Kirui, O.K., Mirzabaev, A. & von Braun, J. 2021 Assessment of land degradation 'on the ground' and from 'above'. *SN Appl. Sci.* **3,** 318. https://doi.org/10.1007/s42452-021-04314-z.

Kong, X. L., Y. L. Li, M. Han, L. X. Tian, J. Q. Zhu, and X. R. Niu. 2020. Analysis of land use/cover change and landscape pattern in the Yellow River Delta during 1986–2016. *J. Southwest For. Univ.* 40 (4):122–31.

Lal, K,. Kumar, D., Kumar A., 2017. Spatio-temporal landscape modeling of urban growth patterns in Dhanbad Urban Agglomeration, India using geoinformatics techniques, *Egypt. J. Remote Sens. Space. Sci.*, Volume 20, Issue 1, Pages 91-102, ISSN 1110-9823, https://doi.org/10.1016/j.ejrs.2017.01.003.

Lambin, E.F., Geist, H.J., Lepers, E., 2003. Dynamics of land-use and land-cover change in tropical regions. *Annu. Rev. Environ. Resour.* 28:205–241. https://doi.org/10.1146/annurev.energy.28.050302.105459.

Lin, C., Wu, C., Tsogt, K., Ouyang, Y., Chang, C., 2015. Effects of atmospheric correction and pan sharpening on LULC classification accuracy using WorldView-2 imagery. *Inf. Process. Agric.* 2:25–36. https://doi.org/10.1016/j.inpa.2015.01.003.

Li Z, Wang S, Song S, Wang Y, Musakwa W., 2021. Detecting land degradation in Southern Africa using Time Series Segment and Residual Trend (TSS-RESTREND). *J. Arid Environ.* [Internet]. 184:104314. https://www.sciencedirect.com/science/article/pii/S0140196320302135.

Lu, D.S., Mausel, P., Brondízio, E.S., and Moran, E. (2004), "Change Detection Techniques," *Int. J. Remote Sens.*, 25, 2365–2407.

Macleod, R.D., Congalton, R.G., 1998, "A Quantitative Comparison of Change Detection Algorithms for Monitoring Eelgrass From Remotely Sensed Data," *Photogramm. Eng. Remote Sens.*, 64(3), 207–216.

Miura, S., Amacher, M., Hofer, T., San-Miguel-Ayanz, J., Ernawati, Thackway, R., 2015. Protective functions and ecosystem services of global forests in the past quarter-century. *For. Ecol. Manage.* 352:35–46. https://doi.org/10.1016/j.foreco.2015.03.039.

Official website of Ministry of Coal, Government of India. https://coal.nic.in/content/historybackground.

Pettorelli, N., Vik, J.O., Mysterud, A., Gaillard, J., Tucker, C. J., Stenseth N. C., 2005. Using the satellite-derived NDVI to assess ecological responses to environmental change. *Trends Ecol. Evol.* **20**: 503–510. https://doi.org/10.1016/j.tree.2005.05.011.

Prăvălie R, Patriche C, Borrelli P, Panagos P, Roşca B, Dumitraşcu M, Nita I-A, Săvulescu I, Birsan M-V, Bandoc G., 2021. Arable lands under the pressure of multiple land degradation processes. A global perspective. *Environ. Res.* 194:110697. https://www.sciencedirect.com/science/article/pii/S0013935120315966.

Rukhovich DI, Koroleva P V, Rukhovich DD, Kalinina N V. 2021. The Use of Deep Machine Learning for the Automated Selection of Remote Sensing Data for the Determination of Areas of Arable Land Degradation Processes Distribution. *Remote Sens.* 13(1). https://www.mdpi.com/2072-4292/13/1/155.

Santra, A., Mitra, S.S., Sinha, S. *et al.* 2020. Performance testing of selected spectral indices in automated extraction of impervious built-up surface features using Resourcesat LISS-III image. *Arab J. Geosci.* **13,** 1229. https://doi.org/10.1007/s12517-020-06183-z.

Sara Abdollahzadeh, Adel Sepehr, Alireza Rashki, 2021. Detecting degraded, prone and transition ecosystems by environmental thresholds and spectral functions, *Remote Sens. Appl.: Soc. Environ.*, Volume 22, 100503, ISSN 2352-9385, https://doi.org/10.1016/j.rsase.2021.100503.

Saranathan, E., Loveson, V. J., Rajamanickam, G. V., 2006. Land suitability study in land degraded area due to mining in Dhanbad District, Jharkhand. *Symposium of ISPRS Commission IV.* https://www.isprs.org/proceedings/XXXVI/part4/RS-GEO-17.pdf.

Sellers P., J., 1985. Canopy reflectance, photosynthesis and transpiration. *Int. J. Remote Sens.* 6: 1335–1372. https://doi.org/10.1080/01431168508948283.

Shankar, U., Boral, L., Pandey, H., & Tripathi, R., 1993. Degradation of land due to coal mining and its natural recovery pattern. *Curr. Sci. 65*(9): 680-687. http://www.jstor.org/stable/24095947.

Singh, A.K., Shukla, N.K., Singh, V., Chaudhary, S.K., Srivastava, S. K., 2007. Indian coal: a source of carbon for artefact industries. *Minetech* 28 (1): 55–60.

Singh, R., Singh, M., Singh., S., Pal, D., Tripathi, N., Singh., R., 2018. Land use/land cover change detection analysis using remote sensing and GIS of Dhanbad district, India. *Eurasian J. For. Sci.*, 6 (2). pp. 1-12.

Sinha, N., Tripathy, S., Pathak, K., 2010. Environmental challenges in uranium mining. *Indian Min. Eng. J.* 49 (11): 23.

Sinha, S., Sharma., K., L., Nathawat, S., M., 2015. Improved Land-use/Land-cover classification of semi-arid deciduous forest landscape using thermal remote sensing, *Egypt. J. Remote Sens. Space. Sci.*, Volume 18, Issue 2, Pages 217-233, ISSN 1110-9823, https://doi.org/10.1016/j.ejrs.2015.09.005.

Sinha, S., Santra, A., and Mitra, S. S., 2018. A Method for Built-Up Area Extraction Using Dual Polarimetric Alos Palsar, *ISPRS Ann. Photogramm. Remote Sens. Spatial Inf. Sci.*, IV-5, 455–458, https://doi.org/10.5194.

Sinha, S., Santra, A., and Mitra, S. S., 2020. Semi-automated impervious feature extraction using built-up indices developed from space-borne optical and SAR remotely sensed sensors, *Adv. Space Res.*, Volume 66, Issue 6, 2020, Pages 1372-1385, ISSN 0273-1177, https://doi.org/10.1016/j.asr.2020.05.040.

Steffen, W., Richardson, K., Rockstrom, J., Cornell, S., E., Fetzer, I., Bennett, E.M., Biggs, R.,Carpenter, S.R., De Vries, W., De Wit, C.A., Folke, C., Gerten, D., Heinke, J., Mace,G.M., Persson, L.M., Ramanathan, V., Reyers, B., Sorlin, S., 2015. Planetary boundaries: guiding human development on a changing planet. *Science* (80). 347, 1259855. https://doi.org/10.1126/science.1259855.

Stracher, Glenn B., Prakash, A., Sokol, Ellina V., Coal and Peat Fires: A Global Perspective, Elsevier, 2015, *Chapter 9* - Impact of Mining Activities on Land Use Land Cover in the Jharia Coalfield, India, Editor(s):, Pages 263-279, ISBN 9780444595096, https://doi.org/10.1016/B978-0-444-59509-6.00009-0.

Tucker, C. J., Pinzon, J. E., Brown, M. E., Slayback D. A., Pak, E. W., Mahoney, R., Vermote E. F., El Saleous N., 2005. An extended AVHRR 8-km NDVIdataset compatible with MODIS and SPOT vegetation NDVI data. *Int. J. Remote Sens.*: 26: 4485–4498. DOI: 10.1080/01431160500168686.

Turner, B.L., Lambin, E.F., Reenberg, A., 2007. The emergence of land change science forglobal environmental change and sustainability. *PNAS*. 104:20666–20671. https://doi.org/10.1073/pnas.0704119104.

UNESCO. 1985: *Living in the Environment*. UNESCO/UNEP.

Y. Pan, X. Li, P. Gong, C. He, P. Shi & R. Pu., 2003. An integrative classification of vegetation in China based on NOAA AVHRR and vegetation-climate indices of the Holdridge life zone, *Int. J. Remote Sens.*, 24:5: 1009-1027. DOI: 10.1080/01431160110115816.

Zhang, Z., Liu, S., Wei, J., Xu, J., Guo, W., Bao, W., Jiang, Z., 2016. Mass change of glaciers in Muztag Ata–Kongur Tagh, Eastern Pamir, China from 1971/76 to 2013/14 as derived from remote sensing data. *PLoS One* 11(1):e0147327. https://doi.org/10.1371/journal.pone.0147327.

Zhou, L., Dickinson, R., Tian, Y., Vose, R, Dai, Y. 2007. Impact of vegetation removal and soil aridation on diurnal temperature range in a semiarid region: application to the Sahel. PNAS. *USA* **104**: 17937–17942. https://doi.org/10.1073/pnas.0700290104.

Zvoleff, A., Wandersee, S., An, L., López-Carr, D., 2014. Land Use and Cover Change. *Oxford Bibliographics*. https://doi.org/10.1093/obo/9780199874002-0105.

16

Assessing Land Susceptibility to Degradation and Sustainability Challenges for Siddharthnagar District, Uttar Pradesh

Rajesh Kumar Abhay[1]*, Shweta Rani[1] and Satish Kumar Saini[2]

[1]Department of Geography, Dyal Singh College, University of Delhi, New Delhi, India
[2]Special Centre for Disaster Research, Jawaharlal Nehru University, New Delhi, India

Abstract
Land is the most basic resource on Earth that makes life possible. The assessment of land susceptibility to degradation from an agricultural sustainability perspective, is very important, as it is directly dependent on the quality of land. Higher land susceptibility may lead to land degradation, and ultimately, agricultural productivity. The objective of this study was to examine the reasons for and variations in the degree of land susceptibility to degradation in the Siddharthnagar district, Uttar Pradesh, based on selected variables. These indicators are rainfall, proximity to the river, soil erosion, drainage density, stream frequency, land use land cover, and cropping intensity. The study was founded on secondary data collected from different sources. Using the multi-criterion method, the level of susceptibility was assessed using different GIS techniques and statistical methods to analyze the data. The study concludes that one-sixth of the district is highly susceptible to degradation, which requires immediate attention from planners. Furthermore, the study identifies the challenges related to the sustainability of the region. It is expected that this study will be helpful for decision-makers and policymakers to take suitable measures for the land sustainability of the region.

***Keywords*:** Land, susceptibility, degradation, multi-criterion, RS, GIS

**Corresponding author*: rkabhay@dsc.du.ac.in

Shruti Kanga, Gowhar Meraj, Suraj Kumar Singh, Majid Farooq and M. S. Nathawat (eds.) Geospatial Technology for Natural Resource Management, (393–410) © 2024 Scrivener Publishing LLC

16.1 Introduction

One of the most important resources is land, which makes life on Earth possible. It is the basis of all economic activities, as it provides raw materials for other economic activities. It is also used as a place of habitation by a majority of living beings, including humans. The United Nations Convention to Combat Desertification (UNCCD), in Article 1 of the Convention, defines land as a terrestrial bio-productive system that consists of soil, plants, and biota that drive ecological and hydrological processes. Land is even more important in rural areas because a large share of the population is engaged in agriculture. More than 54% of India's population works in agriculture and related industries (Bai *et al.*, 2009), contributing 19.9% to India's Gross Domestic Product (GDP) (Shagun *et al.*, 2021). Land quality has been decreasing in many parts of the world for many years (Basuki *et al.*, 2016). This is evident from the downward trends in the production of many crops over the years. The major reasons for this include the fragmentation of land, continuous sowing of the same crop, unwise use of chemical fertilizers and pesticides, low awareness among farmers, and lack of finance. This leads to an increase in the susceptibility of land to undergo a negative change which threatens food security (Barbier *et al.*, 2016) on one hand and on the other hand it also creates hurdles for the livelihood security of many people (Kawy *et al.*, 2019).

Susceptibility is a scale used to estimate the vulnerability of a land to degradation. Susceptibility is the degree to which land changes due to natural forces, man-made interventions, or an aggregate impact of both (Mashame *et al.*, 2016). In other words, land susceptibility refers to how easily it is degraded through the contributions of various variables. The higher the susceptibility, the more easily it would be degraded, and vice versa (Meshesha *et al.*, 2012). Degradation is a loss of quality and quantity. Long-term declines in ecosystem productivity and function are caused by land degradation. Land degradation is a global issue, although its severity varies across regions. Land degradation causes financial losses correlated with reduced land and soil productivity and has an inverse impact on the environment (Safriel *et al.*, 2007). Soil erosion and nutrient depletion are generally the major land deterioration factors, and utilization of land more than its capability also contributes to land degradation (Abhay *et al.*, 2015). Therefore, land susceptibility mapping, improvement of degraded land, and protection of susceptible land are necessary to fulfil the needs of the food, timber, and biomass energy fiber of the increasing population to refine economic output in developing countries.

The quality and characteristics of land resources are affected by agricultural practices, geology and geomorphology, slope of the area, rainfall, drainage system of the area, and deforestation. Periodic land use with the aid of remote sensing technologies and techniques, land cover change can be detected, and the factors impacting land susceptibility can be identified. It can also be assessed with scenario analysis, aerial photographs can be used for the present LULC mapping, and modeling can be performed using the RUSLE approach. Land degradation is positively correlated with susceptibility and negatively correlated with resilience; therefore, it is not only determined by susceptibility but also by the level of resilience (resilience is the capacity to restore its quality following a change towards degradation). Natural disasters (such as floods and climate change) and human activities (such as overgrazing, small- and large-scale mining, deforestation, farming on steep slopes, and clearance of vegetation) influence the ecosystem and degrade land on a large scale. The usefulness of the world's land area has dropped by 23%, and pollinator loss might threaten annual global crops worth $577 billion (UNCCD). Soil erosion accelerates land degradation and vice versa, and soil productivity and its environmental moderating capacity are also reduced in the long term due to soil degradation. Regions with significant terrain alteration, high relief, and slopes with significant intensity and duration of heavy precipitation have the highest soil erosion vulnerability.

Several studies have been conducted on land degradation mechanisms worldwide and in India. Only six attempts have been made over the last 30 years to assess the global trends of land degradation. Various investigations have been conducted to analyze seasonal LULC variations, extent and land deterioration factors, their impact, and to develop a model for land deterioration in semi-arid regions, including South America, Africa, and Asia. The reasons for land susceptibility to degradation, extension, and measures for its improvement have been studied in the watershed zones of Central Java Province and the river basins of India. GIS and remote sensing have also been utilized to evaluate the potential for land degradation (Lal *et al.*, 2001). Although few studies have been conducted on land susceptibility to degradation in Siddharthnagar district, even land is susceptible to degradation due to floods and improper and excessive use of land for agriculture.

The research region lies in the Himalayan foothills of Uttar Pradesh's Siddharthnagar District. It shares an international border with Nepal. The Siddharthnagar district's economy is based on agricultural activities. More than 93% of people live in rural areas and work in agricultural activities (DCHB, Census of India, 2011). Therefore, it is important to determine the reasons for susceptibility to land deterioration in the area to safeguard the livelihoods of local people. Thus, the quality of the land and agricultural production can increase. Planning and policies should be designed in such a way that they can minimize land degradation and its effect on agricultural productivity. The objectives of this study are twofold. The first was to assess land susceptibility to degradation using the multi-criterion method, and the second was to identify susceptibility challenges related to land degradation in the Siddharthnagar district.

16.2 Material and Methods

16.2.1 Geographical Profile of the Study Area

Siddharthnagar is one of the 75 districts in Uttar Pradesh and extends from 27°0' to 27°28' North latitude to 82°45' to 83°10' East longitude. The district covers a total size of 2,895 km^2. It shares an international border with Nepal in the north and shares a district boundary with Maharajganj and Gorakhpur in the east; Gonda, Basti, and Sant Kabir Nagar in the south; and Balrampur district in the west (Census of India, 2011). Administratively, the district is divided into five tehsils and 14 Community Development (CD) blocks (locally known as Vikas Khand) (Figure 16.1).

The Siddharthnagar district is in the foothills of the Himalayas known as the "Tarai region." The physiography of the district has been divided into three categories based on geology, soils, topography, climate, and natural vegetation: the Nangarh Tarai, Rapti flood plain, and Kadirabad-Betwa plain. The topography of the district is plain and the soil is fertile. The average elevation of the district is 290 ft. The slope of the district is gentle. It is in the south and east from the north and west. Many rivers flow in Siddharth Nagar such as Rapti (207 km), Budhi Rapti (88.5 km), Baan Ganga (29.50 km), Parasi, Kooda (50 km), Ghoongi (24 km), Burni, Kunhra, Siswa, and Jamwar (50 km) (Census of India, 2011; Patrika News Paper). The Rapti River originates in the Himalayan hills of Nepal. It flows west or northwest to the east. Rivers often follow a changeable course and cause floods in nearby areas during rainy season.

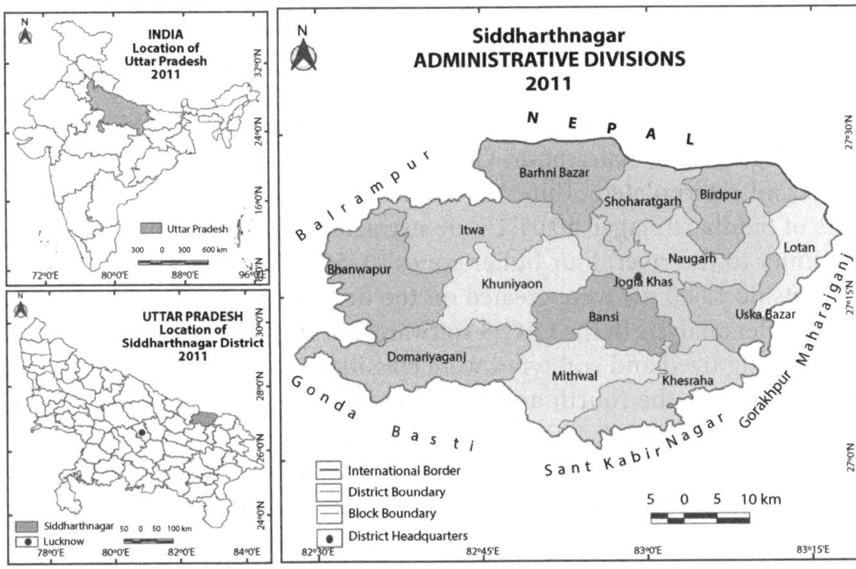

Figure 16.1 Location of the study area. Source: Prepared by the authors based on the Census of India, 2011.

The population of Siddharthnagar district is 25,59,297, of which 93.72% are rural and the rest are urban (Census of India, 2011). The population density in the study area was 884 persons per km^2 (Census of India, 2011). The economy of the district is based on agriculture as more than 90% of the population lives in rural areas. Agriculture is practiced during all three seasons: Rabi, Kharif, and Zaid.

16.2.2 Data Source and Research Methodology

16.2.2.1 Data Sources and Selected Indicators

The goal of this study was based on secondary data gathered from numerous sources. Data were collected from the Rainfall Statistics of India, India Metrological Department (2018), Agriculture Contingency Plan (ACP) Report, Siddharthnagar, Uttar-Pradesh, District Census Handbook (DCHB, Census of India (2011), United Nations Geological Survey (USGS, 2020), and Sankhikiya Patrika Internet-Based Data Entry and Retrieval System (SPIDER). To measure land susceptibility to degradation, seven indicators (five physical and two human activity-related indicators) were selected based on the literature review and the study area was retained. These seven indicators are rainfall, proximity to the river, soil erosion, drainage density (DD), stream frequency (FU), LULC, and cropping intensity (CI) (Figure 16.2).

To extract the rainfall pattern in Siddharthnagar district, the rainfall data for Siddharthnagar and six neighboring districts (Gorakhpur, Sant Kabir Nagar, Basti, Balrampur, Gonda, and Maharajganj) have been collected and interpolated. Thus, the spatial patterns of rainfall in different parts of Siddharthnagar district were extracted. For the second indicator, proximity to the river, four buffer zones (0–500, 500–1,000, 1,000–1,500, and outside 1,500 m) were created on the drainage map of the study area using buffer analysis in the Q-GIS software. For the third indicator, the soil map was digitized and soil types were classified into two groups based on soil texture. For the fourth and fifth, the indicator drainage system map was digitized, and the district was divided into 49 equal-sized grids to measure drainage density and stream frequency. The ratio of the length of a stream channel to its area is known as the drainage density (Eq. (16.1)). The ratio of the number of streams to the drainage basin area is known as stream frequency (Eq. (16.2)). The sixth indicator was based on land use prepared using LANDSAT 8 satellite imagery with a spatial resolution of 30 m. A super-wise classification scheme is applied to analyze land use in the Siddharthnagar district. The last indicator was cropping intensity,

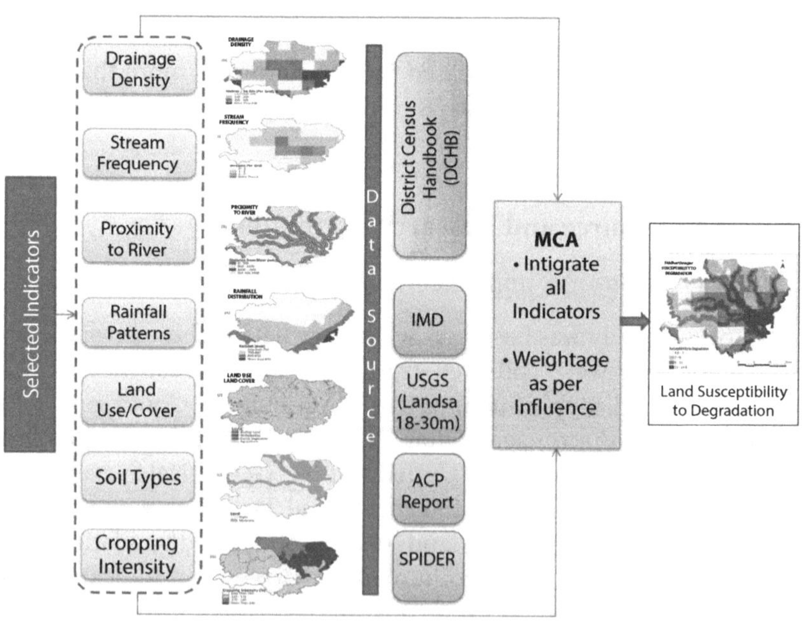

Figure 16.2 Data source and methodology. Source: Prepared by the authors.

which was categorized into four categories: low, moderate, high, and very high cropping intensity.

$$Dd = L / A \tag{16.1}$$

$$Fu = N / A \tag{16.2}$$

Where, Dd = Drainage density; Fu = Stream frequency; L = Length of stream channel; A is the area of the drainage basin, and N = Number of streams.

Finally, a Multi-Criterion Analysis (MCA) method was used to integrate all indicators. The layers of the selected indicators were intersected using Q-GIS software. A weightage was given to each indicator according to its influence on land susceptibility. Higher weightage is given to the indicator that intensely affects the land and vice versa (weightage is given as per the positive relation with susceptibility). A higher MCA value reflects higher susceptibility to land degradation. Finally, the study area was categorized into four levels of susceptibility based on MCA values: very high, high, moderate, and low land susceptibility to degradation.

16.3 Results and Discussion

All seven selected indicators affected the land susceptibility in different parts of the district at different levels. The indicator-wise analysis is explained below.

16.3.1 Rainfall Pattern

Rainfall is a crucial indicator that causes soil erosion and makes land susceptible to degradation. High intensity rainfall along steep slopes, soil permeability, and less vegetation create ideal conditions for soil erosion. However, gentle slopes and forests exhibit lower rates of erosion and land susceptibility. The district was divided into four regions based on the average rainfall. The first was the northern part, where the average rainfall was less than 750 mm (low rainfall). Below this region, moderate rainfall (750–800 mm) was recorded. High rainfall (800–850) was received in the southwest and southeast regions. Very high (more than 850 mm) mean rainfall was received in the southeast region of the district (Figure 16.3a). As mentioned, higher rainfall is the reason for higher susceptibility to land

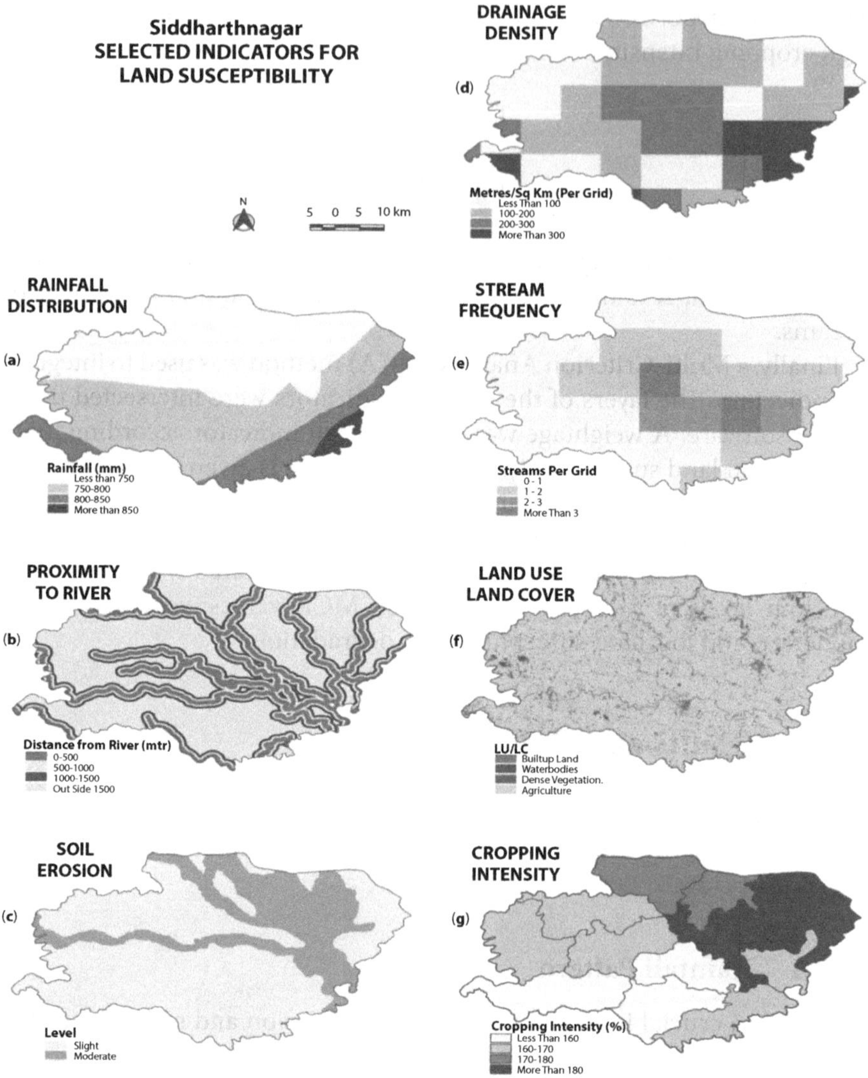

Figure 16.3 Land susceptibility indicators and level of susceptibility. Source: Prepared by the authors.

degradation. The percentages of areas with very high, high, moderate, and low rainfall susceptibility are 1.7, 12.4, 39.8, and 46.1, respectively (Table 16.1). Therefore, the rainfall susceptibility was determined accordingly. Extremely high, high, moderate, and low rainfall receiving locations have very high, high, moderate, and low rainfall vulnerability, respectively.

Table 16.1 Different indicators, their susceptibility and area under different susceptibility class.

Parameter	Range	Susceptibility	Area (km^2)	Percent
Rainfall (cm)	Less Than 750	Low	1,334.6	46.1
	750–800	Moderate	1,152.2	39.8
	800–850	High	359	12.4
	More Than 850	Very High	49.2	1.7
Buffer zone (m)	0–500	Very High	444.6	15.4
	1,000	High	398.3	13.7
	1,500	Moderate	352.2	12.2
	1,500 and above	Low	1,699.9	58.7
Soil	Slight	Slight	2,043.9	70.6
	Moderate	Moderate	851.1	29.4
Length of Rivers/Grid (m)	Less Than 100	Low	888.8	30.7
	100–200	Moderate	1,166.7	40.3
	200–300	High	558.7	19.3
	More Than 300	Very High	280.8	9.7
No. of Streams/Grid	0–1	Low	1572	54.3
	1–2	Moderate	914.8	31.6
	2–3	High	243.2	8.4
	More Than 3 (Maximum)	Very High	165	5.7
LU/LC	Agricultural Land	Very High	2,567.9	88.7
	Dense Vegetation	High	63.7	2.2
	Water bodies	Moderate	55	1.9
	Build-up Land	Low	208.4	7.2
Cropping Intensity	Less Than 160	Low	639.8	22.1
	160–170	Moderate	399.5	13.8
	170–180	High	1,126.2	38.9
	More Than 180	Very High	729.5	25.2

16.3.2 Proximity to River

Proximity to the river is an important variable that increases susceptibility to degradation. Close proximity to a river makes an area highly susceptible to degradation. It has been observed that fast-moving water along the river erodes soil at the bank. Furthermore, the flood also damages standing crops, which leads to huge economic losses to farmers. To analyze this, four buffer zones were created (0–500, 500–1,000, 1,000–1,500, and outside 1,500 m) on both sides of the rivers in Siddharthnagar. Therefore, areas near (0–500 m) are highly susceptible to land degradation. The proximity to the river susceptibility decreased away from the river (Figure 16.3b). Thus, the susceptibility is low outside the 1,500 m buffer zone. The area of the district in each proximity to the river susceptibility class is listed in Table 16.1.

16.3.3 Soil Erosion

The intensity of soil erosion had a significant impact on the rate of land degradation. The extent of soil erosion greatly depends on the soil texture. Because water infiltration is mostly dependent on soil texture, this factor has the greatest impact on soil erosion. Infiltration is often low in fine-textured materials (such as clay), which favors runoff and erosional processes. Fundamental soil characteristics, such as texture, structure, soil organic matter concentration, clay minerals, exchangeable cations, and water retention and transmission capabilities, all affect the susceptibility of soils to erosive agents. Soil erosion increases soil degradation. Drought conditions, either natural or man-made, contribute to a reduction in vegetation cover. Additionally, poor agricultural methods cause the loss of soil organic matter, and inadequate irrigation techniques cause salinization. All these factors contribute to an increase in soil erosivity and, ultimately, desertification. Therefore, erosion appears to be the most significant indicator of processes leading to land degradation and desertification. The soil of the district has been categorized into two classes based on erodibility factors: slightly eroded and moderately eroded (Table 16.1). Moderately eroded soils were found on the riverbanks. Low erodible soil was found away from the riverbanks (Figure 16.3c). Soil erosion susceptibility was moderate in moderately soil-eroded regions of the district, and vice versa.

16.3.4 Drainage Density

One of the most significant causes of land deterioration is drainage density. This is the ratio of the aggregate length of the drainage to the overall drainage area. This affected the features of erosional development. It displays the region's meteorological conditions, vegetation cover, runoff potential, and infiltration capacity of the land. The drainage density is affected by the geology, climate, and characteristics of the drainage area. It was assumed that a higher drainage density in an area makes it more susceptible to degradation. Statistically, it can be expressed as follows: Drainage density in the district is measured for which the entire study area is divided into 49 grids. Subsequently, the average drainage length was measured. A very high (>300 m/km^2) drainage density was observed, especially in the southeast region (Figure 16.3d). Most rivers converge due to the northwest to southeast slopes. A high density (200–300 m/km^2) is mostly found in the central region of the district. It covers 19.7% of the district in the high drainage density susceptibility class (Table 16.1). Moderate drainage density (100–200 m/km^2) is found in the north, west, east, and a small area towards the south. The drainage density was low in grids where the length/grid was less than 100 m. Areas with low drainage densities were scattered throughout the study area in patches. In areas with high drainage density, there is a higher risk of land degradation and vice versa.

16.3.5 Stream Frequency

The total number of streams per square meter is known as stream frequency. High-stream frequency locations offer a significant potential for erosion. The ratio of the total number of streams to the area of the drainage basin was used to compute stream frequency.

The area was divided into 49 grids and classified into four categories based on the stream frequency. When there are more than three streams, the highest stream frequency is obtained. Only two grids, one in the district's center and the other in its northeastern portions, had extremely high stream frequencies (Figure 16.3e). High stream frequency ranges were located near very high stream frequency areas. Moderate stream frequency areas are located between very high, high, and low stream frequency regions, covering 31.6% of the geographical area (Table 16.1). The stream frequency is low in grids where the number of streams is either zero or one. All border regions have low stream frequencies, except the southeast.

16.3.6 Land Use/Land Cover

The land is covered by natural and man-made structures. In the past decades, LULC has rapidly changed with increasing population, extensification of cultivation, urbanization, and industrialization. Such LULC changes have accelerated desertification and land degradation, increasing the risk of mass failures, flooding, erosion, reservoir siltation, and other related problems. LULC categories also play a significant role in infiltration, evapotranspiration, runoff, sediment generation, and stability of slopes. Land use intensification leads to land degradation.

Siddharthnagar is divided into four major LULC categories: agriculture, built-up land, dense vegetation, and water bodies. Land susceptibility is very high for agricultural land because it is the most utilized region because the district's economy is dependent on agriculture. The use of excess fertilizers, over-irrigation, and high cropping intensity make it highly susceptible to degradation. Dense vegetation is also available, including forests and orchards. Vegetation is an important part of land use. Different vegetation types were observed in the study area. It is found in patches. Their role in controlling soil erosion and checking for land degradation is important. However, it is highly susceptible to the use of chemical fertilizers in orchards and expanding agricultural land in dense vegetation areas. Some dense vegetation patches were visible in the northeastern region. Waterbodies are a major part of the land cover. Water bodies in the study were found in the form of rivers, ponds, johads, lakes, and canals (Figure 16.3f). The rivers flow from the west, north, and northeast towards the southeast. Water bodies are moderately susceptible to erosion due to the intensity and volume of water, which makes land eroded and marshy. The built-up area, which includes settlements and other constructed structures, is distributed all over the district but is dense in Naugarh (district headquarters), the urban region of Itwa, Domariyaganj, and Mithwal. There is low susceptibility because the area is covered with concrete, and no direct activity is linked to the land there.

16.3.7 Cropping Intensity

The number of times a field is used throughout a cropping year is referred to as cropping intensity. It also serves as a gauge of land efficiency, which is the proportion of the sown net area that is planted or harvested. This can also be described as the amount of cultivated land planted with crops. Small subsistence and tenant farms tend to have more intensive cropping, which necessitates a greater quantity of irrigation resources. Land-use intensity, crop intensity, and input intensity are the three elements that contribute to land degradation. Cropping intensity varies according to the size of landholdings. Large landowners frequently over-irrigate cash crops, which contributes to salinity and water logging, which degrades the land. Similarly, smallholders and tenant farmers extensively use their tiny holdings to pay for self-subsistence, cost of production, and land rent, which also contributes to soil degradation.

Based on cropping intensity, the study area was grouped into four categories. Susceptibility to cropping intensity was positively correlated with cropping intensity (Table 16.1). Birdpur, Lotan, Naugarh, and Jogia Khas blocks have very high cropping intensities. It is high in only two blocks, Barhni Bazar and Shoharatgarh. These were located along the border of Nepal (Figure 16.3g). Moderate cropping intensity is found in five blocks of the district (Bhanwapur, Itwa, and Khuniyaon in the west and Uska Bazar and Khesraha in the southeast). It is low in the remaining three blocks, Domariyaganj, Mithwal, and Bansi, which are in the southwest, south, and central regions of the district in a series, respectively.

16.4 Spatial Disparities in Land Susceptibility to Degradation

In the Siddharthnagar district, multi-criteria analysis techniques have been applied to integrate all variables. Weight was given based on the level of susceptibility. The higher the susceptibility, the greater is the weight. Thus, the overall aggregate weights were divided by the number of variables to obtain the final MCA values. The results showed that MCA values range from 4.86 to 14.86. Based on this, the study area was divided into four categories of land susceptibility to degradation. This is discussed in detail below.

Very high land susceptibility values, higher than 10.99. A major part of this category is in the southerneastern parts of the district and along rivers. It covers 463.09 km^2 (16% of the district) (Table 16.2). Most areas of the Uska Bazar and Jogia Khas blocks are highly susceptible to land degradation. The reasons for very high land susceptibility are attributed to very high drainage density, stream frequency, high to very high rainfall, availability of moderately erodible soil, proximity to rivers, and dominance of agricultural land use. Regions of the study area that have a susceptibility index value between 9 and 10.99 are included in the highly susceptible category. 985.44 km^2 (34.04%) was considered highly susceptible to degradation. Highly susceptible areas were located next to the highly susceptible regions (Figure 16.4). The value for moderately susceptible regions was between 7

Table 16.2 Land susceptibility to degradation in Siddharthnagar District.

S. no.	Susceptibility values	Area (km²)	Percent	Land susceptibility
1	Less Than 7	440.99	15.23	Low
2	7.00 to 8.99	1,005.48	34.73	Moderate
3	9.00–10.99	985.44	34.04	High
4	More Than 10.99	463.09	16.00	Very High

Source: Calculated by the authors.

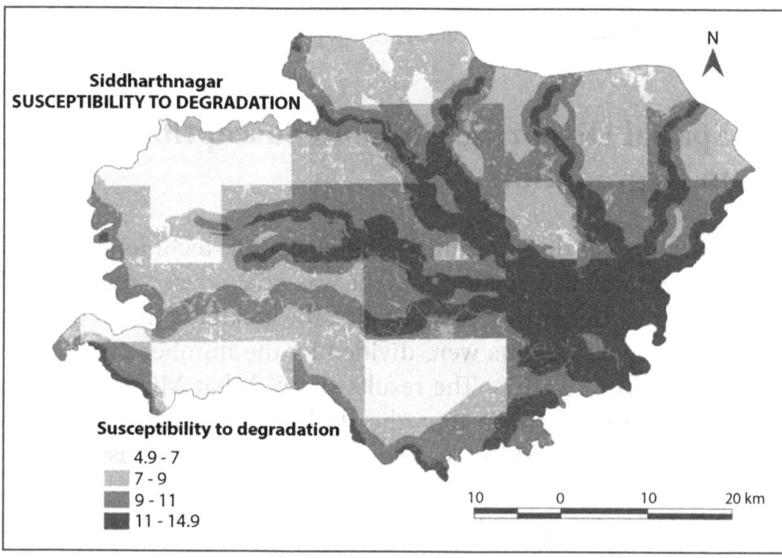

Figure 16.4 Land susceptibility to degradation in Siddharthnagar District. Source: Prepared by the authors.

and 8.99. It includes 1,005.48 km² (34.73%) of Siddharthnagar. Moderately susceptible areas were in the northwestern, southern, and southwestern parts of the district. Study areas with have susceptibility values less than 7 indicated low susceptibility. Approximately 440.99 km² (15.23%) of the district was associated with this category. The major low-susceptibility regions are in the northwestern, southwestern, and southern parts of the district. Some small patches of low susceptibility were located throughout the district. These regions of the study area are less susceptible to degradation due to low cropping intensity, low rainfall, low drainage density, and stream frequency, and the soil available in these regions is slightly erodible.

16.4.1 Land Degradation and Sustainability Challenges

Land degradation has become a significant global issue over the last few decades and is a major threat to food security and sustainability. A variety of natural and human activities, including deforestation, overgrazing cattle, mismanagement of agricultural land, industrial development, soil erosion, and changing weather patterns, have been blamed. Demand for food, settlement, wood, and water has increased with the rapidly increasing population. A large proportion of the population lives and works in rural areas in many developing nations. Agriculture dominates the economy in rural areas, and land is a major source of income for residents.

The implications of desertification and land degradation on human livelihoods, the environment, global economic growth, and migration are becoming increasingly obvious. Farming and rural communities are the worst affected by loss of land productivity. As land degrades, production costs increase. Even large-scale commercial farmers may discover that farming is unfeasible due to increased costs. Small-scale subsistence farmers are often unable to cover these expenditures. Farm workers and others are compelled to relocate to urban areas as a result, only to experience unemployment and poverty. Thus, land degradation has multiple and multiplier effects on agricultural production and sustainable development.

The 2030 agenda for Sustainable Development's strategic framework is the SDGs process, which includes its targets and indicators. Land degradation is an obstacle to achieving many SDGs either directly or indirectly. It is directly related to eradicating hunger and achieving sustainable agriculture (SDG 2). Good health and well-being due to reduced nutrition in agricultural production (SDG 3); safeguarding and restoring water-related ecosystems, such as hills, woodlands, wetlands, streams, reservoirs, and lakes (SDG 6); no poverty, decent work, and economic growth as the economy of rural population is purely dependent on land (SDG 1 and SDG 8); encouraging the sustainable use of land and safeguarding terrestrial ecosystems (SDG 15); and preventing and mitigating the effects of climate change (SDG 13) are SDGs that will also be difficult to achieve if land degradation is not controlled. The emphasis is on establishing tools and improving nations' capacities to collect regular data and focusing on analysis using suitable techniques in the context of SDG 15 (Life on Land), and especially target 15.3 "Protect and restore terrestrial ecosystems and promote sustainable usage". Therefore, it is very important to take measures to minimize land degradation and proceed with developmental activities in a sustainable manner.

World leaders vowed to work toward eradicating land degradation in the context of sustainable development at the Rio+20 summit. Land Degradation Neutrality (LDN) has now been adopted to define a state in which maintaining the amount and quality of land resources is given more focus to ensure and stabilize ecosystem functions and food security over time. Collection of data to monitor land conditions is important; however, it requires timely availability of data. Despite the uncertainty, there is consensus that there will be a huge cost of land degradation if not stopped. The global economy, particularly the economies of less-developed nations, will suffer significantly as a result. Thus, it is crucial for humans to rationalize their economic activities while keeping in mind their future requirements and sustainability paths.

16.5 Conclusion

Land degradation is one of the most important local, regional, national, and global environmental issues. The purpose of this study was to determine the spatial patterns of land susceptibility to degradation in Siddharthnagar District. In this study, seven indicators were used to analyze the level and spatial patterns of land susceptibility to degradation. These are the intensity of rainfall, proximity to the river, soil erosion, drainage density, stream frequency, land use land cover and cropping intensity. It was found that 16%, 34.04%, 34.73%, and 15.23% of the area of the district had very high, high, moderate, and low susceptibility to degradation, respectively as per the result of the land susceptibility to degradation index. It is concluded that drainage density, stream frequency, water bodies, and proximity to river indicators are most important factor affecting land degradation. Other selected indicators also impact land degradation, but they are less effective in plain regions. The MCA method used in this study was suitable and easy to calculate using Q-GIS software. The results obtained using this methodology were satisfactory. The results are feasible for crosschecking through primary field observations. The secondary data used in this chapter are easily available from various sources.

The government, other bodies, and decision makers will benefit from this study's findings in developing their plans and policies. Farmers in the study region can make better decisions regarding their agricultural and allied activities. The results of this study will be helpful for further studies on agricultural, land, and rural development-related research.

The following steps can be taken to manage and stop land deterioration: lowering soil erosion and raising soil organic matter levels; promoting nutrient cycles and balance in the soil; preventing, minimizing, and mitigating soil salinization and alkalinization; preventing and minimizing soil contamination; safeguarding and promoting soil biodiversity; improving soil water management; reducing soil sealing; and preventing and mitigating of soil compaction.

Acknowledgments

The authors are grateful to the authorities of the Indian Council of Social Science Research (ICSSR), New Delhi, and MHRD (IMPRESS Scheme) for their financial support (F. No. IMPRESS/P2171/644/SC/2018-19/ICSSR), and the head of the institution, Dyal Singh College, for providing the necessary facilities. Additionally, the authors are grateful to the reviewers for their insightful feedback on the paper.

References

Abhay, R. K. and Patra, P. (2015). Land susceptibility to degradation in the Kendujhar plateau, Odisha. In Sharma, V. R. (ed), *Contemporary environmental issues.* (pp. 3-25). New Delhi: Research India Press.

Agriculture Contingency Plan. (undated). Siddharthnagar district, Uttar Pradesh.

Annual Report (2017-18). Department of Agriculture, Cooperation & Farmers Welfare, Ministry of Agriculture & Farmers Welfare, Government of India.

Bai, Z., & Dent, D. (2009). Recent land degradation and improvement in China. *Ambio*, 150-156.

Barbier, E. B., & Hochard, J. P. (2016). Does land degradation increase poverty in developing countries?. *PLoS One*, 11(5), e0152973.

Basuki, T. M., Wijaya, W. W., & Wahyuningrum, N. (2016). Spatial distribution of land susceptibility to degradation and recommendation for its improvement: a case study in the upper Solo Sub-Watershed. *J. Degraded Min. Lands Manage.*, 4(1), 689-696.

District Census Hand Book (DCHB). (2011). Siddharthnagar district. Census of India.

Eureta, R., (2022). Land Degradation. In *Enviropaedia Rethinking Reality*, accessed through http://www.enviropaedia.com/topic/default.php?topic_id=147.

Kawy, W. A. M., & Darwish, K. M. (2019). Assessment of land degradation and implications on agricultural land in Qalyubia Governorate, Egypt. *Bull. Natl. Res. Cent.*, 43(1), 1-14.

Lal, R. (2001). Soil degradation by erosion. *Land Degrad. Dev.*, 12 (6), 519-539.

Mahala, A. (2020). Land degradation processes of Silabati river basin, West Bengal, India: A physical perspective. In *Gully erosion studies from India and surrounding regions* (pp. 265-278). Springer, Cham.

Mashame, G., & Akinyemi, F. (2016). Towards a remote sensing based assessment of land susceptibility to degradation: Examining seasonal variation in land use-land cover for modelling land degradation in a semi-arid context. *ISPRS Ann. Photogramm. Remote Sens. Spatial Inf. Sci.*, 3, 137.

Meadows, M. E., & Hoffman, M. T. (2002). The nature, extent and causes of land degradation in South Africa: legacy of the past, lessons for the future?. *Area*, 34(4), 428-437.

Meshesha, D. T., Tsunekawa, A., & Tsubo, M. (2012). Continuing land degradation: cause–effect in Ethiopia's Central Rift Valley. *Land Degrad. Dev.*, 23(2), 130-143.

Safriel, Uriel N. (2007). The assessment of global trends in land degradation. *Climate and land degradation*. Springer, Berlin, Heidelberg, 1-38.

17

Examining Socio-Economic Realities and Challenges in West Bengal's Labor Force: A Case Study

Arindam Mondal[1*], Naina Sambher[2] and D.D. Sharma[2]

[1]*Department of Geography, Visva-Bharati University, West Bengal, India*
[2]*Department of Geography, Himachal Pradesh University, Himachal Pradesh, India*

Abstract

The working-class population, that is, the labor force, is of great significance for the progressive growth of the economy. There are various factors that determine the productivity of the labor force, among which socioeconomic factors dominate. For this purpose, this study highlights workers' economic and social conditions, the problems they face in their respective sectors, and the level of support they receive from the concerned authorities. Based on secondary data sources, a correlation matrix was used for the various indicators in West Bengal. The study found a strong linear correlation between urban and rural populations. This is primarily because West Bengal is one of the most prominent states in the country, from an agricultural perspective. The correlation coefficient between the state's total population and slum population is moderate, indicating the migration of individuals from rural to urban areas in search of employment opportunities. The ratio of the total population to workers in the domestic industry was below the average. As the proportion of daily wage-based workers continued to decline, there was a negative correlation between these two variables. This factor-based analysis demonstrates that the industrial sector has experienced a significant reduction in its share in recent years, whereas the agricultural sector has played a crucial role in maintaining overall development.

Keywords: Labor force, primary sector, secondary sector, industry

**Corresponding author*: arindammondalari95@gmail.com

Shruti Kanga, Gowhar Meraj, Suraj Kumar Singh, Majid Farooq and M. S. Nathawat (eds.) Geospatial Technology for Natural Resource Management, (411–424) © 2024 Scrivener Publishing LLC

17.1 Introduction

Human activities that lead to the flow of production, opportunities, income, and wealth are known as economic activities. Economic activities are classified into three major groups: Primary, Secondary, and Tertiary activities. Primary activities include agriculture, mining, fishing, forestry, pasturing and manufacturing industries, construction, transport, communication, and trade, which are secondary activities. Manufacturing has emerged as a high-growth sector in India. In the next 12 years, there will be 175 million additions to the country's labor force. Over the last 60 years of Indian independence, the primary sector's composition in gross domestic product (GDP) has been replaced by the secondary and service sectors. From a contribution of 40% of the GDP in 1950–51, the secondary and service sectors (2006–07) account for 80% of the country's GDP. This sectoral shift has gained impetus since 1991 with the introduction of new economic and industrial policies, termed "liberalization" (Ghosh, 2011). During the process of growth from 1950–51 to 1999–2000, the Indian economy experienced a change in the production structure with a shift away from agriculture towards industry and the tertiary sector. The share of the industry has increased from 16% to 27.12% during the same period (Joshi, 2004). West Bengal witnessed the highest growth in non-agricultural employment from 2004 to 2010 among all the states in India. The state also witnessed the highest growth in manufacturing employment between 2004–05 and 2011–12 (Chowdhury and Chakraborty, 2016). A.G.B Fisher said 'In every progressive economy there has been a steady shift of employment and investment from the essential primary activities to secondary activities of all kinds and a still greater extent into tertiary production. It is true that in the case of West Bengal, industrialisation and urbanisation are imparting new values to rural areas of West Bengal' (Das, 2013). India is predicted to be the world's second largest country by 2050. The Indian government has always placed a high weight on industrial development. The secondary sector, which contributes one-fourth of the Indian Gross Domestic Product depends mainly on the primary sector for the need for raw materials to be used in the production process. Available natural raw materials are used to create goods and services for consumption. To boost industries in the economy to achieve self-sufficiency in production, which guards the economy of the country against foreign competition, various governmental plans, policies, and strategies have been made (Sharma *et al.*, 2021). This sector can easily be attributed to the backbone of the economy, with major developing and rising companies in the secondary sector, such as transportation, construction, and manufacturing.

17.2 Review of Literature

To gain insight into the studies conducted on socio-economic development and its impact on workers, several studies have been reviewed. A few selected studies are discussed below.

Tiwary *et al.* (2012) in their work focused on industrialization with the introduction of newly developed technologies, which plays an important role in the development of a country. Construction is a rapidly growing industrial sector in India. It has a wide range of activities with the employment of a reasonable number of workers, who are mainly working in the informal sector. In India, approximately 340 million (92%) workers are in the unorganized sector, and approximately half of them are in the construction industry. The government of India and State governments presented acts and rules for the regulation of working conditions and workers in the industry. Due to the uncertainty of their work days and wages, they face regular socioeconomic challenges. Karak (2017) conducted a study on a secular relative decline in terms of employment and value-added in manufacturing industries compared to other states, an ancillarization and flexibilization of production into small-scale factories with very less workers, and a differential impact of this ancillarization on basic goods and consumer goods industries—these three trends were very much visible in West Bengal in the time of 1980–91. During the period of 1980s high agricultural growth, mass poverty among the rural poor and the poor growth of consumer goods industries all these issues had reasons like 'The Green Revolution's impact on West Bengal's agriculture, the nature and effect of the Left Front's land reforms, and the role of rural commercial capital. According to a study conducted by Chattopadhyay (2004), West Bengal lost its earlier status as one of the highly industrialized states of the country. Its share of all-India net value-added, the share of employment, and the number of factories has drastically decreased. Overall, the profitability of the manufacturing sector has also decreased. The productivity of capital in manufacturing has also declined, whereas labor productivity has increased. The Total Factor Productivity (TFP) of the manufacturing sector in the state has been declining, whereas it has been increasing in India. The TFP of the six industry groups that played a dominant role during the early 1960s decreased, except for the jute industry, which itself is a dying industry because of a lack of demand. The proportion of the population dependent on agriculture in the state is 65%, which is well above the national average of 56%. In short, industrial development in a State such as West Bengal under the rule of a political combine that is qualitatively different from the usual political outfits of a market economy democracy, especially one with a well-articulated and globally known political and policy position different in all the major aspects from that of liberalizers, has to show its character as a crucial a matter as the path and pattern of industrial development. This must be different in several critical respects, such as agency, approach, content, outcome, and spread effects. It has also to be such that in a reckonable, foreseeable period the new industrial growth approach can make a visible positive structural impact on the shape and size of the economy of West Bengal (Kabra, 2007). While there was a clear policy shift towards large-scale industrialization in the state of West Bengal during the early 1990s, not much improvement can be discerned in the performance of manufacturing output. A steady imbalance in labor management, economic viability, production, and administrative management is leading West Bengal towards a deteriorated organized manufacturing sector. The industrial policy of 1994–95 made West Bengal a worthy destination for industrial investment by improving infrastructure and simplifying regulatory processes. The functioning of the manufacturing sector in an organized manner has not been affected by moderately developed manpower. Until very recently, small entrepreneurs in the unorganized sector typically produced more than half of the state's manufacturing output. Physical infrastructure in the state was still desired, particularly in the domains of power, water, and ports. A multi-layered power structure exists in the state, where each local unit of the party wields enormous power over the local people through illegal processes and clientelism promotion (Chakravarty and Bose, 2010). West Bengal witnessed the highest growth in non-agricultural employment between 2004–05 and 2009–10 among all the states in India. The state also witnessed the highest growth in manufacturing employment between 2004–05 and 2011–12. Employment in West Bengal is generally characterized by massive informalization of labor, stagnant real wages, and minimal social security. While there has been an increase in employment in the manufacturing sector, its share in the West Bengal GSDP has been declining. Even within the top manufacturing sectors, employment, informalization, and the lack of social security are widespread. Moreover, a significant proportion of workers in the manufacturing sector earn wages lower than the national floor-level minimum wage. Since labor productivity in West Bengal is less than that in India, the employment-generating potential of West Bengal is greater than India for any level of the GDP growth rate. However, such low-productivity employment is neither remunerative nor rewarding in terms of health and mental satisfaction. The employment growth that has taken place in West Bengal, while being a better situation than most other states, is not a model that can be replicated across the country (Chowdhury and Chakraborty, 2016). India holds the second position in tea production all over the world after China, and West Bengal is one of the major states producing tea, which is a secondary sector product. The tea industry plays a vital role in India's socioeconomic conditions. Approximately 70% of workers in the tea industry are women, who represent a highly skilled labor force, trained by previous generations. Despite the introduction of plucking machines and pruning knives in the tea industry, their skill sets remain unmatched. The rights of female tea garden workers are still neglected, and they are facing several problems in their daily lives (Sarkar *et al.* 2016). Another important industry this state is the silk industry. As the Industrial Revolution in the West began, the silk industry in West Bengal started to decline rapidly. In the post-independence period, attempts were made to revive the home-based artisanal industry in India to expand employment and income opportunities in the rural economy. However, during 2002–2012, there was an exodus of more than 23,000 silk farmers and 10,000 silk weavers from the industry. With certain governmental help, only then is it expected to have a higher level of income from this unorganized silk-producing sector and possibly hold back the exodus of silk artisans in the districts of West Bengal (Roy & Dey, 2017). According to the data, West Bengal received investment proposals of Rs 1.12 trillion, with the potential to create 314,000 jobs in May 2011. However, 2012 was particularly rugged on the industrial front of West Bengal (Acharya, 2013).

17.3 Aims and Objectives

The aims and objectives of this study are as follows:

- To determine the economic and social conditions of workers.
- To analyze problems in the working sectors faced by workers.

17.4 Study Area

West Bengal is a state of India located in the eastern part of the country. The number of districts increased from 19 (2001 census) to 23 (Figure 17.1). It has a population of approximately 91 million, which is the

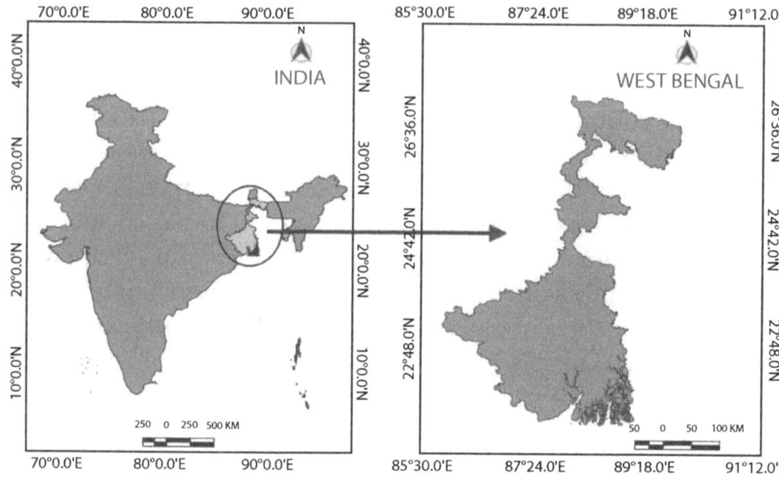

Figure 17.1 Location map. Source: Prepared by the authors.

fourth largest state-wise population in India. It covers an area of approximately 34,000 sq. mi. The location of the state ranges from 27°13'15" to 21°25'24" North latitudes and from 85°48'20" to 89°53'04" East longitude. Bangladesh borders it in the eastern part, Nepal and Bhutan borders are present in the northern section, Odisha, Jharkhand, and Bihar are present in the west, and the Bay of Bengal is in the south. Kolkata, the state capital, is the third-largest metropolitan city in the country. Other major cities are Durgapur, Asansol, Siliguri, Bardhhaman, Kharagpur, Krishnagar, and Haldia. According to 2015 data, West Bengal ranks sixth in the GSDP list of the country. Recently, a steady decline has been observed in the agricultural sector of West Bengal. The industrial sectors of the state are present in different parts of the state, such as Kolkata, Haldia, Durgapur-Asansol, and Darjeeling. Engineering products, electronics, steel, leather, textiles, automobiles, railway coaches and wagons, frigates, chemicals, and fertilizers, jute, and tea are major manufacturing industries in the state.

17.5 Data Source and Methodology

This study was based on secondary data collected from various government departments. Secondary data were collected from the District Census Handbook, Census of India 2011, Statistical Handbook 2013, and Database on Labour and Employment Statistics of West Bengal 2014-15. The data were portrayed using a correlation matrix of various socioeconomic indicators in all the districts of West Bengal, such as male literacy, female literacy, male workers, female workers, marginal workers, main workers, daily workers, and non-workers.

17.6 Analysis and Discussions

The correlation of various factors is analyzed and interpreted as follows:

1. High Correlation Matrix

There was a strong linear correlation between the total population and rural populations (Table 17.1). The main reason for this is that West Bengal is one of the predominant states in the country, from an agricultural perspective. Agriculture is the principal occupation of West Bengal. It leads the country in rice, potato, and jute production, and has significantly contributed to tea, sugarcane, and oilseed production. All crops are labor intensive and require timely maintenance. Most workers are engaged in the primary sector, whereas the others are daily workers working in nearby towns.

Approximately one-third of the state's population is urban in nature. The urban population has increased by four points over the past decade (2001–2011) (Census, 2011). Kolkata is the state capital and third-most populous city of the country, constantly attracting people around the country from the beginning because it is an economic, political, and cultural hub. People residing in urban areas engage in secondary activities and part-time work. However, compared to these large towns and cities, small towns with populations below 20,000 have shown higher growth rates. Based on the physiographical division of the state, nearly half of the urban population belongs to the southern plains region.

2. Moderate Correlation Matrix

The correlation coefficient between the total population and slum population of the state is moderate, which indicates the migration of people from rural to urban areas for opportunities (Table 17.1). West Bengal ranks third among all the states in terms of the total slum population as per the Census of India, 2011, while it ranks fifth in terms of slum percentage. The percentage of the slum population increased from 5.13% to 7.03%. Economic pull factors such as more job opportunities, desire for higher living standards, and diversification in work opportunities have led people to shift to urban areas, but poor urban governance, overpopulation, the affordability crisis, and social factors in urban areas have led to the growth of the slum population.

Table 17.1 Correlation matrix of socioeconomic indicators of West Bengal.

	Total population	Rural population	Urban population	Slum population	Main male workers	Main female workers	Marginal male workers	Marginal female workers	Total male workers	Total female workers	Household industry workers	Male literacy	Female literacy	Average literacy	Daily workers (2015)	Non-workers
Total Population	1	.803	.701	.460	.676	.679	.421	.384	.687	.664	.398	.079	.024	.050	−.076	.675
Rural Population	.803	1	.268	.210	.354	.409	.564	.644	.440	.685	.332	−.109	−.237	−.189	−.195	.425
Urban Population	.701	.268	1	.837	.675	.687	.048	−.137	.587	.310	.171	.348	.439	.415	.045	.556
Slum Population	.460	.210	.837	1	.643	.678	.114	−.144	.569	.314	.049	.367	.552	.490	.074	.537
Main Male workers	.676	.354	.675	.643	1	.899	.478	.233	.974	.672	.534	.208	.408	.333	.537	.968
Main Female Workers	.679	.409	.687	.678	.899	1	.386	.265	.869	.753	.523	.085	.301	.215	.281	.867
Marginal Male Workers	.421	.564	.048	.114	.478	.386	1	.845	.663	.795	.385	.237	.104	.215	.626	.646
Marginal Female Workers	.384	.644	−.137	−.114	.233	.265	.845	1	.418	.834	.270	−.008	−.229	−.138	.307	.395

(*Continued*)

Table 17.1 Correlation matrix of socioeconomic indicators of West Bengal. (*Continued*)

	Total population	Rural population	Urban population	Slum population	Main male workers	Main female workers	Marginal male workers	Marginal female workers	Total male workers	Total female workers	Household industry workers	Male literacy	Female literacy	Average literacy	Daily workers (2015)	Non-workers
Total Male Workers	.687	.440	.587	.569	.974	.869	.663	.418	1	.780	.556	.233	.367	.321	.616	.992
Total Female Workers	.664	.685	.310	.314	.672	.753	.795	.834	.780	1	.483	.041	.007	–.105	.353	.764
Household Industry Workers	.398	.332	.171	.049	.534	.523	.385	.270	.556	.483	1	–.231	.001	–.105	.330	.585
Male Literacy	.079	–.109	.348	.367	.208	.085	.237	–.008	.233	.041	–.231	1	.849	.951	.275	.159
Female Literacy	.024	–.237	.439	.552	.408	.301	.104	–.229	.367	.007	.001	.849	1	.971	.248	.311
Average Literacy	.050	–.189	.415	.490	.333	.215	.215	–.138	.321	–.105	–.105	.951	.971	1	.275	.254
Daily Workers (2015)	–.076	–.195	.045	.074	.537	.281	.626	.307	.616	.353	.330	.275	.248	.275	1	.627
Non-Workers	.675	.425	.556	.537	.968	.867	.646	.395	.992	.764	.585	.159	.311	.254	.627	1

Source: Calculated by the authors based on District Census Handbook data 2011, Statistical Abstract, 2015 &Database on labour and employment statistics of West Bengal-2014–15.

Here, there was a moderate to good correlation. The proportion of female main workers was alarmingly low compared to the male percentage. The percentage ratio of males to females (84.4% and 15.6%, respectively) was lower than the national ratio. As West Bengal is one of the dominant agricultural product-yielding states, most of its workforce (Table 17.2) is engaged in agricultural activities. Gender inequality in the state labor market is a pervasive problem, especially in rural areas. Women's labor force participation is affected by several issues, such as proper access to decent work, working conditions, wage differentials, domestic responsibilities, workplace discrimination, and the availability of various other benefits such as social security and good working conditions.

Workers who might have worked for a short period of the whole year were considered marginal workers. Below is the below-average relationship between both male and female marginal workers and total population size. Among all the districts, Dakshin Dinajpur and Darjeeling had the lowest, while South 24 Pargana and Paschim Medinipur had the highest number of male and female marginal workers, respectively, where males occupied more than half of the total number and females occupied more than one-third of the total number (Economics and Statistical Abstract, 2011). However, there has been a significant change observed in the 2001 and 2011 census data of the male–female marginal population, which shows more than a one-eighth increase in male workers, while female workers have decreased by the same points as an increase in males. However, the number of total marginal workers has increased by approximately 1.8 million.

Table 17.2 Main workers engaged in various fields in the study area.

	Main workers in West Bengal		
Types of fields	**Total**	**Male**	**Female**
Cultivators	4,203,767	3,940,399	263,368
Agricultural Labours	5,869,498	4,943,086	926,412
Household Industry Workers	1,518,128	869,039	649,089
Other Workers	14,095,237	11,925,755	2,169,482
Total	**25,686,630**	**21,678,279**	**4,008,351**

Source: Census of India, 2011.

3. Low Correlation Matrix

The relationship between the total population and household industry workers was below the average. Household industries are mostly run by a single family or together within a community, and a negligible amount of capital investment takes place. With the rapid growth in urbanization and industrialization in India, these family based or community-based industries are declining at a disturbing rate, which has led to a decrease in the number of household industry workers in the present day. The decline in these industries is relatively more obvious in the rural sector than in the urban sector, as a large section belongs to rural areas.

The state's literacy rate is higher than that of India's. West Bengal's literacy rate has increased by eight points in a decade from 2001 to 2011 (calculated excluding the 0–6 years population), where the male population dominates when compared with the female literacy rate. These improvements can be attributed to the expansion of newly developed educational services, a reduction in inequality and disparity, and an increase in opportunities. The number of registered factories has increased over the years, as shown in Figure 17.2. Although there is a disparity in literacy rates among the districts of West Bengal, Uttar Dinajpur had the lowest percentage, and Purba Medinipur ranked the highest. In the context of male and female literacy rates, Purba Medinipur and Kolkata districts have the highest, and Uttar Dinajpur and Purulia have the lowest male and female literacy rates, respectively. Factors such as poverty, diversified rural–urban

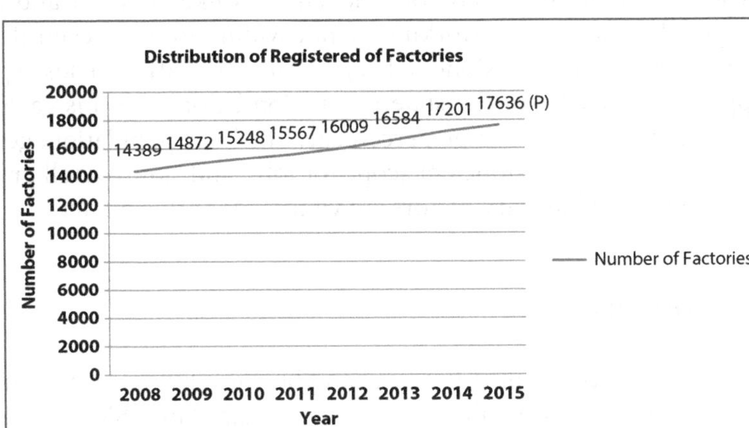

Figure 17.2 Source: Statistical Abstract of West Bengal, 2015 (25 Defense factories excluded).

facilities, large population size, and the primitive mindset of rural parents are slowing down the advancement of literacy in the state.

4. Negative Correlation Matrix

There is a negative relationship between daily workers as the percentage of daily wage-based workers decreases slowly. Different wage amounts for male and female workers, decreasing opportunities due to an increase in more machinery-based approaches, and bare wage amounts lead to uncountable household issues, all of which add more fuel. Different studies show upsetting information that the suicide rate of daily wage workers has been extreme in the contemporary period. Several unions (CITU, AITUC, and INTUC) have formed a Joint Action Committee (JAC) in the state to address issues about workers in certain occupations (e.g., beedi and brickmaking); meetings are held every two months and awareness campaigns are organized. In the brick kiln industry, the JAC has helped brick workers unionize; it has also promoted the implementation of minimum wages and the provision of other welfare benefits (Sen, 2012). West Bengal has introduced several schemes to benefit informal workers, both generally and in specific areas, such as transport, beedi manufacturing, and construction. Females with high levels of education prefer to work as main workers rather than marginal workers, thus showing a negative correlation between the two as they acquire more skills and are self-reliant.

The group of people who do not belong to working age is not involved in any kind of work or work for durations less than what they do, such population is considered as non-workers, such as students, persons who are involved in household activities, dependents like children and aged people, people who are not working or not willing to work even if they belong to working age, pensioners, beggars, etc. The 2011 census reports that approximately 62% of the state population belongs to this category, surpassing the country's average percentage. The huge population, gender disparity in education, insufficient scope for jobs, and slow overall growth of the secondary and tertiary sectors are vital seeds of this matter.

17.7 Conclusions

West Bengal is the controlling financial and business hub of eastern India. Moreover, approximately 9 million people are engaged in MSMEs, which is the second highest in the country. The number of non-workers is high due to inadequacy in the availability of jobs, and gender disparity in jobs, while most of the population belongs to the pensioner class and lacks willingness to work. More workers are moving towards daily wages as they receive various incentives for the same and are covered under various benefits provided by the government. The vital factors contributing to non-workers are non-willingness to work, gender disparity in jobs, low scope in job sectors, lack of skills, low educational levels, and slow growth in tertiary and secondary sectors. Factors such as poverty, diversified rural–urban facilities, large population size, and the primitive mindset of rural parents are slowing down the advancement of literacy in the state. People residing in urban areas engage in secondary activities and part-time work. This factor-based analysis shows that the manufacturing sector has witnessed a vast decline in recent years in its share when again the agricultural sector played a pivotal role in holding onto the overall growth, which still failed to control the unstable situation. However, if a proper connection is created between political transition and policy influence, it will take the condition towards a better place.

Acknowledgments

The authors thank the anonymous reviewers and editor for their comments and suggestions.

Declaration of Conflicting Interests

The authors declare no potential conflicts of interest with respect to the research, authorship, or publication of this article.

Funding

The authors received no financial support for the research, authorship, or publication of this article.

References

Acharya, N. (2013, December 28). West Bengal saw 97% decline in industries since 2010. *Business Standard India*. https://www.business-standard.com/article/economy-policy/west-bengal-saw-97-decline-in-industries-since-2010-113122800521_1.html

Census tables | Government of India. (n.d.). Retrieved December 23, 2022, from https://censusindia.gov.in/census.website/data/census-tables

Chakravarty, D., & Bose, I. (2010). Industrializing West Bengal? The case of institutional stickiness. *Governance Working Papers.*

Chattopadhyay, S. K. (2004). Trends in Total Factor Productivity of Manufacturing Sector in West Bengal: A Sectoral and Temporal Analysis. *Reserve Bank of India Occasional Papers, 25*, 75–103.

Chowdhury, S., & Chakraborty, S. (2016). *Employment Growth in West Bengal: An Assessment* [Working Paper]. eSocialSciences. https://econpapers.repec.org/paper/esswpaper/id_3a10935.htm

Das, S. (2013). Sectoral Transformation of Working Population and Status of Employment-A Case Study of Chanditala CD Block–I, Hugli District, West Bengal. *J. Eng. Sci., 2*(4), 1–7.

Ghosh, A. (2011). Physical infrastructure and development of secondary sector:an econometric analysis for six states in India. *J. Dev. Areas, 44*(2), 207–216.

Joshi, S. (2004). Tertiary Sector-Driven Growth in India: Impact on Employment and Poverty. *Econ. Polit. Wkly., 39*(37), 4175–4178.

Kabra, K. (2007). Some Issues Concerning Industrialisation of West Bengal. *Mainstream, 45*(20), 5.

Karak, A. (2017). Pattern of Industrial Growth in West Bengal during 1980–1991: Structural Demand and Agriculture–Industry Relations. *J. South Asian Dev., 12*(1), 65–88. https://doi.org/10.1177/0973174117700468

Roy, C., & Dey, A. (2017). *Murshidabad Silk Industry in West Bengal: A Study of its Glorious Past & Present Crisis* [MPRA Paper]. University Library of Munich, Germany. https://econpapers.repec.org/paper/pramprapa/92847.htm

Sarkar, S., Chowdhury, A., Roy, P., & Chowdhury, M. (2016). Socio-economic status of the tea garden women workers: A case study from Sub-Himalayan regions of West Bengal, India. *Int. J. Appl. Res., 2*(10), 681–684.

Sharma, V. R., Maurya, A., Sachar, P., Kumar, M., Rani, U., & Sanu, S. K. (2021). Mapping the Regional Imbalances of Demographic Profile: Case Study of Awadh and Middle Doab, 2011. *Horizon J. Soc. Sci., XII*, 12–24.

Statistical Abstract of West Bengal. (2015). Bureau of Applied Economics and Statistics Government of West Bengal. http://wbpspm.gov.in/SiteFiles/Publications/10_18052017120546.pdf

Tiwary, G., Gangopadhyay, P. K., Biswas, S., Nayak, K., Chatterjee, M. K., Chakraborty, D., & Mukherjee, S. (2012). Socio-economic status of workers of building construction industry. *Indian J. Occup. Environ. Med., 16*(2), 66–71. https://doi.org/10.4103/0019-5278.107072

18

Reviving the Spiritual Heartland: Enhancing Cultural Tourism in Kushinagar Through SWOT Analysis

Ankit Singh* **and Anjali Shukla**

Department of Geography, Deen Dayal Upadhyaya Gorakhpur University, Gorakhpur, India

Abstract

This study explores the diverse fields of cultural tourism, with an emphasis on Kushinagar, an important spiritual and cultural site in Uttar Pradesh, India. The purpose of this study is to conduct a thorough SWOT analysis of the area and examine current tourism trends in Kushinagar. It was found that tourism initially grew until 2018–19, then declined sharply owing to the COVID–19 pandemic in 2019–20, before recovering in 2021–22. SWOT analysis identifies Kushinagar's cultural tourism strengths, weaknesses, opportunities, and threats. This study makes recommendations to enhance Kushinagar's tourist industry through the development of theme-based tourism circuits, infrastructure improvement, heritage site preservation, and a local-to-global approach. These initiatives can provide a comprehensive cultural experience for visitors, encourage intercultural dialogue, and boost the economy. The research is qualitative in nature and based on secondary data sources.

***Keywords*:** Cultural tourism, Kushinagar, SWOT analysis, tourism circuit, local to global

**Corresponding author*: singh.ankit019@gmail.com

Shruti Kanga, Gowhar Meraj, Suraj Kumar Singh, Majid Farooq and M. S. Nathawat (eds.) Geospatial Technology for Natural Resource Management, (425–448) © 2024 Scrivener Publishing LLC

18.1 Introduction

Cultural constructs, intrinsically complex and evolving, serve as the cornerstone of human civilizations, embodying the shared beliefs, values, customs, artistic expressions, and social practices that define our collective ethos. This culturally fashioned lens mediates our interactions with the world, influencing our interpretations, engagements, and self-awareness. Be it perceived as an intellectual state, a marker of societal development, an artistic domain of excellence, or an all-encompassing lifestyle, culture ceaselessly shapes individuals and communities.

Cultural tourism represents a distinct form of travel that prioritizes authentic engagement with the local heritage and artistic expression of a destination. It provides an avenue for tourists to delve into the unique traditions, history, architecture, and lifestyle of a locale, thereby facilitating a profound understanding of its people and cultural frameworks. Cultural tourism goes beyond the superficial tourist experience and enables travellers to forge a deeper connection with local communities; partake in traditional festivities; explore historical sites, museums, and art galleries; and savor the quintessential culinary delights of a region.

Renowned scholars have expounded on the concept of cultural tourism, framing it as a travel discipline centered on the discovery and appreciation of a destination's cultural heritage and artistic manifestations. Cultural tourism is defined as immersion in customs, traditions, arts, and historical narratives that are indigenous to a specific locale (Richards, 2011). Corroborating this view, cultural tourism is a quest for distinct experiences embedded in the cultural matrix of a destination, incorporating activities such as visiting museums and historical sites, participating in festivals, and interacting with local communities (Urry, 1990).

A tangible manifestation of these conceptual interpretations of cultural tourism can be observed in Kushinagar, a destination steeped in spiritual and cultural significance, situated in the Indian state of Uttar Pradesh. Renowned as the site where Lord Buddha achieved Mahaparinirvana and was cremated, Kushinagar is a seminal landmark in the annals of Buddhism, Jainism, and other religions. Home to ancient stupas, monuments, and archaeological sites, this sacred land attracts a global influx of visitors eager to engage with the spiritual heritage and cultural riches of Buddhism. Kushinagar, with its deep-rooted historical and religious relevance, exemplifies how cultural tourism serves as a portal for travellers to navigate through the intricate tapestry of traditions, rituals, and teachings that underpin the identity of a destination.

18.2 Study Area

The Kushinagar District is located in the northeastern corner of the state. It lies between latitudes of 26°33' and 27°18' north and longitudes of 83°31' and 84°25' east. In the West, it borders the Gorakhpur and Maharajganj districts, while the Maharajganj district extends into its northern portion. The state of Bihar borders the district to its east and southeast and shares its southern border with the Deoria District. The Burhi Gandak and Chhoti Gandak rivers drain this area. It is exclusively in the Marsh or Terai region. It is 2,906.0 km^2 of the area. Figure 18.1 Shows the location map of the study area.

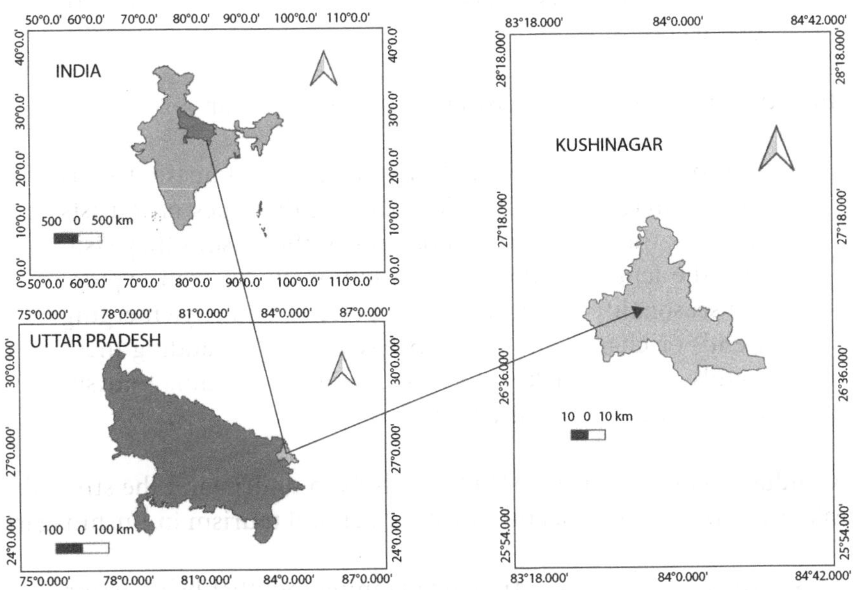

Figure 18.1 Location map of Kushinagar district.

18.3 Objectives

This study aims to:

1) Identifying major cultural places in Kushinagar.
2) Investigate the prevailing tourism trends in Kushinagar.
3) Conduct a comprehensive SWOT analysis to understand the strengths, weaknesses, opportunities, and threats of cultural tourism in Kushinagar.

The objectives outlined in your study are focused on cultural tourism in Kushinagar and provide a clear direction for your research. These are elaborate upon in each of these objectives:

1. Identification of major cultural places in Kushinagar:

- This objective was to conduct a comprehensive survey or study to identify and document significant cultural places and landmarks in Kushinagar. These include historical and religious sites, museums, and other locations that have cultural significance. The goal was to create a list of key places that attract tourists and contribute to the cultural identity of the region.

2. Investigating prevailing tourism trends in Kushinagar:

- This objective involves analyzing the current tourism trends in Kushinagar. This includes studying the types of tourists that visit the area, their demographics, the reasons they visit, and the length of their stay. Additionally, we can explore the seasonality of tourism, peak periods, and any emerging trends or changes in tourist behavior. Understanding these trends can help make informed decisions regarding tourism development and marketing.

3. Conduct a comprehensive SWOT analysis to understand the strengths, weaknesses, opportunities, and threats of cultural tourism in Kushinagar:

- SWOT analysis is a strategic planning tool that helps assess an organization, destination, or project. In this case, you want to apply it to cultural tourism in Kushinagar. Breakdown this objective:
- Strengths: Identify the internal factors that contribute to the success of cultural tourism in Kushinagar. These may include cultural heritage, infrastructure, local support, and unique attractions.
- Weaknesses: Examine the internal factors that hinder cultural tourism in Kushinagar. These could include infrastructure limitations, a lack of marketing, or any internal challenges that need to be addressed.

- Opportunities: Exploring external factors that can be leveraged to enhance cultural tourism. This may involve emerging tourism trends, partnerships with other stakeholders, or new marketing strategies.
- Threats: Identify external factors that pose challenges to cultural tourism in Kushinagar. This could include competition from other destinations, political instability, environmental concerns, or other potential threats.

By accomplishing these three objectives, This study provides a comprehensive understanding of cultural tourism in Kushinagar. This information can be valuable for local authorities, businesses, and other stakeholders in the tourism industry in making informed decisions regarding the development and promotion of cultural tourism in the region.

18.4 Methodology

This study was based on secondary data sources. The data were acquired from a variety of official sources, including the District Statistical Abstract, District Census Handbook (2011), Ministry of Tourism Annual Report, the U.P. Tourist Strategy, and several periodicals. GIS and cartographic techniques were used to create maps and graphs. To achieve the objectives of this study, a structured methodology was employed. First, to identify the major cultural places in Kushinagar, a two-fold approach was adopted. A comprehensive literature review was conducted to compile an initial list of cultural sites, drawing from historical records, online resources, and existing research. Subsequently, on-site surveys will be conducted in Kushinagar to validate and augment the list. These surveys involve visits to cultural places, direct observations, and interviews with local experts and residents. To investigate prevailing tourism trends, surveys and questionnaires are designed and administered to tourists, local businesses, and tourism agencies. Additionally, secondary data from government tourism departments and travel reports were also collected. The collected data will then be analyzed to identify key trends and patterns in tourist demographics, preferences, and behavior. Finally, a comprehensive SWOT analysis is conducted, drawing on the identified cultural places and tourism trends. This analysis will help understand the strengths, weaknesses, opportunities, and threats pertaining to cultural tourism in Kushinagar. The findings provide recommendations for the development and enhancement of cultural tourism in the region.

18.5 Results and Discussion

18.5.1 Major Cultural Places in Kushinagar District

18.5.1.1 Mahaparinirvana Temple

The Parinirvana Temple in Kushinagar is a significant Buddhist site with great religious and historical importance. It is believed to be where Gautama Buddha attained Parinirvana, his final liberation from the cycle of birth and death. The temple stands as a memorial to Buddha's passing away and serves as a place of pilgrimage and veneration for Buddhists around the world. Cultural places in Kushinagar district are shown in Figure 18.2.

The history of the Parinirvana Temple can be traced back to ancient times. Accounts from Chinese pilgrims, such as Hiuen-Tsang and I-tsing, mention the existence of a temple at the site during their visits in the 7th and 8th centuries. In the 19th century, archaeological explorations and excavations were undertaken in Kushinagar, leading to the rediscovery of the temple. Archaeologist Alexander Cunningham played an important role in uncovering and documenting the remains of Parinirvana Temple.

The central shrine of the Parinirvana Temple, which contained a 6.10-meter-long reclining statue of the Buddha, served as the building's principal attraction. From three distinct angles, Lord Buddha appears to be sleeping, thinking, and smiling. Buddha was depicted in this statue lying

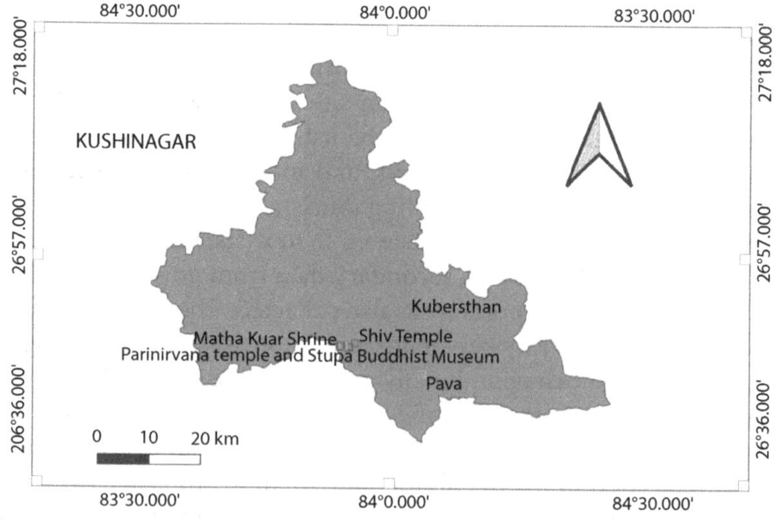

Figure 18.2 Map showing cultural places in Kushinagar district.

on his right side in the position of Nirvana. This symbolized the final liberation and attainment of Nirvana. The three mourners depicted on the stone slab connected to the front of the building are thought to be Ananda, Subhadda, and Mallika.

Adjacent to the temple, other structures, such as monastic cells, prayer halls, and ancillary buildings catered to the needs of the Buddhist community and visiting pilgrims. These structures contributed to the overall architectural ensemble of the temple complex.

18.5.1.2 Mahaparinirvana or Nirvana Stupa

Mahaparinirvana Stupa is a replica of the main stupa discovered by Carlleyle in 1876, and is situated next to the Mahaparinirvana Temple. It was excavated to a depth of 4.3 m, which uncovered a copper plate and other items from the Gupta Era. On the plate, it was written in Sanskrit that the monk Haribala deposited these things at Nirvana Stupa. The Nirvana Stupa was initially constructed by King Asoka, according to Hsüan Tsang, who travelled to Kusinara in AD 637. Additionally, he noticed a stone pillar in front of it, which documented Tathagata's nirvana. It is thought that the Mallas built the Nirvana Stupa at the beginning to retain the Buddha's relics. Later, King Asoka expanded it, and during the Gupta era, it was further developed. The 23-meter-tall Nirvana Stupa was refurbished in 1927 with funding from Myanmar.

18.5.1.3 Matha Kuwara Shrine

Lord Buddha took a brief respite at the Matha Kuar temple before entering Mahaparinirvana. He gave his final sermons to his followers in this revered place, emphasizing the transient nature of life and the necessity of spiritual practice. After this significant event, the Matha Kuar shrine was adorned with a colossal Buddha image in the Bhumisparsha mudra, representing the moment of enlightenment. This magnificent statue, which stands 3.05 m tall and was expertly carved from a single block of blue stone from the Gaya region, is said to be approximately 1,000 years old. Matha Kuwara, the name of the shrine, means simply "forehead prostration," which is what visitors to the shrine perform. The temple was built in 1927 from the donations of two Myanmar devotees.

18.5.1.4 Ramabhar Stupa

The Mallas, the dominant clan in Kushinagar, took the Buddha's remains to Makuta-bandhana, the customary location where their chieftains were crowned, after six days of veneration. They performed the sacred cremation ceremony, as stated in Part VI of Mahaparinirvana Sutta. The story also describes how Brahmin Dona divided the Buddha's relics. The Mallas initially built the Cremation Stupa, which still stands today, sometime after the Buddha was cremated. King Asoka renovated it in the third century BC, and Kumaragupta, the Gupta king, renovated it again in the fifth century AD. When archaeologist Alexander Cunningham visited the location in 1866–1862, he discovered a massive mound known as Ramabhar Tila, built with substantial bricks and rising approximately 15 m above the nearby fields.

The Cremation Stupa features a circular drum with a diameter of 34 m, resting on a platform with a diameter of 47 m. The mound had already undergone substantial excavation before Cunningham's arrival, including the drilling of a massive circular well that was over 3 m in diameter and 15 m deep. Additionally, three trenches were cut through the mound, with one trench reaching nearly ground level and the other two smaller in size. Further excavations confirmed the site's identification as Buddha's cremation site by exhibiting a sizable quantity of clay seals bearing Buddhist verses.

18.5.1.5 Pava, Where the Buddha had his Last Meal

18.5.1.5.1 Religious Significance

The Mahaparinirvana Sutra recounts an incident in which the Buddha, accompanied by monks, stayed at the mango grove of his disciple Cunda in Pava. Cunda prepared a special dish called "sukara maddava," which only the Buddha ate. Unfortunately, this meal caused severe illness and pain in the Buddha. Despite his condition, he continued his journey to Kusinara, resting 25 times along the way. To comfort Cunda, the Buddha explained that his offering had great merit comparable to the offering of milk and rice when he attained enlightenment. The Buddha assured Cunda that his act had brought about positive results, such as long life, beauty, good health, and heavenly rewards.

18.5.1.5.2 Stupa of the Mallas of Pava

The Mallas of Pava claimed some of the Buddha's relics after his Parinibbana and built a stupa to enshrine them. Recently, the Archaeological Survey of India (ASI) has excavated a large mound believed to be the location of Cunda's mango grove, where the Mallas of Pava built a stupa. Hsuan Tsang, a Chinese pilgrim who visited Kusinara in the 7th century AD, mentioned the presence of a stupa built by King Asoka. which acts as a memorial for the Buddha's final meal and his eventual illness and demise.

18.5.1.5.3 Pava and Jain Religion

Pava holds significance not only in the context of Buddhism, but also in relation to Jainism. Figure 18.3 shows cultural places in Kushinagar district related to both Buddhism and Jainism. According to Jain tradition, it is believed to be the site where Mahavira, the 24th and last Tirthankara of Jainism, obtained his ultimate liberation, also known as moksha or nirvana detail list of cultural places in Kushinagar district have been showed in Table 18.1.

Figure 18.3 Cultural places in Kushinagar district.

Table 18.1 Cultural places in Kushinagar district.

Cultural places	Description
Aniruddhawa Village	This is the site where the remains of Lord Buddha were distributed to eight kings.
Buddha Ghat of Kushinagar	The Buddha Ghat, which is situated alongside the Hiranyavati River, was built as a memorial to Gautam Buddha's cremation. Buddhists see it as sacred, same as Hindus regard the Ganges River as sacred.
Japanese Meditation Park	This park, which is next to the Nirvana temple, is perfect for yoga and meditation sessions. Buddhist monks frequently chant and meditate here, and it is surrounded by tall Sal trees and flowery plants.
Vietnamese Temple of Kushinagar	A Buddhist temple built by the Vietnamese community.
Myanmar Stupa & Temple (Burmese Temple)	Also known as Burma Temple, this is a Buddhist temple built by the Burmese community.
Wat Thai Kusinara Chalermraj Buddhist Temple and Monastery	A Thai Buddhist temple and monastery.
Kuber Asthan	Kuber Asthan was constructed by Lord Kuber, a devotee of Shiv Bhagwan and a god of wealth.
Sun Temple at Turkpatti Mahuaa	Kushinagar's Sun Temple is a Gupta-era relic. The Surya Bhagwan idol dates to the fourth or fifth century C.E.

18.5.1.6 Buddhist Museum

The museum is home to a wide variety of artifacts and displays that highlight the teachings and life of Gautama Buddha. It has a diverse collection of artifacts, such as statues, sculptures, inscriptions, coins, ceramics, and pieces of ancient buildings. These artifacts, which are from diverse eras of Buddhist history, provide information about the development of Buddhist culture and art. Ancient stone carvings of pictures from the life of the Buddha, tranquil statues of the Buddha in various poses, finely carved reliefs displaying Buddhist symbols and motifs, and manuscripts carrying Buddhist scriptures.

The data depicts how many tourists, both domestic and foreign, visited the Kushinagar district of Uttar Pradesh, India, between 2012–13 and 2021–22. Table 18.2 shows the year-wise list of total tourist arrival in Kushinagar district of Uttar Praesh, India as well as Figures 18.4 and 18.5 shows Year-wise tourist arrival and total tourist arrival in Kushinagar district of Uttar Pradesh, India, between 2012–13 and 2021–22 respectively.

Trends in Tourism in Kushinagar (2012–2022):

Growth from 2012–13 to 2018–19:

- Between 2012–13 and 2018–19, Kushinagar witnessed a consistent increase in tourist footfall.
- The number of Indian tourists grew from 750,171 in 2012–13 to 1,006,764 in 2018–19.
- The influx of foreign travellers also saw an upward trajectory, rising from 62,292 in 2012–13 to 89,693 in 2018–19.

Table 18.2 Year-wise tourist arrival. Source: U.P. Tourism.

Year	Indian	Foreigner	Total
2012–13	750,171	62,292	812,463
2013–14	774,872	65,435	840,307
2014–15	805,553	68,830	874,383
2015–16	825,218	73,514	898,732
2016–17	861,760	76,221	937,981
2017–18	897,546	80,147	977,693
2018–19	1,006,764	89,693	1,096,457
2019–20	301,424	22,465	323,889
2020–21	177,002	0	177,002
2021–22	848,631	41,058	889,689
TOTAL	**7,248,941**	**579,655**	**7,828,596**

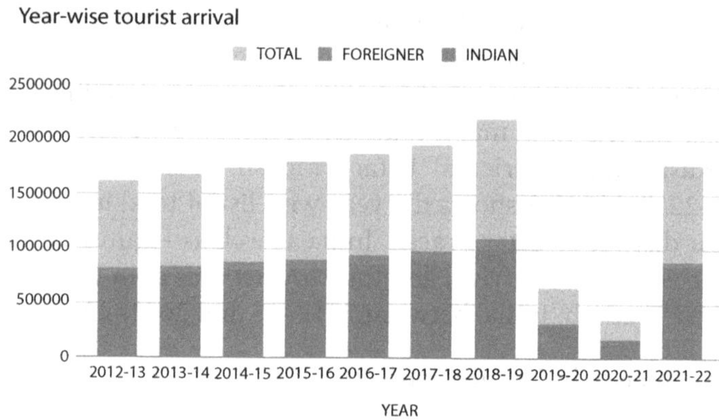

Figure 18.4 Year-wise tourist arrival.

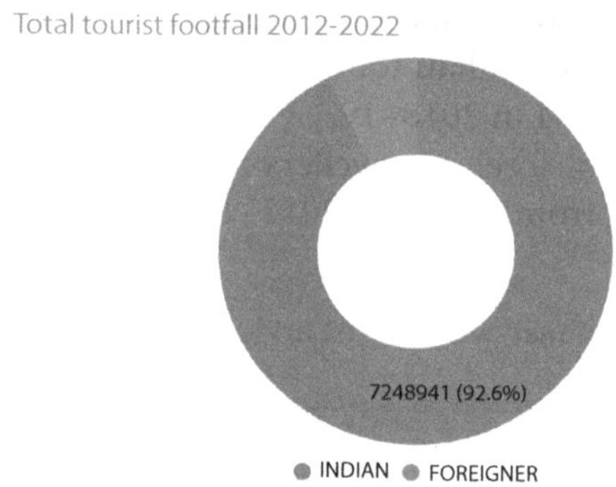

Figure 18.5 Total tourist arrival.

- The total number of tourists visiting Kushinagar surged from 812,463 in 2012–13 to 1,096,457 in 2018–19.

Sharp Decline in 2019–20:

- However, in 2019–20, there was a substantial decline in tourism.
- The overall number of visitors sharply decreased to 323,889, accounting for less than one-third of the previous year's figures.
- The number of Indian tourists decreased to 301,424 and international visitors dropped to 22,465.

Impact of COVID-19 in 2020–21:

- The years 2020–21 witnessed a further drop in tourism, mainly attributed to the effects of the COVID-19 pandemic and related travel restrictions.
- Foreign visitors were virtually nonexistent, indicating severe restrictions on international travel.
- Indian tourist numbers also declined significantly, with 177,002 recorded visitors, a considerable drop from previous years.

Recovery in 2021–22:

- In 2021-22, there was a resurgence of tourist visits to Kushinagar.
- The number of tourists from other countries increased by 41,058, and Indian tourists showed a substantial increase, with 848,631 more visitors.
- The total number of visitors in 2021–22 reached 889,689, marking a recovery from the previous year but still falling short of the pre-2019–20 figures.

This scientific presentation provides a structured and objective account of tourism trends in Kushinagar over specified years. It emphasizes statistical data and trends while maintaining a neutral and factual tone.

The general pattern shows that tourism started rising until 2018–19, started falling quickly in 2019–20, and then kept falling in 2020–21, most likely as a result of COVID–19. Although the numbers have not yet reached their pre-2019 levels, there is a comeback in 2021–2022. This analysis provides an overview of the trends in tourist arrivals in Kushinagar district over the years. The data reflect the impact of various factors on tourism, including the COVID-19 pandemic.

18.5.1.7 SWOT Analysis of Cultural Tourism in Kushinagar District

A SWOT analysis of cultural places provides a comprehensive examination of their internal strengths and weaknesses as well as external opportunities and threats, illuminating their overall position in the tourism sector. Table 18.3 shows the SWOT matrix to shows the internal strengths and weaknesses in the tourism sector.

Table 18.3 SWOT matrix.

STRENGTHS • Cultural and historical importance. • Architectural and archaeological wonders. • Multiple religious traditions. • Geographical location. • Hub of ancient trade routes and highways. • Connectivity in terms of transportation • Local art and crafts.	**WEAKNESSES** • Potential lack of proper preservation and maintenance of cultural sites and monuments. • Lack of effective marketing and promotion • Absence of diversified tourism offerings beyond the cultural aspects. • Lesser international recognition compared to other prominent Buddhist sites. • Tourists concentrated in certain seasons.
OPPORTUNITIES • Development of Tourism Infrastructure. • Collaboration with International Buddhist Communities. • Cultural Festivals and Events. • Eco-tourism.	**THREATS** • Competition with Other Cultural Sites. • Environmental Degradation. • Cultural Commodification.

Strengths:

1. **Cultural and historical importance:** Kushinagar is the site where Lord Buddha delivered his last sermon, attained Mahaparinirvana, and was cremated. This connection to Buddhism attracts tourists from around the world and contributes to the district's cultural vibrancy.
2. **Architectural and archaeological wonders:** The reclining statue of Lord Buddha in the famous Buddhist temple stands as a remarkable piece of artistry. Ancient stupas, monuments, and archaeological remains scattered throughout Kushinagar provide insights into the rich cultural heritage of the region. In addition to Buddhist sites such as Ramabhar Stupa and Mahaparinirvana Temple, there are also other sites like Parshwanath Jain Temple.
3. **Multiple religious traditions:** Besides Buddhism, the district is associated with Jainism and is believed to be the

place where Lord Mahavir, the 24th Tirthankar, attained Parinirvana. This intermingling of religious influences creates a unique cultural landscape and offers visitors an opportunity to explore and appreciate multiple faiths. In addition to Buddhist sites such as Ramabhar Stupa and Mahaparinirvana Temple, other religious sites such as Parshwanath Jain Temple and Shiv Temple are also present.

4. **Geographic location:** This contributes to the cultural significance. Situated on the banks of the Gandak River and near the Terai region of the Himalayas, Kushinagar's natural surroundings have provided a serene and tranquil backdrop for meditation, attracting sages, saints, and spiritual practitioners throughout history. The region's pious and charming ambiance adds to its appeal as a cultural destination.
5. **Hub of ancient trade routes and highways:** The district promoted cultural interaction and trade between many regions by acting as a bridge between key routes, such as Ayodhya-Janakpur and Rajgir-Vaishali-Shravasti. Additionally, ancient Pawa Nagar City in Kushinagar was a vital center for trade and commerce in ancient India. Its strategic location at the intersection of important trade routes encourages the flow of people, ideas, and goods.
6. **Connectivity in terms of transportation:** The region is well connected to important cities and towns through highways and railroads. A comfortable road trip for tourists is made possible by its proximity to the National Highway 27.

According to "India Tourism Statistics 2022," air travel is the most preferred mode of transport for foreign tourists arriving in India. Despite the pandemic in 2020 and 2021, the shares of air travel were 79.2% and 87.5%, respectively. Therefore, it is important to improve the Kushinagar International Airport, which will enable global connectivity, making it simpler for Buddhist pilgrims from nations such as Thailand, Myanmar, and Sri Lanka. In addition, Kushinagar has an advanced local transit system mode of transport (2021) have been showed in Figure 18.6. according to Free Trade Agreements (FTAs).

7. **Accommodation:** To meet the various demands and price ranges of visitors, a variety of hotels, guesthouses, and resorts are available. This availability of accommodations ensures that tourists have convenient options for staying.

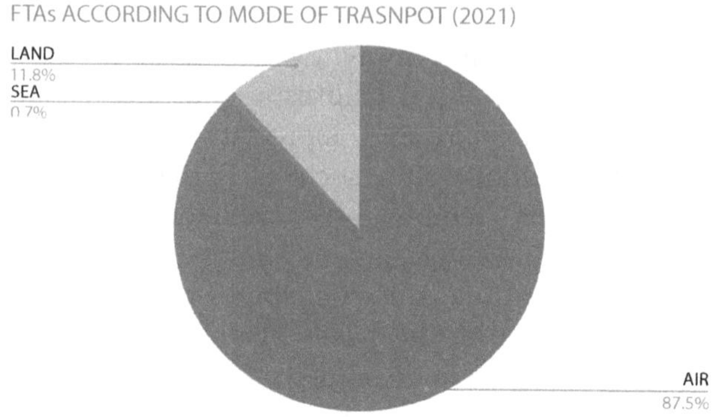

Figure 18.6 FTAs according to mode of transport (2021). Source: India tourism statistics.

8. **Local art and crafts:** Kushinagar is known for its bamboo and cane crafts, where artisans skillfully weave bamboo and canes into a variety of products. This craft includes baskets, mats, furniture, and decorative items. Other local arts and crafts include puppetry, wood carvings, embroidery, and handicrafts.

Weaknesses:

1. **Potential lack of proper preservation and maintenance of cultural sites and monuments:** Some historical and archaeological sites in Kushinagar may not have received adequate attention or funding for their upkeep. This can lead to issues such as the deterioration of structures, insufficient signage or information, and limited accessibility for visitors. Such shortcomings can hinder the overall cultural experience and prevent tourists from fully appreciating the destination's historical and cultural significance.
2. **Lack of effective marketing and promotion:** Insufficient awareness and visibility of the district's cultural attractions may result in fewer tourists visiting the area. Limited promotional efforts can prevent Kushinagar from reaching its full potential as a hotspot for cultural tourism. Despite its significance, Kushinagar is not as prominently featured in India's Incredible India campaign as other Buddhist sites such as Bodh Gaya, Sarnath, or Lumbini.
3. **Absence of diversified tourism offerings beyond cultural aspects:** While Kushinagar's cultural heritage is undoubtedly

significant, the district may lack other complementary attractions or activities to enhance the overall tourist experience. Apart from religious sites, there are limited recreational activities or attractions for tourists, such as adventure sports, entertainment, and shopping centers.

4. **Lesser international recognition compared to other prominent Buddhist sites:** While Kushinagar holds immense historical and religious significance as the place where Lord Buddha attained Mahaparinirvan, it may not enjoy the same level of global recognition as destinations like Bodh Gaya or Varanasi. Bodh Gaya, where Buddha attained enlightenment, is more internationally recognized and receives a significantly higher number of foreign tourists compared to Kushinagar.
5. **Tourists concentrated in certain seasons:** Kushinagar may experience a surge in tourist arrivals during specific times of the year, such as religious festivals (Budhpurnima), holiday periods, or favorable weather conditions.

Opportunities:

1. **Development of Tourism Infrastructure:** There is potential for the development of hotels, restaurants, and transport facilities in Kushinagar. For example, attracting investment and partnerships to build luxury hotels, fine-dining restaurants, and efficient transportation systems can enhance the overall tourism experience for visitors.
2. **Collaboration with International Buddhist Communities:** Engaging with Buddhist communities worldwide can promote Kushinagar as a pilgrimage site. For instance, establishing partnerships with Buddhist organizations and temples abroad can help spread awareness of Kushinagar's cultural significance and attract a global audience of Buddhist travellers.
3. **Cultural Festivals and Events:** Hosting cultural festivals and events can be a major attraction for tourists interested in experiencing the local culture. For instance, hosting annual cultural events that highlight the traditional music, art, and customs of Kushinagar can foster a lively atmosphere and attract tourists interested in engaging in cultural experiences.
4. **Eco-tourism:** Promoting the biodiversity and scenic surroundings of Kushinagar can draw eco-friendly tourists and

people who like nature. Creating eco-friendly excursions and lodgings such as narrated nature walks, animal safaris, and eco-lodges, may provide visitors with immersive and sustainable experiences while protecting the local ecosystem.

Threats:

1. **Competition from Other Cultural Sites:** Other cultural destinations in India and around the world compete with Kushinagar, which may deter travellers from looking for comparable cultural experiences. Kushinagar must create unique selling propositions (USPs) that set them apart from other travel destinations to counter this challenge. This may involve emphasizing the particular characteristics of Kushinagar's cultural history that make it distinctive, such as its connection to Lord Buddha's final moments and its historical importance.
2. **Environmental Degradation:** As Kushinagar's tourism grows, there is a chance that the environment could deteriorate, endangering both natural and cultural landmarks. The adoption of sustainable tourism practices is essential for reducing this hazard. This can incorporate programs that promote eco-friendly transportation choices, appropriate water use, and waste-management systems.
3. **Cultural Commodification:** The over-commercialization of cultural customs and heritage may cause cultural experiences in Kushinagar to lose their originality and become less authentic. To counter this threat, guidelines should be developed to preserve the authenticity of cultural practice. This can include setting limits on commercial activities at cultural sites, promoting community involvement in preserving traditions, and educating tourists about respectful behavior and cultural sensitivity.

18.6 Conclusion and Recommendations

From the aforementioned analysis, it can be concluded that Kushinagar, Uttar Pradesh, has great potential for cultural tourism because of its extensive historical and cultural history, particularly in Buddhism. This paper highlights how tourism increased until 2018–19, then saw a downturn because of the COVID-19 pandemic, and then experienced a resurgence in 2021–22. The SWOT analysis shows that the Kushinagar district has great potential as a destination for cultural tourism because of its profound historical and religious significance, outstanding architecture and archaeology, and thriving cultural heritage influenced by many different religious traditions. These strengths may be utilized to address weaknesses, such as inadequate infrastructure and services for tourists, poor maintenance of cultural monuments, and insufficient marketing efforts. There are opportunities to encourage Buddhist pilgrimages, promote cross-cultural interaction, and ensure historical preservation. Additionally, the inclusion of regional craft and the use of digital platforms for global outreach can increase Kushinagar's attractiveness as a unique cultural tourism destination. However, effective management, is necessary to reduce vulnerabilities, including competition from other cultural places, potential environmental damage, and risks related to cultural commodification. To realize its full potential, it is suggested to focus on identifying circuits more similar to Buddhism (Buddhist circuit, Ramayana circuit, etc.), improving infrastructure and amenities, restoring tourist sites, developing a network of cultural sites within Kushinagar, and connecting them to regional and sub-continental cultural sites (local to global approach). Therefore, to ensure Kushinagar's sustained growth and development as a prominent cultural tourist hub, strategic planning, and sustainable practices, and inclusion of local populations in tourism efforts are crucial.

Based on the conclusions drawn from the analysis, here are some recommendations for realizing the full potential of Kushinagar as a cultural tourism destination:

1. **Identifying and Developing Tourism Circuits**:

    - Focus on developing and promoting specific tourism circuits that align with Kushinagar's cultural and historical strengths. For example, the creation of a dedicated Buddhist circuit, Ramayana circuit, or other thematic routes can enhance visitor experience.

2. **Infrastructure Improvements**:

    - Invest in infrastructure development and improvement to provide better amenities and services for tourists. This should include upgrading transportation facilities, accommodation, and other essential infrastructure elements.

3. **Heritage Site Restoration**:

    - Prioritize the restoration and maintenance of cultural and historical sites. Ensuring that these sites are well preserved and accessible to visitors is crucial for the overall tourism experience.

4. **Networking Cultural Sites**:

    - Develop a network of cultural sites within Kushinagar and connect them to regional and sub-continental cultural destinations. This approach, often referred to as a "local to global" strategy, can promote cross-cultural interactions and enhance the overall appeal of the region.

5. **Promotion and Marketing**:

    - Invest in effective marketing strategies to promote Kushinagar as a cultural tourism destination. Use digital platforms and collaborate with tourism agencies to reach a global audience. Emphasize the unique cultural heritage and religious significance of the area.

6. **Encouraging Pilgrimage Tourism**:

    - Leverage the profound historical and religious significance of Kushinagar in attracting more pilgrims. It should be promoted as a significant destination for Buddhist pilgrimages and facilitate the necessary infrastructure to support religious tourism.

7. **Preservation of Historical and Cultural Heritage**:

    - Ensure the preservation of historical and cultural heritage through sustainable practices. Implement conservation measures to protect architectural and archaeological treasures for future generations.

8. **Incorporation of Local Crafts**:

   - Integrate local crafts and artisanal products into the tourism experience. This not only supports local economies but also adds to the authenticity of cultural tourism experiences.

9. **Effective Management**:

   - Implementing effective management practices to mitigate vulnerabilities. This includes monitoring and addressing issues related to competition from other cultural destinations, potential environmental damage, and the risks of cultural commodification.

10. **Sustainability and Community Involvement**:

    - Emphasize sustainable tourism practices and ensure that local communities are actively involved in tourism efforts. Their participation and benefits from tourism can lead to better local support and the preservation of cultural heritage.

To fully harness Kushinagar's potential as a cultural tourism hub, a strategic, holistic, and sustainable approach is necessary. It should encompass infrastructure development, heritage preservation, marketing efforts, thematic circuit development, and community involvement. With proper planning and execution, Kushinagar can continue to grow as a prominent cultural tourism destination while preserving its rich cultural heritage.

References

Agrawal, M., Choudhary, H., & Tripathi, G. (2010, November 16). Enhancing Buddhist tourism in India: an exploratory study. *Worldw. Hosp. Tour. Themes*, *2*(5), 477–493. https://doi.org/10.1108/17554211011090102

Agrawal, Vineeta. (2016). A Review of Indian Tourism Industry with SWOT Analysis. *Journal of Tourism & Hospitality*, 05. 10.4172/2167-0269.1000196.

Asadi, R. (2011). Strategies for development of Iran cultural tourism. *Aust. J. Basic Appl. Sci.*, *5*(8), 1021-1034.

Buddhist Monuments by Debala Mitra. Published by Shishu Sahitya Samsad Pvt. Ltd., 32A Acharya Prafulla Chandra Road, Calcutta 9.

Chandel, R.S., Kanga, S. (2018). Use of Geospatial Techniques to Manage Tourists & Administration: A Case Study of Mount Abu, Rajasthan. *Environ. & We An Int. J. Sci. Technol.*, 13(1), 63-78.

Chandel, R.S., Kanga, S. (2019). Ecotourism Potential in Western Rajasthan: A Case Study of Jaisalmer District. *J. Clim. Change Water*, 6(1), 8-15.

Chandel, R.S., Kanga, S. (2020). Sustainable Management of Ecotourism in Western Rajasthan, India: A Geospatial Approach. *Geoj. Tourism Geosites*, 29(2), 521-533.

Chandel, R.S., Kanga, S., Singh, S.K. (2021). Impact of COVID-19 on the Tourism Sector: A Case Study of Rajasthan, India. *AIMS Geosci.*, 7(2), 224-243. DOI: 10.3934/geosci.2021014.

Chandel, R.S., Rai, P.K., Kanga, S., Singh, R. (2023). An Assessment and Management of Ecotourism Based on Water and LULC: A Geospatial Approach of Jodhpur, Rajasthan, India. In: Rai, P.K. (eds) River Conservation and Water Resource Management. *Advances in Geographical and Environmental Sciences.* Springer, Singapore. https://doi.org/10.1007/978-981-99-2605-3_12

Chandel, R.S.; Kanga, S. (2023). GIS and Management Planning for Sustainable Ecotourism Development: A Case Study of Jodhpur, Rajasthan, India. *Sustain. Agri Food Environ. Res.*, 11(1), https://doi.org/10.7770/safer-V11N1-art2505.

Chandel, R.S.; Kanga, S.; Singh, S.K.; Đurin, B.; Oršulić, O.B.; Dogančić, D.; Hunt, J.D. Assessing Sustainable Ecotourism Opportunities in Western Rajasthan, India, through Advanced Geospatial Technologies. *Sustainability* 2023, *15*, 11473. https://doi.org/10.3390/su151411473

Chhabra, D. (2011, October). Cultural Tourism Research Methods. *Ann. Tour. Res.*, *38*(4), 1669–1671. https://doi.org/10.1016/j.annals.2011.07.014

India Tourism Statistics, 2022, Government of India.

Kanga, S., Sharma, L. K., Pandey, P. C., Nathawat, M. S., & Sinha, S. (2011). Geospatial approach for allocation of potential tourism gradient sites in a part of Shimla District in Himachal Pradesh, India. *J. GIS Trends*, 2(1), 1-6.

Kanga, S., Sharma, L.K., Pandey, P.C., Nathawat, M.S., Sharma, S.K. (2013). Forest Fire Modeling to Evaluate Potential Hazard to Tourism Sites Using Geospatial Approach. *J. Geomat.*, 7(1), 93-99.

Last Days of the Buddha. *A Translation of the Mahaparinibbana Sutta by Sister Vajira.* Buddhist Publication Society, Kandy, Sri Lanka 1964.

Nega D (2018) An Investigating the SWOT Analysis of Cultural Tourism Heritage for Sustainable Tourism Development: The Case of LALIBELA Rock Hewn Churches, World Heritage Site, Ethiopia. *J. Tour. Hospit.,* 7: 395. doi: 10.4172/2167-0269.1000395

Nezha, M., Rossi, A., El Khalidi, K., Pavel, A. B., El Ouaty, O., & Fekri, A. (2021). A SWOT Analysis to understand the impact of tourism industry on the Three pillars social Economy and Environment. In *SHS Web of Conferences* (Vol. 119, p. 04004). EDP Sciences

Research on the Development Strategies of Rural Tourism in Suzhou Based on SWOT Analysis. (2012, March 16). Research on the Development Strategies of Rural

Tourism in Suzhou Based on SWOT Analysis - ScienceDirect. https://doi.org/10.1016/j.egypro.2012.01.20

Richards, G. (ed.) (2001): *Cultural Attractions and European Tourism*, CAB International, Wallingford, UK

San, C. K. (2001). Buddhist pilgrimage. *Subdng Jaya Buddhist Association, Malaysia*.

Sharma, Anukrati (2013); A SWOT analysis of Rajasthan tourism; *Spectr.: J. Multidiscip. Res.* Vol. 2 Issue 6.

Subramoniam, S., Al-Essai, S. A. N., Al-Marashadi, A. A. M., & Al-Kindi, A. M. A. (2010). SWOT analysis on Oman tourism: A case study. *J. Econ. Dev. Manage. IT Finance Marketing*, *2*(2), 1.

Urosevic, N. (2012). Cultural identity and cultural tourism: Between the local and the global (a case study of Pula, Croatia). *Singidunum J. Appl. Sci.*, *9*(1), 67–76. https://doi.org/10.5937/sjas1201067u.

Yadav, A.K., Kanga, S., Kumar, R. (2018). Web-Enabled GIS-Based Tourism Information System for Shimla Municipality (H.P), India. *i-manager's J. Inf. Technol.*, 7(1), 18-23.

Index

Also of Interest

Check out these other related titles from Scrivener Publishing

Mathematical Modeling in Agriculture, Edited by Sabyasachi Pramanik, Niranjanamurthy M., Ankur Gupta, and Ahmed J. Obaid, ISBN: 9781394233717. The main goal of the book is to explore the idea behind data modeling in smart agriculture using information and communication technologies and tools to make agricultural practices more functional, fruitful and profitable.

Agroforestry, Edited by Abhishek Raj, Manoj Kumar Jhariya, Arnab Banerjee, Ramesh Kumar Jha, and Krishan Pal Singh, ISBN: 9781394231133. This book offers comprehensive insights into the management of agroforestry for livelihood security and sustainable development in the tropics, addressing ecological interactions, productivity, and the monetization of carbon credits, while also outlining a future roadmap and policy challenges.

Smart Agritech: Robotics, AI, and Internet of Things (IoT) in Agriculture, Edited by Santosh Kumar Srivastava, Durgesh Srivastava, Korhan Cengiz, and Pramod Gaur, ISBN:9781394302959. The main goal of *Smart Agritech: Robotics, AI, and the Internet of Things (IoT) in Agriculture* is to explore how emerging technologies such as robotics, artificial intelligence (AI), and IoT can be leveraged to improve efficiency, sustainability, and productivity in agriculture.

DIGITAL AGRICULTURAL ECOSYSTEM: Revolutionary Advancements in Agriculture, Edited by Kuldeep Singh and Prasanna Kolar, ISBN: 9781394242931. The book comprehensively explores the dynamic synergy between modern technology and agriculture, showcasing how advancements such as artificial intelligence, data analytics, and smart farming practices are reshaping the landscape to ensure food security in the era of climate change, as well as bridging the gap between cutting-edge research and practical implementation.

THE DIGITAL AGRICULTURAL REVOLUTION: Innovations and Challenges in Agriculture through Technology Disruptions, Edited by Roheet Bhatnagar, Nitin Kumar Tripathi, Nitu Bhatnagar and Chandan Kumar Panda, ISBN: 9781119823339. The book integrates computational intelligence, applied artificial intelligence, and modern agricultural practices and will appeal to scientists, agriculturists, and those in plant and crop science and management.

HANDBOOK OF AGRICULTURAL BIOTECHNOLOGY: Volume I Nanopesticides, Edited by Charles Oluwaseun Adetunji and Julius Kola Oloke, ISBN: 9781119836148. The book provides detailed information on the application of nanopesticides for the management of numerous agricultural pests and pathogens, as well as to foster an innovative pathway toward future sustainable agriculture and food systems.

HANDBOOK OF AGRICULTURAL BIOTECHNOLOGY: Volume II Nanobioherbicides, Edited by Charles Oluwaseun Adetunji, Julius Kola Oloke, ISBN: 9781119836155. The book provides a detailed examination of the application of nanobioherbicides that come from plants including information on the different metabolites derived from numerous plants that could become bioherbicides.

HANDBOOK OF AGRICULTURAL BIOTECHNOLOGY: Volume III Nanofungicides, Edited by Charles Oluwaseun Adetunji and Julius Kola Oloke, ISBN: 9781119836162. This book provides state-of-the-art information about recent advancements in the application of nanobiofungicides for effective management of post-harvest diseases and agricultural fungal diseases, including blights, mildews, molds, and rusts.

HANDBOOK OF AGRICULTURAL BIOTECHNOLOGY: Volume IV Nanoinsecticides, Edited by Charles Oluwaseun Adetunji and Julius Kola Oloke, ISBN: 9781119836179. The book provides detailed information about the application of repellent products that contain plant-based ingredients known as nanobioinsecticides, including the pesticide evaluation scheme guidelines for repellent testing, relevant information about the procedures to evaluate several repellent compounds, the development of new products that offer high repellency, and guidelines for consumer safety.

Books in the series, "Bioprocessing in Food Science"

NUTRACEUTICALS FROM FRUIT AND VEGETABLE WASTE, Edited by Vidisha Tomer, Navnidhi Chhikara, Ashwani Kumar, and Anil Panghal, ISBN: 9781119803508. Nutraceuticals from Fruit and Vegetable Waste, the latest volume in the series, "Bioprocessing in Food Science," is an essential reference on the valorization of fruit and vegetable waste and the state of the art for fruit and vegetable processing.

NUTRITIONAL SCIENCE AND TECHNOLOGY: Concept to Application, Edited by Tejpal Dhewa, Anil Kumar Puniya, and Anil Panghal, ISBN: 97811198089. This new volume in the groundbreaking series, "Bioprocessing in Food Science," Nutritional Science and Technology, is a comprehensive resource covering topics such as human nutrition, food safety concerns, and emerging technologies like nanotechnology, and it also addresses the need for innovative food products, contamination, and regulatory challenges in the global food market.

HARVESTING FOOD FROM WEEDS, Edited by Prerna Gupta, Navnidhi Chhikara, and Anil Panghal, ISBN: 9781119791973. An exciting new volume in the groundbreaking new series, "Bioprocessing in Food Science," Harvesting Food from Weeds focuses on the latest processes, industrial applications, and leading research on this important, evolving new area, for engineers, scientists, students, and other industry professionals.

MICROBES IN THE FOOD INDUSTRY, Edited by Navnidhi Chhikara, Anil Panghal, and Gaurav Chaudhary, ISBN: 9781119775584. This newest volume in the groundbreaking new series, "Bioprocessing in Food Science," focuses on the latest processes, industrial applications, and leading research on microbes in the food industry, for engineers, scientists, students, and other industry professionals.

NOVEL TECHNOLOGIES IN FOOD SCIENCE, Edited by Navnidhi Chhikara, Anil Panghal, and Gaurav Chaudhary, ISBN: 9781119775584. The third book in the series, "Bioprocessing in Food Science," covers the latest and greatest new technologies in food science.

FUNCTIONAL FOODS, Edited by Navnidhi Chhikara, Anil Panghal, and Gaurav Chaudhary, ISBN: 9781119775560. Presenting cutting-edge information on new and emerging food engineering processes, *Functional Foods*, the second volume in the groundbreaking new series, "Bioprocessing in Food Science," is an essential reference on the modeling, quality, safety, and technologies associated with food processing operations today.

THERMAL FOOD ENGINEERING OPERATIONS, Edited by Nitin Kumar, Anil Panghal, and M. K. Garg, ISBN: 9781119775591. Presenting cutting-edge information on new and emerging food engineering processes, Thermal Food Engineering Operations, the first volume in the new series, "Bioprocessing in Food Science," is an essential reference on the modeling, quality, safety, and technologies associated with food processing operations today.

Printed in the USA/Agawam, MA
September 16, 2026

910661.152